Advances in Coatings Deposition and Characterization

Advances in Coatings Deposition and Characterization

Selected Articles Published by MDPI

MDPI • Basel • Beijing • Wuhan • Barcelona • Belgrade

For citation purposes, cite each article independently as indicated on the article page online and as indicated below:

LastName, A.A.; LastName, B.B.; LastName, C.C. Article Title. *Journal Name* **Year**, *Article Number*, Page Range.

Contents

About the Selection Editors

Dr. Alessandro Lavacchi
Istituto di Chimica dei Composti OrganoMetallici (ICCOM-CNR), Via Madonna del Piano 10, 50019 Sesto Fiorentino, Firenze, Italy

Prof. Dr. Massimo Innocenti
Department of Chemistry, Università di Firenze, Via della Lastruccia 3-13, 50019 Sesto Fiorentino, Firenze, Italy

Prof. Dr. Steve Bull
School of Chemical Engineering and Advanced Materials, Newcastle University, Bedson Building, Newcastle upon Tyne, NE1 7RU, UK

Prof. Dr. Philippe Dubois
Center of Innovation and Research in Materials & Polymers, University of Mons, Place du Parc, 23 7000 Mons, Belgium

Dr. Michele Fedel
Department of Industrial Engineering, University of Trento, Via Sommarive n. 9, 38123 Trento (TN), Italy

Preface to "Advances in Coatings Deposition and Characterization"

Coatings offer the unique opportunity to create architectures that combine the functionality of two or more materials, conferring unique properties to objects with an extremely large palette of solutions. For this flexibility, thick and thin films have terrific impacts on the most relevant societal challenges. Computers, food packaging, airplanes, and cars, to mention a few familiar objects from everyday life, rely heavily on coatings.

To celebrate the key role that coatings have in society, and in science and technology, this book collects a selection of relevant reviews and original research articles published in "Coatings" in 2017 and 2018. Papers have been selected based on their broad impact and balancing between the two major aspects of coatings science and technology: deposition and characterization.

Alessandro Lavacchi

Article

Development of YSZ Thermal Barrier Coatings Using Axial Suspension Plasma Spraying

Dapeng Zhou [1,*], Olivier Guillon [1,2] and Robert Vaßen [1]

[1]   Forschungszentrum Jülich GmbH, Institute of Energy and Climate Research, Materials Synthesis and Processing (IEK-1), Jülich 52425, Germany; o.guillon@fz-juelich.de (O.G.); r.vassen@fz-juelich.de (R.V.)
[2]   JARA-ENERGY, Jülich Aachen Research Alliance (JARA), Jülich 52425, Germany
*   Correspondence: da.zhou@fz-juelich.de; Tel.: +49-2461-61-96841

Received: 12 July 2017; Accepted: 7 August 2017; Published: 10 August 2017

Abstract: The axial injection of the suspension in the atmospheric plasma spraying process (here called axial suspension plasma spraying) is an attractive and advanced thermal spraying technology especially for the deposition of thermal barrier coatings (TBCs). It enables the growth of columnar-like structures and, hence, combines advantages of electron beam-physical vapor deposition (EB-PVD) technology with the considerably cheaper atmospheric plasma spraying (APS). In the first part of this study, the effects of spraying conditions on the microstructure of yttria partially-stabilized zirconia (YSZ) top coats and the deposition efficiency were investigated. YSZ coatings deposited on as-sprayed bond coats with 5 wt % solid content suspension appeared to have nicely-developed columnar structures. Based on the preliminary results, the nicely developed columnar coatings with variations of the stand-off distances and yttria content were subjected to thermal cycling tests in a gas burner rig. In these tests, all columnar structured TBCs showed relatively short lifetimes compared with porous APS coatings. Indentation measurements for Young's modulus and fracture toughness on the columns of the SPS coatings indicated a correlation between mechanical properties and lifetime for the SPS samples. A simplified model is presented which correlates mechanical properties and lifetime of SPS coatings.

Keywords: axial suspension plasma spraying; thermal barrier coating; yttria-stabilized zirconia; microstructure; lifetime; indentation fracture toughness

1. Introduction

Thermal barrier coatings (TBCs) are widely used in aircraft and industrial gas-turbine engines to improve the durability and efficiency of engines [1,2]. TBCs are complex multilayer systems composed of an oxidation-resistant metallic bond coat (BC) and a thermal insulating ceramic top coat (TC) [3]. The ceramic top coat is typically made of 7–8 wt % yttria partially-stabilized zirconia (YSZ). Due to the refractory nature of YSZ with a melting point of ~2700 °C, high temperature materials processing technology is required [4]. The two primary widely-used methods for depositing TBCs are electron beam-physical vapor deposition (EB-PVD) and atmospheric plasma spraying (APS) [5]. During the EB-PVD process, a high-energy electron beam is used to melt and evaporate ceramic ingots in a vacuum chamber. Subsequently, the vapor deposits onto a preheated substrate at a deposition rate of, typically, several μm/min. Due to the vapor phase condensation and shadowing effect [5], columnar-structured TBCs with a high level of strain tolerance can be achieved [6]. However, the high manufacturing cost limits the use of EB-PVD on severe thermo-mechanically loaded parts, such as first-row blades.

Versatility and low deposition cost make APS an attractive technique for depositing TBCs in commercial applications [7,8]. In the APS process, ceramic powder with a particle size of tens of micrometers is injected into an arc plasma jet and deposited onto the substrates. Within the plasma jet, the particles are accelerated and melted, followed by impaction, rapid solidification, and forming a coating on the substrate [3]. A lamellar microstructure is obtained by stacking splats during deposition. Attributing to deposition-induced defects (pores, arrayed micro-cracks and interfaces), typical APS coatings offer lower thermal conductivity than EB-PVD coatings [5,9]. However, under severe thermo-mechanical loading conditions, lamellar TBCs deposited with APS exhibit lower lifetimes compared with EB-PVD coatings [10]. There seems to be an advantage of the columnar EB-PVD compared to the micro-cracked and porous APS microstructure. Hence, the possibility to achieve such a microstructure by a modified APS process, namely suspension plasma spraying (SPS), is highly attractive.

SPS is a spraying technology with which high-performance TBCs can be deposited at low cost [11–13]. In the SPS process, liquids (e.g., ethanol or water) are used as carrier media to inject sub-micron or nano-sized ceramic particles into the plasma jet. With SPS, TBCs with different porosity levels, as well as different microstructures, such as porous, vertically cracked, and columnar structures can be achieved [5,14,15]. Especially, columnar SPS coatings with high porosity, low thermal conductivity, and much finer microstructure than conventional APS coatings have a great potential for industrial applications [16,17]. More recently, a new axially-injected suspension plasma spraying technology has been developed [18]. In this process, the suspension is axially injected into the core of the high-temperature and high-velocity plasma jet. Due to the axial injection method, particles can be well melted and accelerated in the plasma jet [19]. Generally, the porosity bands in SPS coatings between individual spray passages can be often found [20]. These bands are due to not properly injected droplets at the outer fringes of the plasma plume. A central injection is able to reduce this effect considerably.

Even though some work about the effect of bond coat roughness on the microstructure of the top coat layer have been reported, such as [17,21,22], the effect of spraying conditions on the microstructure of top coats sprayed with axial SPS have only be partially investigated so far. The objective of the present work is to investigate the influence of the deposition conditions (bond coat roughness, stand-off distance, input powder, and solid content of suspension) on the microstructures of SPS coatings. Based on the results, appropriate microstructures have been selected for thermal cyclic experiments. The rather moderate thermal cycling lifetime of SPS coatings compared with that of APS coatings are discussed with respect to the mechanical properties of the coatings. Furthermore, a simplified model was used to correlate the mechanical properties and the lifetime of SPS coatings. This study can shed light on improving the lifetime of axial SPS thermal barrier coatings.

2. Materials and Methods

2.1. Materials

Commercially-available 9.7 wt % yttria-stabilized zirconia powder (TZ-5Y, Tosoh Corporation, Tokyo, Japan) and 7.5 wt % yttria-stabilized zirconia powder, which was made by mixing TZ-5Y with TZ-3Y (5.4 wt % YSZ, Tosoh Corporation, Tokyo, Japan) were used in this work. The powder was dispersed in ethanol with the addition of a dispersant (PEI, Ploysciences, Warrington, PA, USA) and zirconia milling balls (d = 3 mm, Sigmund Lindner GmbH, Warmensteinach, Germany). The suspension was milled on a roller cylinder (120 min^{-1}, 24 h) in order to produce a homogeneously-dispersed suspension (30 wt % in solid content). After milling, the suspension was diluted with ethanol to 10 wt % and 5 wt %. After milling, the particle size distribution was measured with a HORIBA LB-550 nanoparticle size analyzer (Retsch Technology GmbH, Haan, Germany). The particle distribution is d_{10} = 0.16 µm, d_{50} = 0.19 µm, and d_{90} = 0.27 µm. The viscosity of 10 wt % and 5 wt % suspension, measured with a viscosimeter (Physica MCR 301, Anton Paar Germany GmbH, Ostfildern, Germany) at a shear rate of 10 s^{-1}, was 1.63 mPa·s and 1.48 mPa·s, respectively.

Stainless steel plates (25 × 25 × 2 mm^3) coated with 200 µm high-velocity oxy fuel (HVOF) bond coat (Amdry 9954, $Co_{32}Ni_{21}Cr_8Al_{0.5}Y$, d_{10} = 11 µm, d_{50} = 20 µm, d_{90} = 50 µm) were used as substrates. Before deposition, the surfaces of THE substrates were carefully treated with: (1) mirror polishing; (2) grinding; (3) mirror polishing followed by grit blasting; and (4) as-sprayed. Additionally, rougher HVOF bond coats were prepared with Amdry 995C powder ($Co_{32}Ni_{21}Cr_8Al_{0.5}Y$, d_{10} = 52 μm, d_{50} = 69 μm, d_{90} = 90 μm) which has a larger particle size than Amdry 9954 [23]. The roughness of the surface-treated bond coats was measured with a double-sided non-contact metrology system (CT 350T, Cyber Technologies GmbH, Eching-Dietersheim, Germany). The roughness was determined as:

- Mirror polishing: R_a = 0.06 μm, R_z = 0.27 μm;
- Grinding: R_a = 0.26 μm, R_z = 1.89 μm;
- Mirror polishing and grit blasting: R_a = 2.82 μm, R_z = 21.6 μm;
- As sprayed HVOF bond coat: R_a = 10.4 μm, R_z = 67.4 μm;
- As sprayed rough HVOF bond coat: R_a = 14.4 μm, R_z = 87.5 μm.

For the thermal cycling tests, button-shaped nickel-based superalloy IN 738 (30 mm in diameter, 3 mm in thickness) was used as the substrate. For the purpose of minimizing the effect of stress generated at the edge, a curvature with a radius of 1.5 mm was machined at the outer edge of the substrates. On the IN 738 substrates, a 150 μm thickness MCrAlY bond coat (Amdry 9954, Oerlikon Metco Company, Wohlen, Switzerland) was sprayed with an F4 torch in a vacuum plasma spray (VPS) facility (Oerlikon Metco Company, Wohlen, Switzerland).

2.2. Plasma Spraying Conditions

Ceramic top coats were sprayed with an Axial III high-power plasma torch (Northwest Mettech Corporation, Vancouver, BC, Canada) which was mounted on a six-axis robot. The Axial III torch contains three cathodes and three anodes which are powered by three independent power sources. The liquid suspension was injected axially into the middle of the three plasma jets which converge within an interchangeable nozzle [19]. A feeding system developed by Forschungszentrum Jülich was used in this work [24]. The feeding rate of the suspension was set to be 30 g/min. The speed of the gun was set to be 1000 mm/s. A mixture of Ar (75 vol %), H_2 (15 vol %), and N_2 (10 vol %) with a flow rate of 245 standard liters per minute (slpm) was used as the working gas. The detailed spraying conditions are listed in Table 1.

Table 1. Spray parameters used for YSZ top coats.

No.	D (mm)	P (kW)	S (wt %)	I (A)	Surface Preparation
A	70	105	10	750	Mirror polishing
B	70	105	10	750	Grinding
C	70	105	10	750	Grit blasting
D	70	105	10	750	As sprayed
E	100	105	10	750	Mirror polishing
F	100	105	10	750	Grinding
G	100	105	10	750	Grit blasting
H	100	105	10	750	As sprayed
I	100	105	10	750	As sprayed (rough)
J	70	84	10	600	As sprayed
K	70	105	5	750	As sprayed
L	100	105	5	750	As sprayed
M	70	105	5	750	As sprayed
N	100	105	5	750	As sprayed
O	70	105	5	750	As sprayed *

D: standoff distance (mm); P: input power (kW); S: solid content in suspension (wt %); I: current (A); *: deposited with 7.5 wt % YSZ powder.

2.3. Microstructure and Porosity Characterization

Metallographic cross-sections of samples were prepared to investigate the microstructure of the top coats with scanning electron microscopes (SEM, Zeiss Ultra 55 FEG-SEM, Carl Zeiss Microscopy

GmbH, Oberkochen, Germany, and Hitachi TM3000, Hitachi High-Technologies Europe GmbH, Krefeld, Germany). The columns and vertical crack density of the coatings were counted from cross-section SEM images (magnification 300×). Vertical cracks and columns intercepting a fixed length line (7 mm in length) which is parallel and a certain distance from the top coat/bond coat interlayer were counted. The crack and column density were calculated by dividing the number of cracks and columns with the cross-section length. Vertical cracks and columns were defined as follows: (1) vertical cracks are cracks running perpendicular to the ceramic/bond coat interlayer and penetrating at least half the thickness of the coating; (2) columns are conical areas with a high density, secluded by linear porous gaps which are perpendicular to the top coat/bond coat interface.

The deposition efficiency of top coats was calculated by dividing the weight change of samples before and after spraying with the weight of the YSZ powder used during spraying. The deposition efficiency η of top coats was obtained with the following equation:

$$\eta = G_c / G_s \times 100\% \tag{1}$$

in which G_c (g) is the weight of top coat deposited on the substrate, G_s (g) is the weight of the YSZ powder injected into the plasma jet during spraying [25].

The porosity of the coatings deposited at different spraying conditions was measured with image analysis (IA). Porosity can be easily detected due to the high degree of contrast between dark pores and high brightness coating material [26]. For each measurement, 10 cross-section SEM images of the top coats with magnification of 300× were used. The defect (pore) size within the columns was also determined with image analysis. The mean equivalent circle diameter (ECD) value of the largest 10% of pores was measured. Assuming all the pores were spheres, the mean diameter of the pores was obtained by multiplying the mean ECD with a constant $(4/\pi)$ [27]. All the image analysis work was conducted with the software AnalySIS (AnalySIS pro, Olympus Soft Imaging Solutions GmbH, Hamburg, Germany).

For the investigation of phases of the top coats, X-ray diffraction was also carried out with a D4 Endeavor (Bruker, Karlsruhe, Germany) using Cu Kα radiation. A scanning range of 2θ from $10°$ to $80°$ with a step size $0.02°$ and a count time 0.5 s/step were used. The Rietveld refinement technique was used to determine the phases and the amount of different phases existing in the coatings [28].

2.4. Mechanical Property Tests

Mechanical property measurements were performed on the metallographic cross-sections of the samples. The hardness and elastic modulus of top coats were measured with a depth-sensing micro-indentation test (H-100 Fischerscope, Helmut Fischer GmbH, Sindelfingen, Germany). The load applied on the indenter was set to be 1 N. Effective Young's modulus was calculated from the initial unloading slope [29]. The elastic modulus of the materials can be obtained with the following equation:

$$E^* = E / 1 - v^2 \tag{2}$$

where E is the elastic modulus (GPa), E^* is effective Young's modulus (GPa), and v is the Poission's ratio; in this work $v = 0.25$ was adopted from [30]. In order to obtain reliable values, 15 indentations were performed on each sample.

For the measurements of fracture toughness, an indentation fracture toughness technique was carried out in this work [31]. A Vicker's indenter with an applied load of 3 N was used to generate proper cracks in metallographic cross-sections of the samples. In order to minimize the effect of column gaps, all the indentations were performed at the central part of the columns. The ratio between surface crack length l (m) and indentation half diagonal length a (m) fell into the following range:

$$0.25 \leq l/a \leq 2.5 \tag{3}$$

This indicates that the generated cracks are Palmqvist cracks [31]. Thus, the following equation was used for calculating indentation fracture toughness of the top coats.

$$K_{IC} = 0.018\left(\frac{E}{H}\right)^{\frac{2}{5}} Ha^{\frac{1}{2}}\left(\frac{a}{l}\right)^{\frac{1}{2}} \tag{4}$$

where K_{IC} is the indentation fracture toughness (MPa·m$^{1/2}$), H is the indentation hardness (MPa), E is the elastic modulus (MPa), a is the indentation half-diagonal length (m), and l is the crack length (m) [31,32]. Ten indentations were made for each fracture toughness determination.

2.5. Thermal Shock Tests

The thermal shock tests were performed in a gas burner rig facility operating with a natural gas and oxygen mixture. The front sides of the samples were periodically heated up to the target temperature and, simultaneously, the reserve sides of the samples were cooled with compressed air to maintain a temperature gradient across the sample. In the aim of getting reliable lifetime results, two specimens were subjected to thermal cycling tests for each kind of microstructure. The detailed information about the test facility and used samples were given in [33]. Thermal cycling lifetime was defined as the number of thermal cycles that a sample survives before the appearance of visible spallation (25 mm^2). During the test, the surface temperature was monitored with an infrared pyrometer, and the substrate temperature was measured with thermocouple which was located in the center of the substrate. In this work, the surface temperature was set to be 1400 ± 30 °C and the substrate temperature was adjusted to 1050 ± 30 °C.

3. Results and Discussion

3.1. Effect of Substrate Roughness on Coating Microstructure

Figure 1 shows the as-sprayed cross-section microstructures of top coats which were sprayed on BCs with different roughness at a stand-off distance of 70 mm. It can be seen that the roughness of the BC layer greatly affects the microstructure of the top coats. Samples A and B, which were deposited onto relatively smooth surfaces, display a microstructure with vertical cracks penetrating through the entire coating thickness. It has been reported that the vertical cracks in the APS TBCs can improve the strain tolerance of TBCs during thermal cycling [34]. It can be expected that vertical cracks in the SPS coatings can increase thermal cycling performance of the coatings, as well. The formation of vertical cracks in traditional APS coatings is related to the cooling and shrinkage of the deposited splats, resulting in large tensile stress and consequent cracking during cooling. This mechanism only works for high depositing temperatures and dense coatings without a large number of micro-cracks for stress relief [35]. For SPS coatings, the splat sizes range from 0.3–2 μm in diameter [15]. Due to the limited size of the splats, the possibility of micro-crack formation during cooling is reduced. Probably, a higher tensile stress level can be built up within the top coat. When the tensile stress level exceeds the strength of the top coat, highly-segmented SPS coatings with vertical cracks are formed.

Branching cracks, which originate from vertical cracks and propagate along porous bands, also exist in top coats (as shown in Figure 1a,b). These branching cracks are detrimental to the thermal lifetime of TBCs. Porous bands existing between successive passes of the spraying gun provide an easy pathway for the propagation of branching cracks under thermal stress. The formation of porous bands is attributed to poorly-heated un-molten particles or resolidified particles travelling in the jet fringes [36]. The extent of porous bands is largely reduced by axial feeding compared with radial feeding. This is also obvious when comparing the present results to former results using radial feeding (see, e.g., [20]). In addition, the porous bands can also be significantly reduced by changing the spraying pattern and suspension solid load content.

When the roughness of the BCs is increased up to R_a = 2.82 μm, as shown in Figure 1c, columns start to grow on the asperities of the bond coat; at the same time, the vertical cracks coexist in the top coat. Thus, the microstructure of Sample C is composed of a mixture of vertical cracks and columns. For sample D, the top coat was deposited on a surface with a roughness about R_a = 10.4 μm. Even though some vertical cracks can be observed, the microstructure of the top coat, as shown in Figure 1d, is a typical columnar structure. All the columns were separated by porous gaps, growing on the asperities of the surface. This observation is consistent with the proposed deposition mechanism by VanEvery et al. [14]. The plasma drag force changes the particles' velocity from normal along the substrate surface. Thus, these particles preferentially impact on asperities of the surface leading to the formation of columns. Although the porous gaps between columns can provide an easy way for ingress of hot gas and corrosive media, they also increase the strain tolerance of the top coats leading to higher thermal cycling lifetimes of the TBCs.

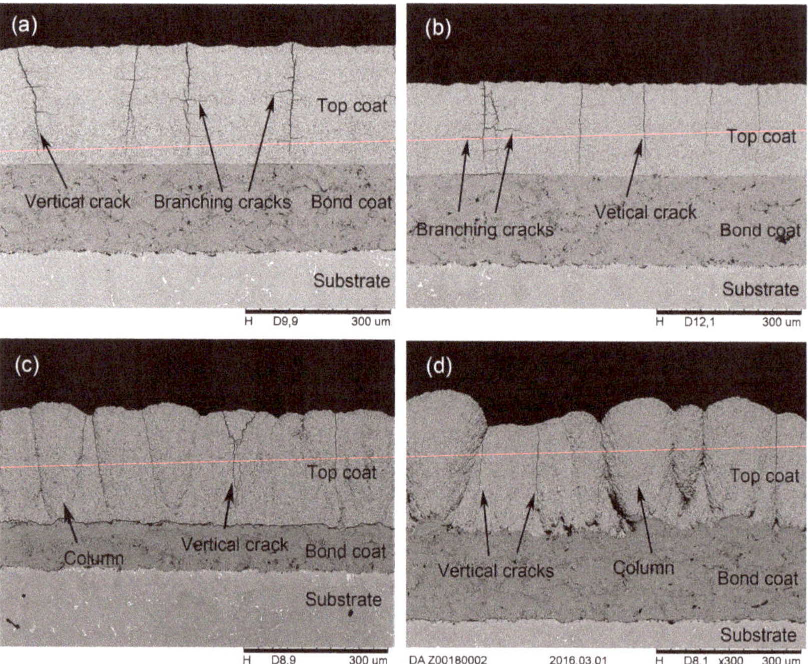

Figure 1. Cross-section SEM image of as sprayed YSZ top coats (Samples A, B, C, and D) deposited at a standoff distance of 70 mm, on BC with different surface treatments: (**a**) mirror polishing; (**b**) grinding; (**c**) grit blasting; and (**d**) as-sprayed.

An evaluation of the crack and column density is shown in Figure 2a. The crack density decreases from Sample B to D, while the column density shows an opposite tendency. It seems that vertical cracks compete with columns and so the crack density is greatly affected by the column density. Probably, columns can release tensile stress within the top coat, which is a prerequisite condition for generating vertical cracks. It should be mentioned here that during the deposition of sample A, a small part of the top coat peeled off from the substrate. The tensile stress built in the top coat was probably partially released leading to the unexpected low crack density of Sample A. The deposition efficiency for Samples A, B, C, and D (as shown in Figure 2b) are roughly constant (45–49%). It seems that the roughness of the bond coat barely affects the deposition efficiency in axial SPS.

The porosity of the top coats was also investigated with image analysis. The results shown in Figure 2c exhibit a very low porosity for Samples A and B, only about 2.9% and 2.6%, respectively. The porosity of the top coats increases with the increase of the bond coat roughness (Samples C and D in Figure 2c). Probably, this is a result of the formation of columns which can introduce additional pores, specifically column gaps, into the top coats. In summary, the bond coat roughness can indirectly influence the porosity of top coats by affecting the microstructure of top coats. It should be mentioned here that very fine porosity in the nanometer range is not accessible by the used low-resolution method.

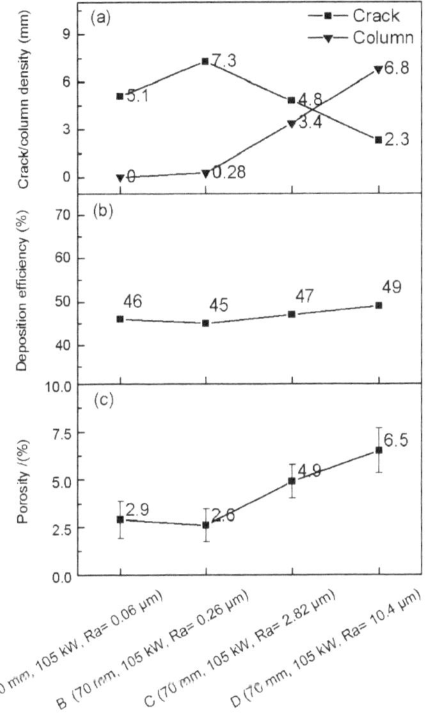

Figure 2. Crack/column density (**a**), deposition efficiency (**b**), and porosity (**c**) for as-sprayed YSZ top coats (Samples A, B, C, and D) deposited on BC with different surface treatments.

3.2. Effect of Stand-Off Distance on the Coating Microstructure

In addition, a set of top coats was deposited on surfaces with different roughness at a longer stand-off distance (100 mm). The cross-section microstructures of the top coats are presented in Figure 3. All top coats show a columnar microstructure. In addition, columnar-structured top coats can be obtained even on smooth surfaces, as shown in Figure 3a,b. This can be explained by considering the Stokes number S_t:

$$S_t = \frac{\rho_p d_p^2 v_p}{\mu_g l_{bl}} \tag{5}$$

in which ρ_p is the particle specific mass (kg/m^3); d_p is the particle diameter (m); v_p is the particle velocity (m/s); μ_p is the plasma gas molecular viscosity (Pa·s); l_{bl} is the thickness of the flow boundary layer (m), which varies as the inverse of the square root of the gas velocity close to the substrate. A longer stand-off distance will lead to lower particle velocities when approaching the substrate, thicker boundary layers and, hence, a reduction of the particle Stokes number. Particles with lower Stokes numbers are more easily able to follow the plasma gas and deposit preferentially on asperities of the BC surface forming columns. In other words, the increase in stand-off distance promotes the formation of columns. It should be mentioned here that columns formed on smooth surfaces (Figure 3a,b) do not directly grow on the surface of the BC, but at a certain distance away from the BC surface instead. This indicates that the previously-deposited ceramic top coat surface can provide asperities on which columns grow. These asperities should have a proper size, which is related with the spraying conditions before the column growth starts.

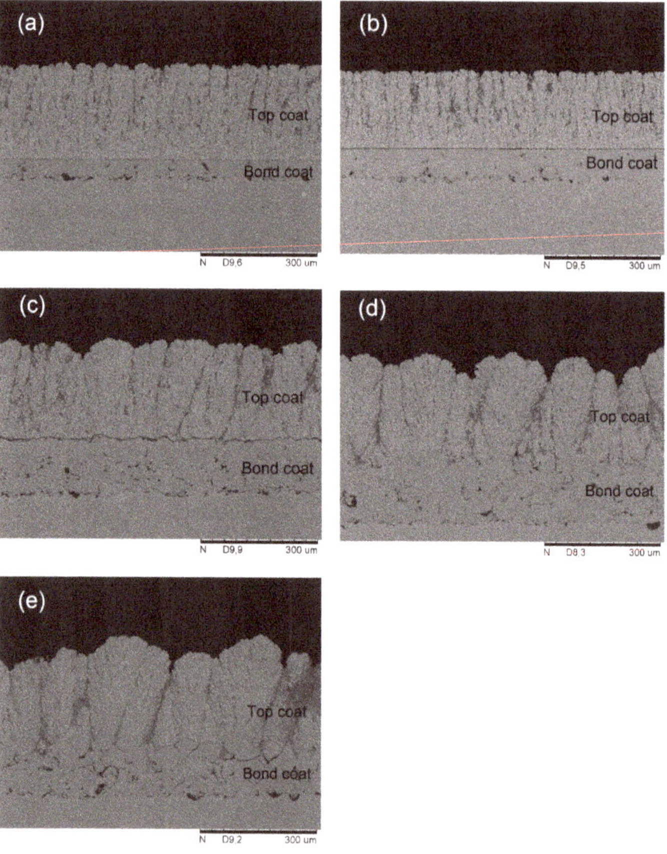

Figure 3. Cross-section SEM images of as-sprayed YSZ top coats (Samples E–I) deposited at a standoff distance of 100 mm, on BC with different surface treatments: (**a**) mirror polishing; (**b**) grinding; (**c**) grit blasting; (**d**) as-sprayed; and (**e**) as-sprayed rough.

The column density of the top coats is presented in Figure 4a. The column density drops gradually with the increase of the BC roughness. This observation is also consistent with deposition mechanism proposed by VanEvery et al. Increasing the roughness will reduce the number of relevant asperities, which promote the growth of columns in a specific area.

The porosity levels of the top coats deposited on surfaces with different roughness values at a 100 mm stand-off distance are presented in Figure 4c. With increasing surface roughness, from R_a = 0.06 µm up to R_a = 2.82 µm, the porosity of top coat increased gradually, and only slightly, increased at higher roughness values. The deposition efficiency of the top coats is considerably reduced at the higher stand-off distance (compare Figures 2b and 4b). A longer stand-off distance increases the number of particles with Stokes numbers below 1 [3]. These particles will follow the gas flow and might never impact on the substrate. In addition, at a longer stand-off distance particles are already cooling down and the substrate temperature is reduced, as well. Both factors lead to a lower sticking probability on the surface. Comparing the porosity level of the top coats deposited at a stand-off distance of 70 mm (Figure 2c) and 100 mm (Figure 4c), it can be seen that coatings deposited at a longer stand-off distance exhibit a higher porosity. This is probably related to the reduced droplet temperature and velocity impacting on the substrates.

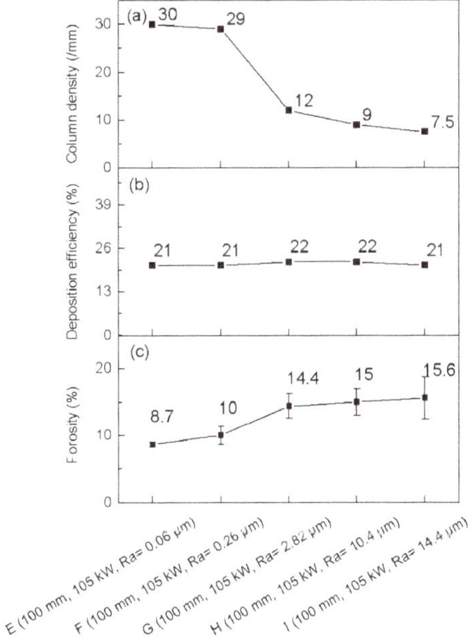

Figure 4. Column density (**a**), deposition efficiency (**b**) porosity (**c**) for as sprayed YSZ top coats (Samples E–I) deposited with 100 mm standoff distance on BC with different surface treatments.

3.3. Effect of the Input Power on the Coating Microstructures

The top coat (Sample J) sprayed with a lower input powder (84 kW instead of 105 kW) was deposited on an as-sprayed HVOF bond coat. The cross-section microstructure of Sample J is presented in Figure 5. It can be seen that Sample J also exhibits a columnar structure. The column density is about 8.8/mm which is higher than the value of Sample D (Figure 2a). Moreover, it has a higher porosity of about 18.5%, surprisingly. It seems that lower input power can greatly increase the porosity of the top coat. The deposition efficiency for Sample J is about 48%, close to the value of Sample D.

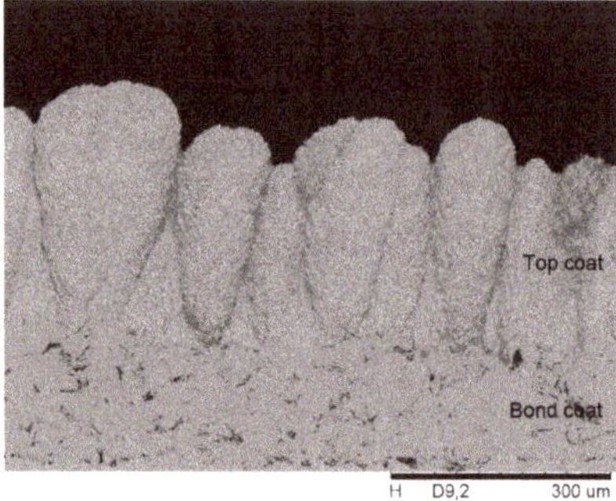

Figure 5. Cross-section image of the as-sprayed YSZ top coat (Sample J) deposited at a 70 mm standoff distance with 84 KW input power on as-sprayed BC.

3.4. Effect of Solid Content on the Coating Microstructures

Cross-section SEM images of Samples K and L, which were sprayed with a 5 wt % solid content suspension, are presented in Figure 6. It can be found that both of the top coats have well-developed columnar structures. No vertical cracks can be observed from the cross-section. The column density for Samples K and L is 10.4 and 11/mm, respectively. Furthermore, the density of the top coats are 19.3% and 28.0%. Compared with Samples D (Figure 2) and H (Figure 4), it seems that lowering the solid content in the suspension from 10 wt % to 5 wt % can promote the growth of columns and decrease the density of the top coats. In addition, the deposition efficiency of Samples K and L were measured to be 60.7% and 27.2%. Surprisingly, the deposition efficiency was also increased. Due to the different injection methods, axial SPS shows higher deposition efficiency than radial SPS. The deposition efficiency of axial injection SPS is close to the efficiency of APS, which can be 60% and above [37].

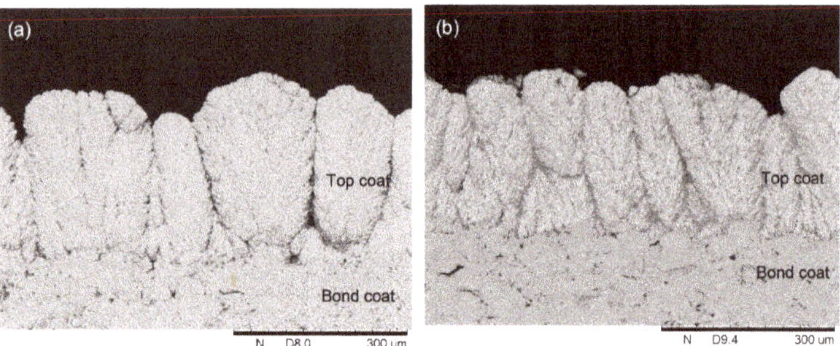

Figure 6. Cross-section images of as-sprayed YSZ top coats (Samples K (**a**) and L (**b**)) sprayed with 5 wt % solid content suspension on as-sprayed BC.

3.5. Thermal Cycling Performance of the Coatings

Based on the investigation of spraying conditions on microstructures, two well-developed columnar structured top coat (Samples M and N) were deposited on VPS bond-coated IN738 substrates for thermal cycling tests. The microstructure of the as-sprayed Samples M and N are shown in Figure 7. The column density for Samples M and N were 10.1 and 12.7/mm, respectively; the porosity determined with image analysis for Samples M and N were up to 22.8% and 29.1%. Sample N, which had a higher column density and porosity, was expected to have a higher strain tolerance and, thus, a longer thermal cycling lifetime under thermal load. However, it turned out that the average thermal cycling lifetime for Samples M and N (as shown in Table 2) were 177 and 76 cycles, respectively. The photographs and cross-section SEM images of thermally-cycled samples are presented in Figure 8. It can be seen on the photographs that a bit of white YSZ residue attached to the bond coat in the delaminated regions. From the cross-section images of the samples after thermal cycling testing, a similar delamination mode is found for both cases.

Table 2. Overview on thermally-cycled samples.

Sample	No.	Thickness (μm)	Bond Coat/Top Coat Interface Temperature (°C)	Top Coat Surface Temperature (°C)	Top Coat Temperature Gradient (°C/μm)	Lifetime (Cycles)
M	1	~255	1119	1346	1.10	193
	2	~257	1098	1284	0.78	160
N	1	~257	1118	1384	1.10	81
	2	~245	1121	1362	1.14	71
O	1	~289	1094	1277	0.63	218
	2	~317	1089	1283	0.61	269

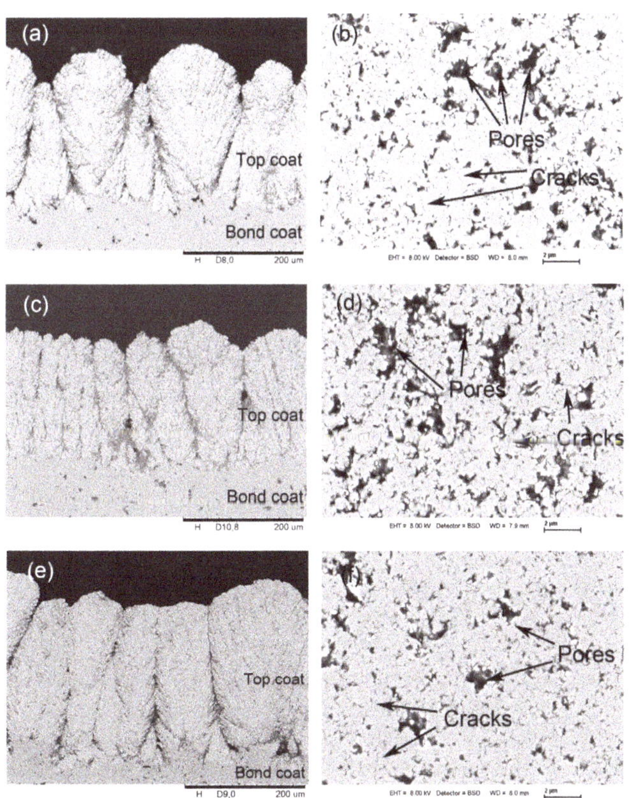

Figure 7. Cross-section SEM images for as-sprayed YSZ top coats: (**a**) Sample M, (**c**) Sample N, (**e**) Sample O; and the detailed microstructure within columns: (**b**) Sample M, (**d**) Sample N, and (**f**) Sample O.

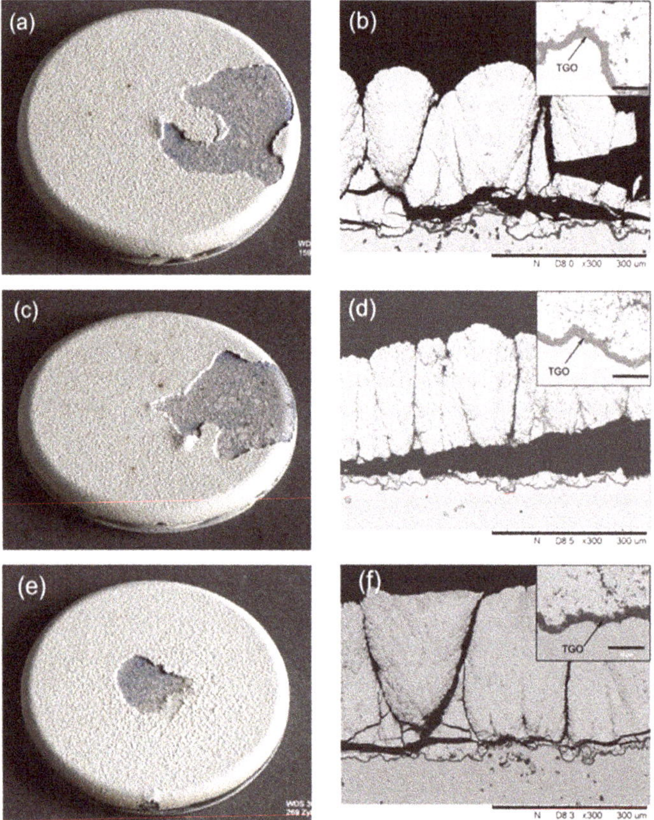

Figure 8. Photographs of thermally-cycled samples: (**a**) Sample M, (**c**) Sample N, and (**e**) Sample O; and cross-section SEM images of thermally-cycled samples: (**b**) Sample M, (**d**) Sample N, and (**f**) Sample O.

The delamination occurred within the top coats, parallel to the top coat/bond coat interface and just above the bond coat. As shown in Figure 8, for both samples the thermal growth oxide (TGO) were formed on the bond coat, but the thickness was limited (2–3 μm). Compared to typical TGO thickness at failure (6–12 μm) [1] the result indicates that the TGO layer growth is not the major reason for the premature failure of the YSZ top coats.

As reported in [38], at a temperature above 1200 °C, YSZ shows an insufficient phase stability. The metastable t' phase, which is formed due to rapid cooling during the deposition process, undergoes a phase transformation into the equilibrium phases (tetragonal and cubic) at high temperature; and upon cooling, the tetragonal phase will transform into the monoclinic phase companied by a volume change. This volume change leads to stresses in the coating and is detrimental for the thermal cycling lifetime of coatings. Thus, an XRD analysis was performed on thermally-cycled samples to investigate the phase evolution during thermal cycling. As shown in Figure 9, all the observed peaks in the XRD patterns of the as-sprayed samples can be attributed to the tetragonal phase. According to the lattice parameter results, the ratio $c/a\sqrt{2}$ for the tetragonal phase falls into the range of 1.010 to 1.000. This indicates that the tetragonal phase in the as-sprayed coating is non-transformable [39]. After thermal cycling, no monoclinic phase was detected from the XRD patterns and the tetragonal phase was non-transformable; but a cubic phase was detected from both thermally-cycled Samples M and N. According to quantitative analysis, the amount of cubic phase in the both coatings is around 12 wt %. The cubic phase with low fracture toughness is detrimental to the lifetime of the coatings. The formation of relatively large amounts of the cubic phase during thermal cycling testing is probably due to the fairly high yttria content in the coatings. In order to exclude the effect of relatively higher yttria content on the lifetime of the coatings, Sample O was produced with 7.5 wt % yttria-stabilized zirconia powder. As a result of using the same spraying parameters, Sample O (as shown in Figure 7e) exhibited similar columnar microstructures as Sample M (Figure 7a). After spraying, Sample O was subjected to thermal cycling testing. It showed that the average lifetime of Sample O is 244 cycles. Compared with Sample M, the lifetime of Sample O was improved slightly, but was still short. From the photograph (Figure 8e) and cross-section SEM image (Figure 8f) of thermally-cycled Sample O, it can be seen that the failure mode is similar to that of Samples M and N. Additionally, the thickness of the TGO layer for Sample O was very thin (about 2–3 μm). The XRD diagrams of as-sprayed and thermally-cycled Sample O are presented in Figure 9. According to the lattice parameter results, the tetragonal phase in as-sprayed and thermally-cycled coatings was non-transformable. In addition, the amount of cubic phase in the thermally-cycled coating was decreased to about 7 wt %. It seems that the relatively high yttria content and the phase transformation of YSZ are not the main reasons for the premature failure of SPS coatings.

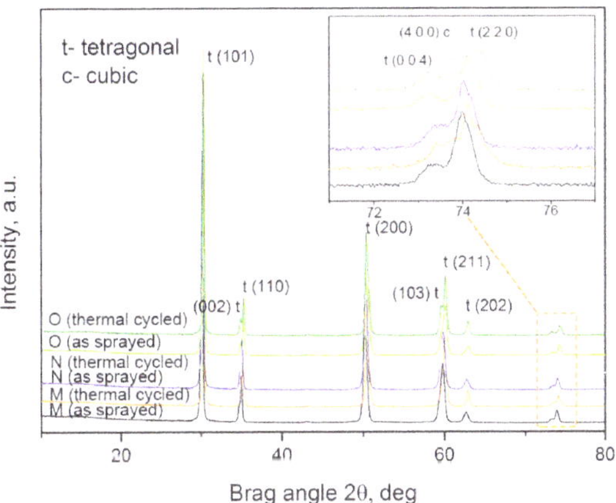

Figure 9. X-ray diffractograms of the as-sprayed and thermally-cycled YSZ samples (M, N, and O).

Another possible explanation for the premature failure of YSZ top coats might be the different mechanical properties, like Young's modulus and fracture toughness, of the SPS coatings. Hence, indentation tests have been conducted to evaluate these properties.

Mechanical properties of as-sprayed and thermally-cycled Samples M, N, and O are listed in Table 3. The as-sprayed fracture toughness for all three samples is lower than the value of typical APS coatings, which is about 2.0–3.3 MPa·m$^{1/2}$ [40]. Sample M, with a longer thermal cycling lifetime, exhibits a higher indentation fracture toughness than Sample N. On the other hand, this sample also shows a higher Young's modulus and, hence, a typically higher stress level was expected during thermal cycling. In order to obtain a better insight into which factor could be the most important one, a simplified evaluation was made.

Table 3. Mechanical properties of as-sprayed and thermally-cycled Samples K, L, and M.

Sample	Condition	Hardness (GPa)	Elastic Modulus (GPa)	Fracture Toughness (MPa·m$^{1/2}$)
M	As sprayed	9.6 ± 1.1	113.7 ± 10.4	1.5 ± 0.3
	Thermal cycled	10.2 ± 1.4	159.5 ± 16.8	2.3 ± 0.3
N	As sprayed	6.2 ± 0.9	81.9 ± 8.5	1.0 ± 0.2
	Thermal cycled	8.4 ± 1.4	127.9 ± 15.3	1.8 ± 0.2
O	As sprayed	8.3 ± 0.6	100.7 ± 7.0	1.7 ± 0.4
	Thermal cycled	10.5 ± 1.0	167.5 ± 13.2	2.5 ± 0.3

Due to thermal expansion mismatch between the YSZ top coat and the metallic substrate, thermal stress is built up during thermal cycling. This stress will be determined by the mismatch strain $\Delta\varepsilon$ which is proportional to the mismatch of the thermal expansion coefficients of coating and substrate (~5 × 10^{-6} /K) multiplied by the temperature change during cycling (~1000 K) and the Young's modulus E of the coating. Of course, the columnar structure leads to an increasing reduction of the stress with the increasing ratio of the coating thickness (for completely penetrating cracks) to crack spacing [41]. On the other hand, stresses will remain in the coating, especially in the center part of the columns close to the top coat/bond coat interface.

Crack propagation and, hence, failure will be governed by the stress intensity factor K_I, proportional to the stress multiplied by the square root of the defect size c and a geometry factor Y, which is $2/\pi^{1/2}$ for spherical pores (and assumed here):

$$K_I = Y \, \sigma \, \sqrt{c} \approx Y \Delta\varepsilon \, E \, \sqrt{c} \tag{6}$$

Generally, the crack growth rate dc/dt is proportional to the ratio of the stress intensity to the critical stress intensity K_{IC} factor (or toughness) with an (often rather high) exponent n; assuming the initial crack size c is much smaller than the critical crack size ac, the lifetime t_f obtained by integration can be approximated by:

$$t_f \propto c \left(\frac{K_{Ic}}{Y \, \Delta\varepsilon \, E \, \sqrt{c}} \right)^n \tag{7}$$

This formula has been used to evaluate the lifetime assuming a rather low strain in the columnar structures (reduced by a factor of two due to the columnar structure) [42]. The results are given in Table 4. The initial defect size c was estimated as being the mean value of the largest 10% of pores. The defect size results are presented in Table 4. As a result of sintering, the elastic modulus and fracture toughness of samples increase during thermal cycling tests. The average elastic modulus and fracture toughness of samples before and after thermal cycling testing were used to calculate the lifetime according to Equation (7).

Table 4. Input parameters and evaluation according to Equation (7) for the SPS coatings.

Sample	Strain	Defect Size (μm)	Average Elastic Modulus (GPa)	Average Fracture Toughness (MPa·m$^{1/2}$)	K_{Ic}/K_I	Lifetime (Cycles)	Lifetime/Pore Size (Cycles/μm)
M	0.0025	0.93	134	1.8	5.06	177	189.2
N	0.0025	1.38	105	1.4	4.05	76	55.1
O	0.0025	0.67	134	2.1	6.87	244	362.4

A plot of the data points plotted in a log-log scale is shown in Figure 10. The rather linear distribution of the data points in this plot indicates that Equation (7) might be an appropriate description of the lifetime data. However, it should be noticed that the exponent n obtained from the slope in Figure 10 is 3.5 and this value is rather small compared to typical results found for ceramics [43]. A more elaborate analysis is necessary. However, the main goal of the lifetime evaluation was to distinguish the relevant reasons for the premature failure of SPS TBCs. Here the correlation between toughness and lifetime shed light on the failure mechanism of SPS TBCs, that the lower fracture toughness of SPS coatings could be the major reason for their early failure.

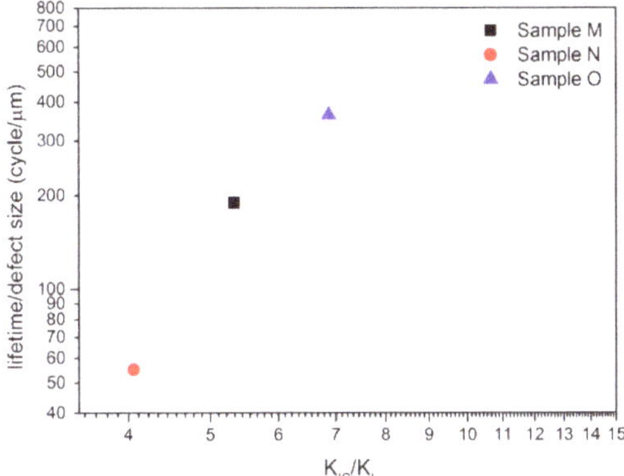

Figure 10. Plot of the lifetime t_f divided by the initial crack length c as a function of the critical stress intensity factor divided by the apparent one (see Equation (7)) for the samples given in Table 3.

4. Conclusions

In this study, thermal barrier coatings with different microstructures were deposited with axial suspension plasma spraying. The effects of spraying conditions on the microstructure of top coats and deposition efficiency were investigated. Additionally, thermal cycling tests were performed on the columnar structured TBCs. The main conclusions are as follows:

- Among the spraying parameters, bond coat roughness and stand-off distance played a dominant role on the microstructure formation of the top coats. Both increasing stand-off distance and reducing bond coat roughness increased the column density. Lowering the solid content in the suspension promotes the growth of columns. Also bond coat roughness, stand-off distance, and input power also had a significant influence on the microstructure and porosity of the top coats. Increasing bond coat roughness and stand-off distance, lowering input power, as well as the solid content in the suspension could enhance the porosity level of the coatings.
- The SPS process with axial injection had a high deposition efficiency of up to 60.7%. The deposition efficiency was mainly influenced by the stand-off distance. With increase of it, the deposition efficiency dropped greatly.
- Columnar-structured SPS TBCs exhibited a moderate thermal cycling performance. An evaluation of the major influencing factors of the lifetime showed that the relatively low fracture toughness is probably the main reason for the premature failure of the SPS TBCs.

Acknowledgments: This work was financially supported by the Helmholtz Association of German Research Centers and the China Scholarship Council (CSC). The authors sincerely acknowledge the contributions of the following colleagues in our institution: Karl-Heinz Rauwald for his invaluable assistance during suspension plasma spraying, Martin Tandler for carrying out the thermal cycling work, Hiltrud Moitroux for taking photographs, Sohn Yoo Jung for quantitative analysis on XRD results, and Sebold Doris for making part of the SEM images. The authors would also like to thank Jürgen Malzbender (IEK-2) for the support of the indentation tests.

Author Contributions: Dapeng Zhou conceived, designed, and performed the experiments, analyzed the data, and co-wrote the article; Olivier Guillon discussed the results and co-wrote the article; and Robert Vaßen directed the experiments, analyzed the data, and co-wrote the article.

Conflicts of Interest: The authors declare no conflict of interest.

References

1. Vaßen, R.; Giesen, S.; Stöver, D. Lifetime of plasma-sprayed thermal barrier coatings: Comparison of numerical and experimental results. *J. Therm. Spray Technol.* **2009**, *18*, 835–845. [CrossRef]
2. Vassen, R.; Stuke, A.; Stöver, D. Recent developments in the field of thermal barrier coatings. *J. Therm. Spray Technol.* **2009**, *18*, 181–186. [CrossRef]
3. Heberlein, J.V.; Fauchais, P.; Boulos, M.I. *Thermal Spray Fundamental*; Springer: New York, NY, USA, 2014.
4. Cao, X.Q.; Vassen, R.; Stoever, D. Ceramic materials for thermal barrier coatings. *J. Eur. Ceram. Soc.* **2004**, *24*, 1–10. [CrossRef]
5. Sampath, S.; Schulz, U.; Jarligo, M.O.; Kuroda, S. Processing science of advanced thermal-barrier systems. *MRS Bull.* **2012**, *37*, 903–910. [CrossRef]
6. Schulz, U.; Saruhan, B.; Fritscher, K.; Leyens, C. Review on advanced EB-PVD ceramic topcoats for TBC applications. *Int. J. Appl. Ceram. Technol.* **2004**, *1*, 302–315. [CrossRef]
7. Mauer, G.; Schlegel, N.; Guignard, A.; Jarligo, M.; Rezanka, S.; Hospach, A.; Vassen, R. Plasma spraying of ceramics with particular difficulties in processing. *J. Therm. Spray Technol.* **2015**, *24*, 30–37. [CrossRef]
8. Nordhorn, C.; Mücke, R.; Mack, D.E.; Vassen, R. Probabilistic lifetime model for atmospherically plasma sprayed thermal barrier coating systems. *Mech. Mater.* **2016**, *93*, 199–208. [CrossRef]
9. Bernard, B.; Bianchi, L.; Malié, A.; Joulia, A.; Rémy, B. Columnar suspension plasma sprayed coating microstructural control for thermal barrier coating application. *J. Eur. Ceram. Soc.* **2015**, *36*, 1081–1089. [CrossRef]
10. Padture, N.P.; Gell, M.; Jordan, E.H. Materials science—Thermal barrier coatings for gas-turbine engine applications. *Science* **2002**, *296*, 280–284. [CrossRef] [PubMed]
11. Schlegel, N.; Ebert, S.; Mauer, G.; Vassen, R. Columnar-structured mg-al-spinel thermal barrier coatings (TBCs) by suspension plasma spraying (SPS). *J. Therm. Spray Technol.* **2015**, *24*, 144–151. [CrossRef]
12. Schlegel, N.; Sebold, D.; Sohn, Y.; Mauer, G.; Vassen, R. Cycling performance of a columnar-structured complex perovskite in a temperature gradient test. *J. Therm. Spray Technol.* **2015**, *24*, 1205–1212. [CrossRef]
13. Ganvir, A.; Curry, N.; Markocsan, N.; Nylén, P.; Toma, F.L. Comparative study of suspension plasma sprayed and suspension high velocity oxy-fuel sprayed YSZ thermal barrier coatings. *Surf. Coat. Technol.* **2015**, *268*, 70–76. [CrossRef]
14. VanEvery, K.; Krane, M.J.; Trice, R.W.; Wang, H.; Porter, W.; Besser, M.; Sordelet, D.; Ilavsky, J.; Almer, J. Column formation in suspension plasma-sprayed coatings and resultant thermal properties. *J. Therm. Spray Technol.* **2011**, *20*, 817–828. [CrossRef]
15. Delbos, C.; Fazilleau, J.; Rat, V.; Coudert, J.F.; Fauchais, P.; Pateyron, B. Phenomena involved in suspension plasma spraying part 2: Zirconia particle treatment and coating formation. *Plasma Chem. Plasma Process.* **2006**, *26*, 393–414. [CrossRef]
16. Ganvir, A.; Curry, N.; Markocsan, N.; Nylén, P.; Joshi, S.; Vilemova, M.; Pala, Z. Influence of microstructure on thermal properties of axial suspension plasma-sprayed YSZ thermal barrier coatings. *J. Therm. Spray Technol.* **2016**, *25*, 202–212. [CrossRef]
17. Curry, N.; Tang, Z.; Markocsan, N.; Nylén, P. Influence of bond coat surface roughness on the structure of axial suspension plasma spray thermal barrier coatings—Thermal and lifetime performance. *Surf. Coat. Technol.* **2015**, *268*, 15–23. [CrossRef]
18. Tang, Z.; Kim, H.; Yaroslavski, I.; Masindo, G.; Celler, Z.; Ellsworth, D. Novel Thermal Barrier Coatings Produced by Axial Suspension Plasma Spray. In Proceedings of the International Thermal Spray Conference, Hamburg, Germany, 27–29 September 2011.
19. Berghaus, J.O.; Legoux, J.-G.; Moreau, C.; Tarasi, F.; Chraska, T. Mechanical and thermal transport properties of suspension thermal-sprayed alumina-zirconia composite coatings. *J. Therm. Spray Technol.* **2008**, *17*, 91–104. [CrossRef]
20. Guignard, A.; Mauer, G.; Vaßen, R.; Stöver, D. Deposition and characteristics of submicrometer-structured thermal barrier coatings by suspension plasma spraying. *J. Therm. Spray Technol.* **2012**, *21*, 416–424. [CrossRef]
21. Sokołowski, P.; Kozerski, S.; Pawłowski, L.; Ambroziak, A. The key process parameters influencing formation of columnar microstructure in suspension plasma sprayed zirconia coatings. *Surf. Coat. Technol.* **2014**, *260*, 97–106. [CrossRef]

22. Joulia, A.; Duarte, W.; Goutier, S.; Vardelle, M.; Vardelle, A.; Rossignol, S. Tailoring the spray conditions for suspension plasma spraying. *J. Therm. Spray Technol.* **2015**, *24*, 24–29. [CrossRef]

23. Rajasekaran, B.; Mauer, G.; Vassen, R. Enhanced characteristics of hvof-sprayed mcraly bond coats for tbc applications. *J. Therm. Spray Technol.* **2011**, *20*, 1209–1216. [CrossRef]

24. Guignard, A. *Development of Thermal Spray Processes with Liquid Feedstocks*; Forschungszentrum Jülich: Jülich, Germany, 2012.

25. Schlegel, N. *Untersuchungen Zu Suspensionsplasmagespritzten Wärmedämmschichtsystemen*; Forschungszentrum, Zentralbibliothek: Jülich, Germany, 2016. (In Germany)

26. Andreola, F.; Leonelli, C.; Romagnoli, M.; Miselli, P. Techniques used to determine porosity. *Am. Ceram. Soc. Bull.* **2000**, *79*, 49–52.

27. Kong, M.; Bhattacharya, R.N.; James, C.; Basu, A. A statistical approach to estimate the 3d size distribution of spheres from 2d size distributions. *Geol. Soc. Am. Bull.* **2005**, *117*, 244–249. [CrossRef]

28. Young, D.S.; Sachais, B.S.; Jefferies, L.C. *The Rietveld Method*; Springer: Berlin, Germany, 1993.

29. Oliver, W.C.; Pharr, G.M. An improved technique for determining hardness and elastic modulus using load and displacement sensing indentation experiments. *J. Mater. Res.* **1992**, *7*, 1564–1583. [CrossRef]

30. Choi, S.R.; Zhu, D.; Miller, R.A. Mechanical properties/database of plasma-sprayed ZrO_2-8wt % Y_2O_3 thermal barrier coatings. *Int. J. Appl. Ceram. Technol.* **2004**, *1*, 330–342. [CrossRef]

31. Niihara, K.; Morena, R.; Hasselman, D. Evaluation of k lc of brittle solids by the indentation method with low crack-to-indent ratios. *J. Mater. Sci. Lett.* **1982**, *1*, 13–16. [CrossRef]

32. Sergejev, F.; Antonov, M. Comparative study on indentation fracture toughness measurements of cemented carbides. *Proc. Estonian Acad. Sci. Eng.* **2006**, *12*, 388–398.

33. Traeger, F.; Vaßen, R.; Rauwald, K.H.; Stöver, D. Thermal cycling setup for testing thermal barrier coatings. *Adv. Eng. Mater.* **2003**, *5*, 429–432. [CrossRef]

34. Karger, M.; Vaßen, R.; Stöver, D. Atmospheric plasma sprayed thermal barrier coatings with high segmentation crack densities: Spraying process, microstructure and thermal cycling behavior. *Surf. Coat. Technol.* **2011**, *206*, 16–23. [CrossRef]

35. Guo, H.; Vaßen, R.; Stöver, D. Atmospheric plasma sprayed thick thermal barrier coatings with high segmentation crack density. *Surf. Coat. Technol.* **2004**, *186*, 353–363. [CrossRef]

36. Fauchais, P.; Montavon, G. Thermal and Cold Spray: Recent Developments. In *Key Engineering Materials*; Trans Tech Publications: Zürich, Switzerland, 2008; pp. 1–59.

37. Moign, A.; Vardelle, A.; Themelis, N.; Legoux, J. Life cycle assessment of using powder and liquid precursors in plasma spraying: The case of yttria-stabilized zirconia. *Surf. Coat. Technol.* **2010**, *205*, 668–673. [CrossRef]

38. Vassen, R.; Cao, X.Q.; Tietz, F.; Basu, D.; Stover, D. Zirconates as new materials for thermal barrier coatings. *J. Am. Ceram. Soc.* **2000**, *83*, 2023–2028. [CrossRef]

39. Viazzi, C.; Bonino, J.P.; Ansart, F.; Barnabé, A. Structural study of metastable tetragonal ysz powders produced via a sol–gel route. *J. Alloy. Compd.* **2008**, *452*, 377–383. [CrossRef]

40. Beshish, G.; Florey, C.; Worzala, F.; Lenling, W. Fracture toughness of thermal spray ceramic coatings determined by the indentation technique. *J. Therm. Spray Technol.* **1993**, *2*, 35–38. [CrossRef]

41. Schulze, G.W.; Erdogan, F. Periodic cracking of elastic coatings. *Int. J. Solids Struct.* **1998**, *35*, 3615–3634. [CrossRef]

42. Vaßen, R.; Kerkhoff, G.; Stöver, D. Development of a micromechanical life prediction model for plasma sprayed thermal barrier coatings. *Mater. Sci. Eng. A* **2001**, *303*, 100–109. [CrossRef]

43. Munz, D.; Fett, T. *Ceramics: Mechanical Properties, Failure Behaviour, Materials Selection*; Springer Science & Business Media: Berlin, Germany, 2013.

Review

Sputtering Physical Vapour Deposition (PVD) Coatings: A Critical Review on Process Improvement and Market Trend Demands

Andresa Baptista [1,2,*], Francisco Silva [2], Jacobo Porteiro [1], José Míguez [1] and Gustavo Pinto [1,2]

[1] School of Industrial Engineering, University of Vigo, Lagoas, 36310-E Marcosende, Spain; porteiro@uvigo.es (J.P.); jmiguez@uvigo.es (J.M.); gflp@isep.ipp.pt (G.P.)

[2] School of Engineering, Polytechnic of Porto (ISEP), 4200-072 Porto, Portugal; fgs@isep.ipp.pt

* Correspondence: absa@isep.ipp.pt; Tel.: +351-228-340-500

Received: 15 September 2018; Accepted: 2 November 2018; Published: 14 November 2018

Abstract: Physical vapour deposition (PVD) is a well-known technology that is widely used for the deposition of thin films regarding many demands, namely tribological behaviour improvement, optical enhancement, visual/esthetic upgrading, and many other fields, with a wide range of applications already being perfectly established. Machining tools are, probably, one of the most common applications of this deposition technique, sometimes used together with chemical vapour deposition (CVD) in order to increase their lifespan, decreasing friction, and improving thermal properties. However, the CVD process is carried out at higher temperatures, inducing higher stresses in the coatings and substrate, being used essentially only when the required coating needs to be deposited using this process. In order to improve this technique, several studies have been carried out optimizing the PVD technique by increasing plasma ionization, decreasing dark areas (zones where there is no deposition into the reactor), improving targets use, enhancing atomic bombardment efficiency, or even increasing the deposition rate and optimizing the selection of gases. These studies reveal a huge potential in changing parameters to improve thin film quality, increasing as well the adhesion to the substrate. However, the process of improving energy efficiency regarding the industrial context has not been studied as deeply as required. This study aims to proceed to a review regarding the improvements already studied in order to optimize the sputtering PVD process, trying to relate these improvements with the industrial requirements as a function of product development and market demand.

Keywords: PVD optimization process; PVD technique; sputtering; magnetron sputtering; deposition improvement; reactors

---

1. Introduction

The physical vapour deposition (PVD) process has been known for over 100 years, and plasma-assisted PVD was patented about 80 years ago [1]. The term "physical vapour deposition" appeared only in the 60s. At that time, the evolution of vacuum coating processes was needed, which was carried out through the development of well-known technologies, such as sputtering, vacuum, plasma technology, magnetic fields, gas chemistry, thermal evaporation, bows, and power sources control, as described in detail in Powell's book [2].

In the last 30 years, plasma assisted PVD (PAPVD) was divided into several different power source technologies such as direct current (DC) diode, triode, radio-frequency (RF), pulsed plasma, ion beam assisted coatings, among others. In the beginning, the process had some difficulties in being understood at a fundamental level and necessary changes were introduced to provide benefits, such as excellent adhesion from the coating to the substrate, structure control, and material deposition at low temperatures [3].

On the other hand, many additional surface treatments have appeared to meet the industrial and commercial needs, developing the performance of a huge number of products. In the last decades, the development of PVD deposition technologies has been focused essentially on the coating of tools, considering the strong evolution of the computer numerical control (CNC) machining processes, since new machining approaches have arisen [4].

PVD technique is a thin film deposition process in which the coating grows on the substrate atom by atom. PVD entails the atomization or vaporization of material from a solid source, usually called target. Thin films usually have layers with thicknesses as thin as some atomic layers to films with several microns. This process causes a change in the properties of the surface and the transition zone between the substrate and the deposited material. On the other hand, the properties of the films can also be affected by the properties of the substrate. The atomic deposition process can be made in a vacuum, gaseous, plasma, or electrolytic environment. Moreover, the vacuum environment in the deposition chamber will reduce the gaseous contamination in the deposition process to a very low level [5].

The last decades showed an evolution of the PVD techniques, aiming to improve coating characteristics and deposition rates without putting aside initial surface cleaning to remove possible contaminations [6,7]. This technique has suffered relevant improvements, mainly in carbides and nanocomposite transition metal nitrides substrates [8–12]. Research has been focused on improving the characteristics of coatings, although the enhancement of the deposition rate effectiveness regarding this process has been the main concern of the industry linked to this kind of techniques [13–15].

The most common surface coating methods in a gaseous state regarding the PVD process are evaporation and sputtering. These techniques allow for particles to be extracted from the target at very low pressure to be conducted and deposited onto the substrate [16].

The reactor that was used in the evaporation process requires high-vacuum pressure values. Generally, these characteristics and parameters have lower atomic energy and less adsorption of gases into the coatings deposition. As a result, a transfer of particles with larger grains leads to a recognized lesser adhesion of the particles to the substrate, compared with the sputtering technique. During deposition, some contaminant particles are released from the melted coating material and moved onto the substrate, thereby reducing the purity of the obtained coatings. Thus, the evaporation process is usually used for thicker films and coatings with lower surface morphological requirements, although this technique presents higher deposition rates when compared with the sputtering process.

Therefore, the sputtering process appears as an alternative for applications that require greater morphological quality of surfaces where roughness, grain size, stoichiometry, and other requirements are more significant than the deposition rate. Due to the stresses generated during the cooling process with the decrease in temperature or the melting temperature of the substrate (polymers), the deposition process presents temperature limitations for certain applications [17–21]. This leads to the Sputtering process becoming more relevant among PVD deposition techniques without forgetting the appearance of new techniques based on the sputtering process to meet the continuous increase in market requirements.

New coating properties, following market and researchers' requirements, have been developed with the emergence of new systems based on conventional processes. Even though the deposition rates that were obtained by the evaporation process are the desired, the truth is that the sputtering deposition techniques made an unquestionable progress in terms of quality and increase in deposition rate, responding to industry and researchers demands interested in this area, even serving as interlayer for further coatings obtained by chemical vapour deposition (CVD) [22].

CVD is another method of deposition under vacuum and is the process of chemically reacting a volatile compound from a material to be deposited with other gases, in order to produce a non-volatile solid that is deposited onto a substrate. This method is sometimes used as pre-coating with the aim of increasing the durability of the substrates, decreasing the friction, and improving the thermal properties—this means that one can combine deposition methods, like layers of PVD and CVD, in the same coating [23–26].

There are also a large number of studies in mathematical modelling and numerical simulation that contribute to the improvement of this process, which may be an advantage over other processes. These studies have a great impact on the improvement of the reactors characteristics that lead in the future to the costs reduction, as well as in the improvement of the mechanical properties of the films [27–32].

This work has as main focus the magnetron sputtering technique since its development will be focused on the improvement of these specific reactors in the future.

2. PVD Coatings

PVD is an excellent vacuum coating process for the improvement of wear and corrosion resistance. It is highly required for functional applications, such as tools, decorative pieces, optical enhancement, moulds, dies, and blades. These are just a few examples of a wide range of already well-established applications [33–35]. The equipment used in this technique requires low maintenance and the process is environmentally friendly. Benefits of PVD coatings are many. PVD can provide real and unique advantages that add durability and value to products. Deposition techniques have an important role in machining processes. Machining tools are probably one of the most exigent applications, which require characteristics, such as hardness at elevated temperatures, high abrasion resistance, chemical stability, toughness, and stiffness [36–45]. In addition, PVD is also able to produce coatings with excellent adhesion, homogeneous layers, designed structures, graduated properties, controlled morphology, high diversity of materials and properties, among others [46–50].

PVD processes allow the deposition in mono-layered, multi-layered and multi-graduated coating systems, as well as special alloy composition and structures. Among other advantages of this process, the variation of coating characteristics continuously throughout the film is undoubtedly one of the most important [32,51,52]. Their flexibility and adaptability to market demands led to the development and the improvement of techniques for the various processes and thus multiple variants have arisen, some of them presented in Figure 1.

These techniques are constantly evolving and continue to be inspiration sources for many studies. Many books and articles spread out the information on these variants, making it difficult to quantify all existing techniques. Sputtering (or cathodic spraying) and Evaporation are the most commonly used PVD methods for thin film deposition.

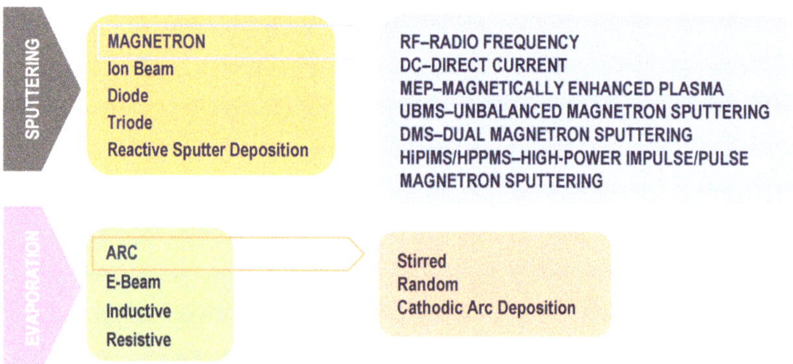

Figure 1. Segmentation of the current physical vapour deposition (PVD) techniques for advanced coatings.

2.1. Evaporation and Sputtering Principles

In PVD techniques, a thermal physical process of releasing or collision transforms the material to be deposited—the target—into atomic particles, which are directed to the substrates in conditions of gaseous plasma in a vacuum environment, generating a physical coating by condensation or the accumulation of projected atoms. A higher flexibility in the types of materials to be deposited and a better composition control of the deposited films are the results of this technique [21,53,54]. Two electrodes connected to a high voltage power supply and a vacuum chamber constitute the PVD reactors, as seen in Figure 2 [21,53,55].

Regarding the sputtering process, fine layers of several materials are applied while using the magnetron sputtering process. The raw material for this vacuum coating process takes the form of a target. A magnetron is placed near the target in sputtering processes. Then, in the vacuum chamber, an inert gas is introduced, which is accelerated by a high voltage being applied between the target and the substrate in the direction of the magnetron, producing the release of atomic size particles from the target. These particles are projected as a result of the kinetic energy transmitted by gas ions whose have reached the target going to the substrate and creating a solid thin film. The technology allows for previous contaminations located on the substrate to be cleaned from the surface—this is by reversing the voltage polarity between the substrate and the target, usually called cathodic cleaning [21].

When considering the technique of e-beam evaporation, this method involves purely physical processes, where the target acts as an evaporation source containing the material to be deposited, which works as a cathode. Note that the system evaporates any material as a function of the e-beam power. The material is heated at high vapour pressure by bombarding electrons in high vacuum, and the particles released. Then, a clashing occurs between the atomic size released particles and gas molecules—inserted into the reactor, with the aim of accelerating the particles, by creating a plasma. This plasma proceeds through the deposition chamber, being stronger in the middle position of the reactor. Successively compressed layers are deposited, increasing the adhesion of the deposited film to the substrate [17–21].

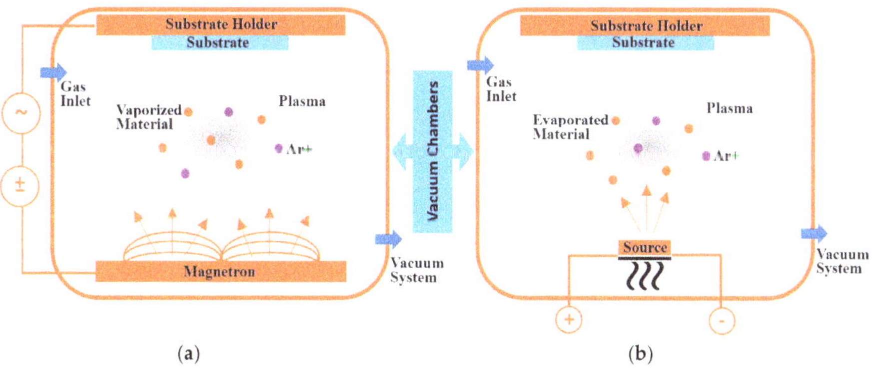

Figure 2. Schematic drawing of two conventional PVD processes: (**a**) sputtering and (**b**) evaporating using ionized Argon (Ar+) gas.

Being a cleaner deposition process, sputtering permits a better film densification, and reduces residual stresses on the substrate as deposition occurs at low or medium temperature [56–58]. Stress and deposition rate are also controlled by power and pressure. The use of targets with larger area facilitates a good uniformity, allowing the control of the thickness by an easy adjustment of the process parameters and deposition time. However, the process may cause some film contamination by the diffusion of evaporated impurities from the source, thus, there are still some limitations in the selection of the materials that were used for the coatings due to their melting temperature.

Furthermore, this process does not allow an accurate control of film thickness. However, it allows for high deposition rates without limit of thickness [59]. For better understanding, a comparison between evaporation and sputtering techniques is summarized in Table 1.

Table 1. Typical features of the PVD [17–21].

Parameters	Sputtering	Evaporation
Vacuum	Low	High
Deposition rate	Low (except for pure metals and dual magnetron)	High (up to 750,000 A min^{-1})
Adhesion	High	Low
Absorption	High	Less absorbed gas into the film
Deposited species energy	High (1–100 eV)	Low ($\sim$0.1–0.5 eV)
Homogeneous film	More	Less
Grain size	Smaller	Bigger
Atomized particles	More Dispersed	Highly directional

2.2. Sputtering Process Steps

To obtain better thin film deposition, it is important to know all process steps regarding the equipment related to the reactor, keeping in attention what takes place in the chamber during the deposition cycle. A preparation process before deposition is necessary, namely cleaning the substrate to achieve a better film adhesion between the coating and substrate. Nonetheless, cleaning the substrates in an ultrasonic bath, outside the vacuum chamber, is also suggested before the substrates are placed on the satellites [60]. An advantage of a vacuum sputtering chamber is the fact that it can be used both for cleaning the substrates, and afterwards, a coating deposition [21,54]. On the other hand, the duration of the cleaning process is considerable, being a disadvantage in terms of industry competition, as it raises final product costs. In order to control the costs, a management of the machine's breakdown times and setups is necessary. As this fact is a drawback to the industry, an optimization of the process' parameters is required to reduce production times. An important parameter to be optimized is the deposition rate, regarding an improvement in the plasma density and energy available in the process. Thus, it is necessary to take into account all of the steps of the process and parameters studied, in order to comply with industry demands [61]. Contamination of the films can be avoided with correct substrate handling and efficient maintenance of the whole vacuum system, as the contamination sources come from bad surface conditions or system related sources. The process cycle time depends mainly on the vacuum chamber size and its pumping system [17].

Following the placement of the substrates on the holders' vacuum chamber, the deposition process takes place regarding the following four important steps, featured in Figure 3:

- The first step—Ramp up—involves the preparation of the vacuum chamber, which consists in a gradual increase of the temperature, induced by a tubular heating and a modular control system; at the same time, the vacuum pumps are activated in order to decrease the pressure inside the chamber. In this type of sputtering reactors, two pumps are used, the first one (primary vacuum) produces a pressure up to 10^{-5} bar, the second one (high vacuum) reaches 10^{-7} bar pressure.
- The second step—Etching—is characterized by cathodic cleaning. The substrate is bombarded by ions from plasma etching to clean contaminations located on the substrate surface. This is an important preparation step for a deposition because it helps to increase adhesion. Indeed, the substrate properties have a direct influence on adhesion, such as substrate material, hardness and surface quality [62,63].
- In the third step—Coating—takes place. The material to be deposited is projected to the substrate surface. Several materials can be used; among these are titanium, zirconium, and chromium nitrides or oxides, among others.
- The last step—Ramp downstage—corresponds to the vacuum chamber returning to room temperature and ambient pressure. In order to achieve this, a specific cooling system is

used—chiller—with two sets of water knockout drums: one is used for the vacuum pumps and the other for cooling targets. Equipment unloading and cooling should not damage coatings' properties. The need for a cooling system is a drawback because it decreases production rate and rises energetic costs.

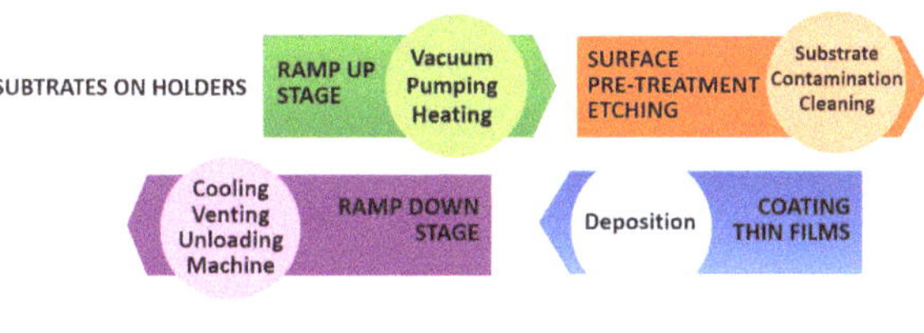

Figure 3. The processing flow for a classic PVD sputtering process.

A global industrial concern is energy consumption to help reduce costs [64]. New policies are expected to drive more innovation, encourage better industry performance, and lead to more energy savings.

The CVD process reveals a higher consumption compared to the PVD when considering all of the process steps. This has been demonstrated by several studies, such as sustainability assessments regarding manufacturing processes, energy consumption, and material flows in hard coating processes [65,66].

A comparative study issued by Gassner et al. [65], revealed a consumption of 112 kWh (process cycle time $\approx$ 5 h; coating thickness $\leq$ 6 µm) regarding the TiN deposition, using Magnetron Sputtering (MS) technique, which can be compared with the 974 kWh consumed in the TiCN/Al$_2$O$_3$ deposition using CVD technique, (process cycle time: 18 h 30 min; coating thickness $\leq$ 30 µm). PVD process, particularly the sputtering process, does not require very high temperatures, such as the high-temperature that was usually developed in the CVD process. Thus, in the CVD technique, the highest energy consumption is centred in the heating step, which is justified since the temperature parameters range between 750–1150 °C in the CVD case, and in the MS deposition, are usually done at lower temperatures, in the range of 350–600 °C. A possible way of reducing energy consumption costs is recovering the residual heat through heating exchange modules. Furthermore, in MS deposition processes, three-quarters of the total energy is usually consumed in the coating step. In order to increase energy coating efficiency, recycling target materials must be considered. Thus, it is possible to conclude that heating, etching, and refrigeration have a much lower contribution to energy consumption. Figure 4 compares the energy consumption for the PVD (using MS) and CVD during the deposition steps [65].

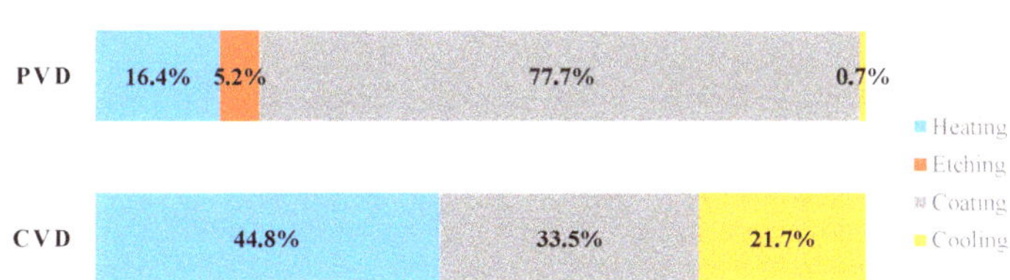

Figure 4. Energy consumption in the different steps of the PVD process: Heating, Etching, Coating, and Cooling. Energy consumption in the steps of the CVD process: Heating, Coating, and Cooling.

The way to sustainability and energy efficiency needs to be thought as an opportunity in reducing costs. The investment in terms of quality improvement of the coatings includes an increasing in adhesion between those and the corresponding substrates, extending their lifespan. This means that the reactor must be deeply studied in terms of parameters and external devices used to improve the deposition process. Beyond the solutions above referred, other possible solutions can be based on the improvement of the cleaning step within the reactor, minimizing the reactor occupation and saving costs in terms of energy and assistant gases. The reduction of dark areas into the reactor through the modelling of the plasma region would increase the useful volume of the chamber, contributing to a more homogeneous thickness of the coating in a large number of coated parts regarding each batch. Moreover, if the quality of the coatings is improved regarding its function, the lifespan of the coating will be enlarged, contributing by this way to energy savings, thus increasing its sustainability.

2.3. Deposition Process Influence Coatings Properties

In the last decades, one has observed an evolution in the approach of researchers regarding the impact of the deposition processes on coating properties.

The quality and variety of thin films has been a focus over the years and has progressed ever since. Currently, due to efficiency and optimization reasons of the industrial processes, new techniques and reactors have emerged with various combinations and possible derivations. However, these new techniques have a great impact on the influence of coating properties. The appearance and evolution of new simulations software also contribute positively to the continuous improvement need.

Having a focus on process improvement, it is important to know the parameters that can be adjusted during coating deposition and substrate cleaning. Some of these parameters can be the number of pumps, the number of targets, type of targets, substrate geometry, reactor occupancy rate, pressure, gas type, gas flow, temperature, current density, bias, among others.

However, parameters changes will have an impact on the film deposition rate and its adhesion. Consequently, one can have changes in the grain size and film thickness that will determine the coating characteristics, namely its hardness, Young's modulus, morphology, microstructure, and chemical composition [67].

The preparation of vacuum chambers for deposition is very important. The presence of oxygen in the vacuum chamber must be removed in order to ensure further vacuum and cleanliness conditions. Reactor cleanliness is also an important factor to keep the coatings free of impurities resulting from other previous materials used in the reactor. It is advised that the pressure must be maintained in the range of 101 and 104 Pa, being the last one the base pressure. Setup conditions will contribute to the creation of homogeneous plasma and an efficient cathodic cleaning. Etching process makes it easier to remove oxides and other contaminants from the substrate surface. The duration of the etching and bias sub-processes is also very relevant. A good plasma etching and excellent substrate surface cleanliness surely provide good adhesion [63]. In addition to adhesion, microstructural and mechanical properties, as well as corrosion properties of thin films have been studied. Gas flow and type are responsible for changes in the microstructural and mechanical properties. To improve the corrosion properties of materials, they have been developed new PVD coating techniques with magnesium alloys [34].

Effect of nitrogen-argon flow ratio on the microstructural and mechanical properties of AlSiN and AlCrSiN coatings that were prepared by high power impulse magnetron sputtering has been studied [68,69]. For AlSiN coatings, the N_2/Ar flow ratios from 5% to 50% had a strong impact on the results obtained. As a result, the hardness of the AlSiN coating increased with increasing nitrogen-argon flow ratio and reached a maximum value of 20.6 GPa [68]. In AlSiN coatings with an increase of N_2/Ar flow ratio, with nitrogen content in the range from 28.2% to 56.3%, prepared by varying the flow ratio from 1/4 to 1/1, resulted in higher hardness and better tribological behaviour with the contribution of the increasing crystallinity [69].

In recent years, in the high-power impulse magnetron sputtering (HiPIMS) process, reactive gases, such as oxygen or nitrogen, have been used. This technique is being applied to improve and adapt the properties of the growing films by their high fraction of ionized sputtered material during the process [70].

Regarding corrosion properties, studies show that this characteristic is tightly influenced by the deposition conditions and coating microstructure. However, it has less influence on the density of the defects [34]. The operating conditions have an effect on the homogeneity of the coating. Since these conditions interfere with the properties of the films, it is important to optimize the substrate and holders' rotation, the number of satellite holders, different initial positions of the substrate face, and take advantage of the chamber area and satellites occupancy space [60,71].

To conclude, the properties of the films are directly related to the deposition process. For this reason, it is unquestionable that progress continues to focus on solving problems that the industry is looking for, with a focus on improving reactors in terms of performance and film properties.

3. Sputtering Depositions Improvements

Process optimization and PVD technique improvements have been the focus of many studies, thus contributing to its success. Recently, the increase in plasma ionization has been the main goal of improvement, increasing the deposition rate. Other attentions are paid to decrease dark areas in the deposition chamber, recycle or improve targets, select gases, and increase atomic bombardment efficiency. The use of responsible energy practices has sensitized users of this technique, even though it is still an area to improve. Enlightening the energy efficiency of the whole process has an impact on costs, but also on the environment [68,72,73].

3.1. Reactors' Parameters and Characteristics

In general, the vacuum chambers that are applied to the coating of tools and components are constantly evolving. However, the industry already presents a wide range of solutions in this field [32,38,39]. The emergence and development of dedicated software that is easy and quick to use through remote control, have contributed to the technological evolution of PVD reactors [74–77]. Manual labour has been replaced by technology because the main purpose is to make the equipment more autonomous and automatic. This will reduce maintenance and management costs, and, on the other hand, increase production by making the investment more profitable.

One of the great advantages of this type of reactors and technology is its ability to deposit a wide range of films into parts with complex geometries of different materials, making the process quite flexible. Loading and unloading workpieces is a simple task since access to the coating areas is extremely easy. Currently, the characteristics of the reactors contribute to its handling. In summary, the main characteristics and parameters of the reactors can be seen in Figure 5.

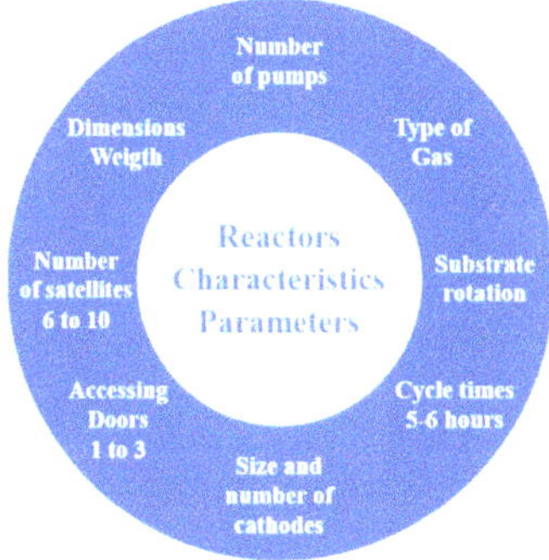

Figure 5. Characteristics and parameters chambers.

It is essential to highlight the importance of the useful diameter of the vacuum chambers, because this parameter will limit the working pressures, and, consequently, the substrates size. The chamber diameter can vary between 400 and 850 mm [74–77] but it can reach up to 2500 mm under customer request [74]. Cycle time is relatively fast, for example for a deposition of 3 μm, usually, takes between 5–6 h. Regarding the number of satellites, this can reach twenty, being more common the use of six or ten satellites.

The rotation systems of the substrates are very important in an industrial context. Its efficiency can reduce costs and improve film properties, with the rotation speed being determinant in the deposition sequence of the layer. Studies have shown that this effect is reflected in the morphologic properties of the coated substrates [60,71,78].

As described in Section 1, in the last years, new pulsed techniques with many potentialities have emerged. The technique with the greatest impact on its development, taking into account all sputtering techniques, is undoubtedly the one using Magnetron. However, it can also be used Diode, Triode, Ion Beam, and Reactive Sputtering systems. The studies on the evolution of the sputtering magnetron technique in DC and RF contribute to the emergence of techniques, such as dual magnetron sputtering (DMS), reactive bipolar pulsed dual magnetron sputtering (BPDMS), modulated pulsed power magnetron sputtering (MPPMS), HiPIMS, dual anode sputtering (DAS), among others.

One of the cheaper power supplies with easy process control is the DC power supply and hence is the most used although the sputter yield is generally much lower [79]. This makes the DC power source the most used in magnetron or pulsed systems. Its major disadvantage is the low rate of ionization. Studies show that only about a fraction of 1% of the target species is sprayed ionized [17]. The DC source only applies when the targets are made of conductive materials. On the other hand, the RF source is only applicable to the use of non-conductive or low conductivity targets. An alternative to using DC and RF sources is the Mid Frequency source (MF). To maintain the plasma in the sputtering process, an alternating high-frequency signal is applied, which allows the current to pass through the target, thus avoiding the accumulation of charges.

The dual magnetron sputtering (DMS) process uses MF power supply and it has been widely used for reactive deposition. It has become increasingly sophisticated, being usually used in systems for industrial applications using magnetron rotation [80,81]. In particular, this method is characterized by a different composition of the targets and way as the film grows. Surface oxidation is one of the sensitive aspects of this technique. To counter oxidation, it is necessary to take into account parameters, such as reactive gas partial pressure, voltage, and sputter rate [80]. Furthermore, in order to receive DC power and apply pulsed-DC power sources in the magnetrons, components need to be configured to switch power. This is possible while using a pulsed power supply [80].

DAS is a technology that allows switching from the commonly used alternating current—mid frequency (AC-MF) mode to a DC power process to reduce the heat load on the substrate.

The BPDMS technique is followed by the DMS technique. The interesting fact about this technique is that it also uses an MF source like DAS. Rizzo et al. [82] used in their study an MF band of 80–350 kHz. Using this technique, it was possible to prevent arc formation and its results showed a high deposition rate of around 0.044 μm min^{-1} using ZrN coating.

The deposition rate is always the focus of improvement when one thinks to upgrade a reactor for industrial purposes. The need of the industry thus obliges it, and, in that sense, in the last years, studies have been conducted also following industrial needs. Some recent investigations have been focused on the increase of spray ionization, on process stability, on new segmented targets, on gas flow optimization of different gases, on the bias influence, on obtaining better absorbers, among others. However, in order to obtain low-cost absorbers as compared to industrial techniques, a laboratory-tested sputtering unit was tested and the results pointed out that the deposition rates were low [83]. On the other hand, in studies regarding the gas flow in sprayed zirconia coatings on flat substrates, the deposition rate results reached 20 μm h^{-1}, which represents a good deposition rate. Other studies regarding the influence of bias voltage and gas flow showed that the temperature increase in the substrate and the application of a bias voltage resulted in a decreased deposition rate. For example, having a substrate temperature of 650 °C and applying a bias voltage at -10 V, it is possible to obtain a deposition rate of 20 μm h^{-1}. However, the deposition rate is reduced to 5 μm h^{-1} if -20 V bias voltage is applied [84,85].

Another approach in an industrial context using the reactor CemeCon® CC800/9 is the study of Weirather et al. [86]. They used the Reactive Pulsed DC magnetron sputtering technique with triangle-like segmented targets. That work contributed to show the potential of this technique in an industrial context, reducing the costs in thin film deposition. $Cr_{1-x}Al_xN$ ($0.21 \leq x \leq 0.74$) was used as a coating material, having obtained low friction values of 0.4 and wear coefficients up to 1.8×10^{-16} m^3 N m^{-1}, in order to obtain good results regarding the tribological properties. The maximum hardness obtained was 25.2 GPa, which proved to be a good result.

A study carried out regarding the plastics industry compared conventional DC, MF pulsed and HiPIMS techniques considering the deposition rate and coating's hardness. It is noteworthy that the complex geometry of the injection moulding tools was an additional challenge in this study, taking into account the three technologies that were used. For the three different technologies, five different targets configurations were used, varying the chemical composition of the $(Cr_{1-x}Al_x)N$ coatings. The HiPIMS technique provided the best results for aluminium deposition rate, which was reflected in an increase from 1.32 to 1.67 μm h^{-1}. In this case, the deposition rates of DC and MF coatings decrease from about 2.45 to 1.30 μm h^{-1}. On the other hand, chromium deposition rate presented the worst results for the HiPIMS technique as compared to DC and MF ones. The morphology, surface, and roughness that were obtained by the HiPIMS technique showed almost constant coatings behaviour [87].

HiPIMS technology allows for combining technologies, such as cathodic arc plasma deposition and ion plating, with this being its greatest advantage [88]. Although this type of reactor appeared in the 1990s, with the evolution of sputtering magnetron technology, just in recent years it has known more interest in its improvement and in exploring its potentialities. Since then, it has been used in the improvement of the spray ionization through the pulsed power that influences the plasma conditions and the coating's properties. When compared to conventional magnetron sputtering, the studies about this technique have shown significant improvements in coating structure, properties, and adhesion [89,90]. On the other hand, the combination of HiPIMS and DC-Pulsed also shows evidence in improving adhesion and morphology while using TiSiN coating [91]. Although versatile, it is necessary to have some care in the process and in the evaluation of results, given the difficulty in obtaining consistent and repeatable results [92].

One variation of the HiPIMS is the power pulses method MPPMS. This technique uses a pulsed high peak target power density for a short period of time and creates high-density plasma with an elevated degree of ionization of the sputtered species [93].

Deep oscillation magnetron sputtering (DOMS) is another variant of HiPIMS. A study that was carried out using this system led to seeking a relation between the ionization of the sputtered species and thin film properties [94]. This investigation had the purpose of identifying the mechanisms which influence the shadowing effect in this technique. To effectively reduce the atomic shadows, it was necessary to accelerate the chromium ions in the substrate sheath in the DOMS, which reduces significantly the high angle component of its collision. A high degree of ionization allows the deposition of dense and compact films without the need for the bombardment of high-energy particles during the coating growing process.

Plasma enhanced magnetron sputtering (PEMS) is an advanced version of conventional DC magnetron sputtering (DCMS). In conventional MS, the discharge plasma is generated in front of the magnetrons, as can be seen in Figure 6a. On the other hand, PEMS assisted deposition has the advantage of generating an independent plasma through impact ionization by accelerating electrons that are emitted from hot filaments in the chamber, which expands through the entire vacuum chamber, as shown in the illustration of Figure 6b. Lin et al. [95] carried out a comparative study between the techniques DCMS and HiPIMS with and without PEMS assistance regarding the deposition of

TiSiCN nanocomposite coatings, concluding that PEMS assistance improves the microstructure and mechanical properties of coatings that are produced by DOMS or DCMS, as well as the reduction of residual stress.

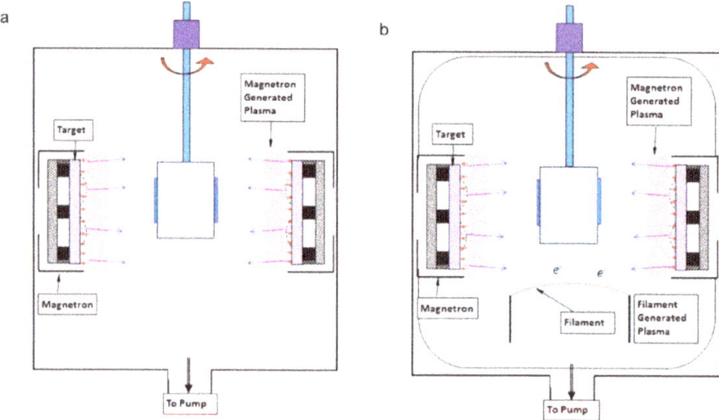

Figure 6. Schematic comparison between (**a**) conventional magnetron sputtering (MS), and the (**b**) plasma enhanced magnetron sputtering (PEMS) assisted process. Reproduced from [95] with permission. Copyright 2018 Elsevier.

The receptivity of the industry to the HiPIMS technique has been very positive bearing in mind the range of reactor power supply. Emerging technologies allow gains around the 30% in the ionization rate and higher charge states of the target ions. This high degree of ionization results in increased advantages of some coating properties, such as improved adhesion and the possibility of consistently covering surfaces with complex geometry [79,96]. Thus, the scientific community has focused on the development of high power magnetic pulsed technologies, since these results are very interesting concerning the industrial context.

3.2. Improvements and Applications of External Devices

Studies show a great interest in using the HiPIMS technique due to its versatility in the production of the PVD coating. This technique has as a disadvantage the deposition rate, which is lower when compared with the conventional sputtering DC. This factor needs to be improved. Some studies have been developed around this concern, trying to overcome the above-mentioned problems by the use of external devices, such as magnetic fields, although the improvements have not been significant yet.

In order to increase the deposition rate of thin films and improve the performance of HiPIMS, Li et al. [97] tested two different vacuum chamber approaches, using five different substrates positions: 0°, 45°, 90°, 135°, and 180°, based in the magnetron cathode in both studies. The first study was focused on the application of an external unbalanced magnetic field. This method indicates that, in the 0° angle substrate position, a substantially higher ion current in the substrate was reported. An increase in plasma density in the substrate region has occurred, showing that this method achieved the expected results. Following the first goal to increase the deposition rate, the second work focuses on more simplified and efficient ion discharge using external electric and magnetic fields with the auxiliary anode. To optimize the magnetic field distribution, the authors used a coaxial electromagnetic coil. This method allowed for a better distribution of the electric field and electric potential in the reactor, increased discharging, plasma intensification, and uni-directionality. The amplitude of the plasma density was five times greater in all positions when compared to the discharging without outer-field HiPIMS [98]. Figure 7 shows the vacuum system during HiPIMS discharge, measuring the ionic current of the substrate in different positions regarding both studies.

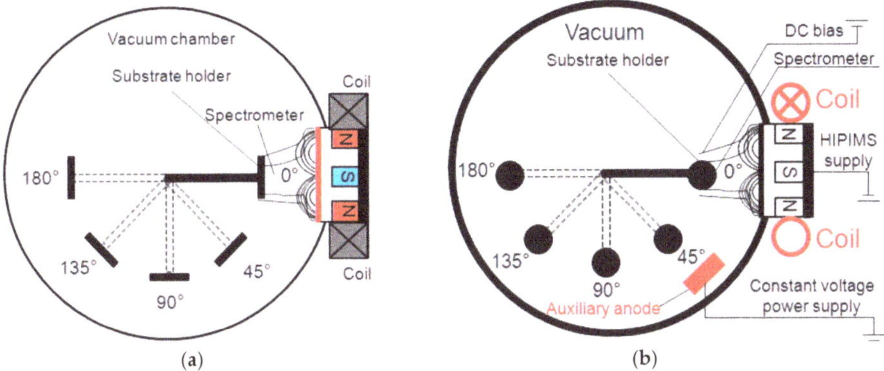

Figure 7. Vacuum system setup using (**a**) external unbalanced magnetic field, (**b**) external electric and magnetic fields, with the auxiliary anode. (**a**) Reproduced from [97] with permission. Copyright 2016 Elsevier. (**b**) Reproduced from [98] with permission. Copyright 2017 Elsevier.

Other examples using an auxiliary magnetic field as external device showed more ambitious results, presenting increased deposition rates between 40% and 140% when considering the inclusion of an external device with different types of targets that were chosen due to their relevance in technological applications. The results were compared with the HiPIMS process without the external device under similar experimental conditions (working gas pressure, average power). Figure 8 shows the configuration of the setup used. However, it is possible to improve the system, as described by the researchers that are involved in that work. It was shown a great potential for deposition improvement in HiPIMS through the control of the magnetic field and pulse configuration [99].

Using magnetic fields, Ganesan et al. [100] showed that it is possible to increase the deposition rate, guiding the ion flux in the direction of the substrate with the application of an external magnetic field using a solenoidal coil, excited with a DC current pulse. This is the scheme that is used in this study, as depicted in Figure 9; it is possible to see in the centre of the chamber the additional solenoidal coil that provides an external magnetic field.

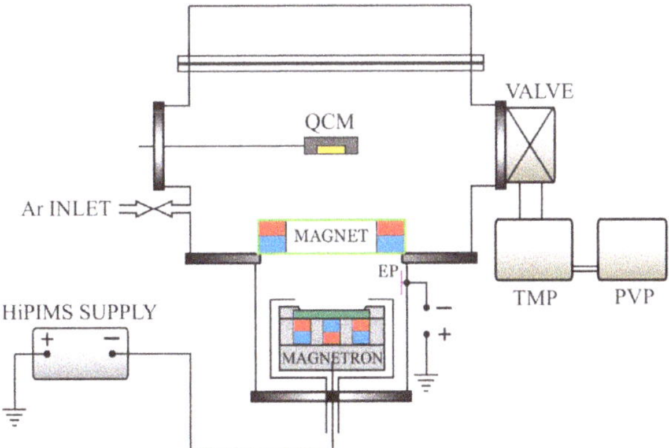

Figure 8. Schematic drawing of the experimental setup used in the work. Reproduced from [99] with permission. Copyright 2018 Elsevier.

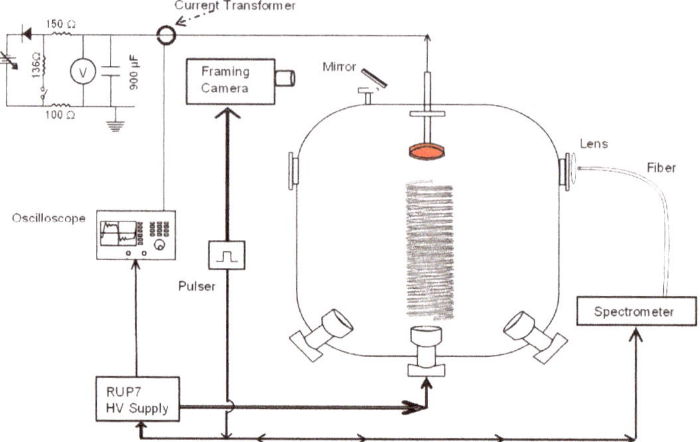

Figure 9. Experimental unit high power impulse magnetron sputtering (HiPIMS) deposition system showing the additional solenoidal coil. Reproduced from [100] with permission. Copyright 2019 Elsevier.

The study shows evidence of intensification in the ionization zone that increases the plasma extension and density, leading to an increase in the deposition rate through the combination of magnetic and reactor magnetron fields. This evidence can be interpreted in Figure 10, where (a) represents a conventional situation of deposition by the generation of transporting ions in neutral (N) and ionized fluxes (I) plasmas, those are deposited on the substrate. The same happens on (b), although applying the magnetic field. The ionization zone will be extended, activating additional ions that will be directed to the substrate, increasing the deposition rate. The results show that an increased maximum peak current and/or in the power density corresponds to a significant improvement in the pulverized ions flow. It has further been found that an increase of about 25% in peak current is seen when a 150 A magnetic field at the start of the HiPIMS pulse is used, inducing a 25% increase in the rate of the target ion emission as compared to the case where no external magnetic field is applied [100].

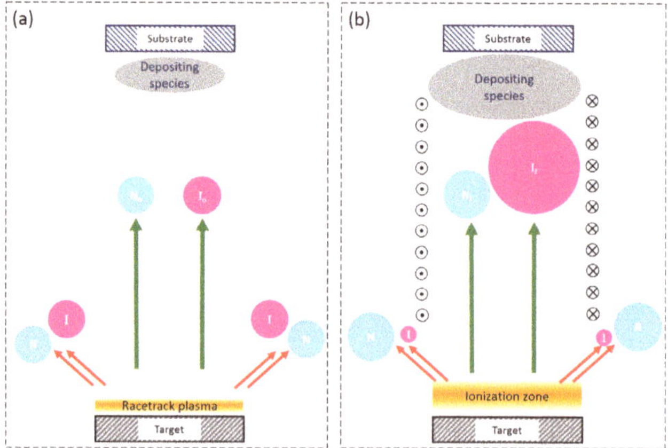

Figure 10. Schematic illustration of a film deposition, (**a**) without the external device and (**b**) with the application of an external magnetic field. Reproduced from [100] with permission. Copyright 2019 Elsevier.

In order to improve Cu films, Wu et al. [101] studied the utilization of a modified HiPIMS system using a positive kick voltage after an initial negative pulse, being possible to control the magnitude and the pulse width of the reverse pulse. This result is interpreted by a bipolar pulsed effect that was studied in detail in this type of deposition. Figure 11 shows the results that were obtained in the deposition of Cu films using three different kick pulses, comparing three different systems: DC magnetron sputtering (DCMS), conventional HiPIMS, and Bipolar Pulsed. It was found that the increase in the voltage amplitude and pulse width of the kick pulse can promote an increase of the deposition rate relatively to the conventional HiPIMS, but even so, the deposition rate that was achieved by the DCMS process showed to be higher. To conclude, the HiPIMS bipolar pulse shows great potential and this new approach can improve Cu film properties such as electronic conductivity and adhesion. However, in order to achieve deposition rates higher than DCMS, the substrates positioning needs to be planned in the centre of the reactor, where the deposition rate is more effective.

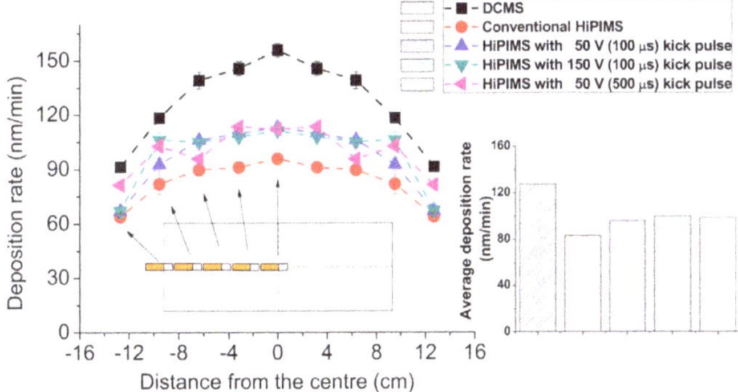

Figure 11. Results relatively to Cu films deposition rate. Average deposition rates for all the samples at different positions in the reactor. Reproduced from [101] with permission. Copyright 2018 Elsevier.

The combination of two or more vacuum chambers in improving the process and its efficiency should be considered. This approach can be seen in the recent work of Bras et al. [102], which simulates an industrial in-line vacuum production of solar cells using as a deposit compound the copper indium gallium selenide (CIGS). The use of the sputtering technique in the industrial context applied in the production of solar cells was demonstrated. In this study, automated arms were used to load and unload the cells. The system presents a process sequence using two vacuum chambers and 25 cathodic spray stations, which has its own heating and helium cooling arrangement. Following the simplified process in Figure 12, it can be seen: (a) in chamber A sputtering stations 1 to 5, where the substrate cleaning and absorber metal layers are carried out; (b) in the transition chamber A to B, the heating station increases the substrate temperature to improve deposition; (c) in chamber B, sputtering stations 6–18 promote the deposition of CIGS layers and the substrate is then rapidly cooled down in the intermediate chamber, and, finally, back to the chamber A; and, (d) sputtering stations 19–25 are producing the buffer layer in an oxygen-containing atmosphere. To finalize the cells, the addition of resistive transparent conductive oxide (TCP) bilayers is needed. The efficiency was demonstrated for cells with a total area of 1 cm^2 and 225 cm^2 with values of 15.1% and 13.2%, respectively.

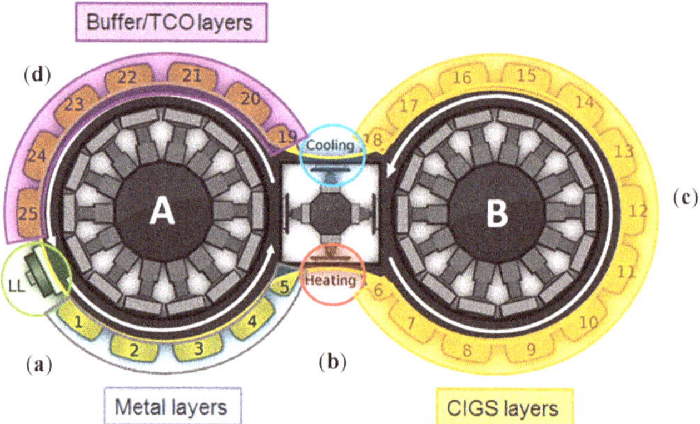

Figure 12. The schematic process sequence of the load-lock DUO solar cell manufacturing system of Midsummer®, (**a**) stations 1 to 5, (**b**) heating station, (**c**) stations 6 to 18 and (**d**) stations 19 to 25. Reproduced from [102] with permission. Copyright 2017 Elsevier.

3.3. Considerations Using CFD Simulation

To study the phenomena that occur in the PVD method, numerical simulation models are usually used. These simulation methods help to solve complex engineering problems in scientific and/or industrial contexts. The most common numerical approaches are finite elements methods (FEM) and computational fluid dynamics (CFD). FEM studies are commonly centred on mechanical properties and CFD is usually focused on process concerns. Initially, studies have been focused mainly on material properties but now the trend is to study the PVD reactor in an industrial context using the simulation, avoiding the cost of stopping the equipment [103,104]. However, the use of numerical methods allows obtaining only an approximate but not exact solution. It is also necessary to have a critical analysis of the results that were obtained through the models because their approximation can introduce errors. Comparing the results that were obtained by the models with experimental results is desirable [105]. FEM helps in studying the phenomenon related to the substrate and coating on their mechanical properties, such as strength, brittleness, adhesion, and performance, among others [31,32,106–108]. CFD is typically focused on the study of fluid flow dynamics, anodic chamber performance prediction, thermal evaluation in a reactor's design, input temperature, the velocity of distribution of the species into the reactor, pressure, and others [109–113]. The quality of the coatings that were obtained in commercial PVD processes is of great importance and therefore its optimization. Thus, it is necessary to take into account the discharge characteristics to know the motion of the neutral gas flow inside the reactor chamber.

Monte Carlo method (DSMC) models are a class of computational algorithms that provide approximate solutions and are widely used in research of thin films [105,114].

Bobzin et al. [28], used direct simulation Monte Carlo method (DSMC) models in CFD analysis to characterized and it assess gas behaviour in the PVD coating process using the Knudsen number (Kn) by means of different approaches: for Kn $\leq$ 0.1 the gas flow is described by the Navier–Stokes equations and for Kn > 0.1 a kinetic approach was used by the Boltzmann equation. In order to validate the model, they used an argon neutral gas flow and molecular nitrogen gas in an industrial scale reactor CemeCon 800/9® typically used for DC-MS and HiPIMS processes. Considering the developed CFD model, they conclude that it presents limitations in the transition flow regime. To accurately predict flow characteristics, only the kinetic model should be considered. The benefits of each model and the comparison between them were studied and showed that the advances in simulation lead to a detailed analysis on the PVD processes of the formation of coatings that are capable of complying with industrial requirements.

Kapopara et al. [29,30], predicted the gases concentration and distribution (argon and nitrogen), density profiles, velocity profiles, and pressure profiles across the sputtering chamber. They conclude that the locations of both gas inlet port and substrate have a crucial influence on the gas distribution inside the chamber. With this study, it was possible to propose a modification of the reactor geometry for a better gas flow over the substrate. This research showed that the CFD simulation has a great potential and its influence is growing over the time on the PVD reactors studies. After the modulation phase, it allows varying parameters, being a strong advantage in an industrial context due to the capacity of predicting the final results regarding the different phenomena that occur in the reactor during the deposition process. The main goal is a reduction in the production time, with a consequent reduction in costs, maintaining the quality standards.

Trieschmann [115] also used the DSMC to study neutral gas simulation on the influence of rotating spokes on gas rarefaction in HiPIMS. This different approach helps to understand the gas dynamics in the harsh discharge condition. It was concluded that the influence of a rotating plasma ionization zone is limited by a segmented time-modulated sputtering inlet distribution [115].

To conclude, the CFD modelling has been carried out to analyze gas flow and its mixing behaviour within the chamber reactor. However, for a better approximation of a real situation, it is important to use different models and compare them. The models defined and studied are important on the advance of geometry and parameters changes in the reactors that can be simulated, also taking into account external devices, in order to improve the process.

4. Concluding Remarks

PVD techniques are in constant evolution, accompanying the appearance of new technologies that are being adapted to the processes. They also meet the increasing demands of the industry. Furthermore, the focus of researchers in the last years has been on improving reactors and the application of external devices to the detriment of improving the properties of films, which has passed to the background, following the needs of industry.

Optimizing energy consumption of PVD processes is an opportunity for improvement. It is in the deposition step that this improvement can be reflected, since it is in this step of the process that PVD shows greater consumption. In the CVD process, it represents 33.5%, whereas in the PVD process, it represents 77.7% of consumption.

New opportunities in the development of techniques have contributed to the appearance of the MF power source that allowed the combination of DC and RF sources. The DC sources remain the most used type while the RF source is the least used. However, the combination of the sources in the DAS technique allowed for reducing the heat load on the substrate, thus improving the film properties, giving to this technique a huge yield for improvement.

External devices have emerged as a result of the PVD techniques enhancement. The HiPIMS method shows evidence of this application and improvement for high-performance thin films. This technology has had a good acceptance by the industry, which contributed to accelerating the researcher's interest. In addition, the good results obtained with the application of external devices has shown an increase in deposition rates due to plasma intensification. The coatings industry has evolved to the HiPIMS reactors due to the facts above-mentioned, and it is believable that this trend remains in the same way in the near future.

With technology evolution, simulations are currently a reality. Softwares, such as FEM and CFD, support this evolution in reducing production costs and adapting external devices. In addition, they respond to solve complex engineering problems in an industrial context. However, the use of CFD in solving coating problems can still grow in the light of its potential. New developments are expected as technology and software advances in deposition systems.

Author Contributions: Conceptualization, A.B. and F.S.; Methodology, A.B. and F.S.; Writing-Original Draft Preparation, A.B.; Writing-Review & Editing, F.S. and G.P.; Supervision, J.P., J.M. and F.S.

Funding: This research received no external funding.

Acknowledgments: Authors Andresa Baptista and Gustavo Pinto thank the support of CIDEM/ISEP.

Conflicts of Interest: The authors declare no conflict of interest.

Nomenclature

AC-MF	Alternating Current—Mid Frequency
BPDMS	Reactive Bipolar Pulsed Dual MS
CFD	Computational Fluid Dynamics
CIGS	Copper Indium Gallium Selenide
CNC	Computer numerical control
CVD	Chemical Vapour Deposition
DAS	Dual Anode Sputtering
DC	Direct Current
DCMS	Direct Current Magnetron Sputtering
DMS	Dual Magnetron Sputtering
DSMC	Direct Simulation Monte Carlo
E-Beam	Electron Beam Gun
FEM	Finite Elements Methods
HiPIMS	High Power Impulse Magnetron Sputtering
HPPMS	High-Power Pulsed Magnetron Sputtering
MEP	Magnetically Enhanced Plasma
MF	Mid Frequency
MPPMS	Modulated Pulsed Power MS
MS	Magnetron Sputtering
PAPVD	PVD Plasma Assisted
PEMS	Plasma enhanced magnetron sputtering
PVD	Physical Vapour Deposition
RF	Radio Frequency
TCP	Transparent Conductive Oxide
UBMS	Unbalanced Magnetron Sputtering

References

1. Berghaus, B. Improvements in and Related to the Coating of Articles by Means of Thermally Vapourized Material. UK Patent 510992, 1938.
2. Powell, C.F.; Oxley, J.H.; Blocher, J.M. *Vapour Deposition*; John Wiley & Sons: New York, NY, USA, 1966.
3. Holmberg, K.; Matthews, A. *Coatings Tribology: Properties, Techniques and Applications in Surface Engineering*; Elsevier Science: Amsterdam, The Netherlands, 1994; p. 440.
4. Tracton, A.A. *Coatings Technology: Fundamentals, Testing, and Processing Techniques*; CRC Press: Boca Raton, FL, USA, 2007; pp. 238–284.
5. Mattox, D.M. *Handbook of Physical Vapor Deposition (PVD) Processing Film Formation, Adhesion, Surface Preparation and Contamination Control*; Knovel: Norwich, NY, USA, 1998.
6. Silva, F.J.G.; Neto, M.A.; Fernandes, A.J.S.; Costa, F.M.; Oliveira, F.J.; Silva, R.F. Adhesion and wear behaviour of NCD coatings on Si_3N_4 by micro-abrasion tests. *J. Nanosci. Nanotechnol.* **2009**, *9*, 3938–3943. [CrossRef] [PubMed]
7. Martinho, R.P.; Silva, F.J.G.; Alexandre, R.J.D.; Baptista, A.P.M. TiB2 Nanostructured coating for GFRP injection moulds. *J. Nanosci. Nanotechnol.* **2011**, *11*, 5374–5382. [CrossRef] [PubMed]
8. Musil, J. Flexible hard nanocomposite coatings. *RSC Adv.* **2015**, *5*, 60482–60495. [CrossRef]
9. Veprek, S.; Veprek-Heijman, M.G.J.; Karvankova, P.; Prochazka, J. Different approaches to super hard coatings and nanocomposites. *Thin Solid Films* **2005**, *476*, 1–29. [CrossRef]

10. Yang, Q.; Zhao, L.R. Microstructure, mechanical and tribological properties of novel multi-component nanolayered nitride coatings. *Surf. Coat. Technol.* **2005**, *200*, 1709–1713. [CrossRef]

11. Martinho, R.P.; Andrade, M.F.C.; Silva, F.J.G.; Alexandre, R.J.D.; Baptista, A.P.M. Microabrasion wear behaviour of TiAlCrSiN nanostructured coatings. *Wear* **2009**, *267*, 1160–1165. [CrossRef]

12. Silva, F.J.G.; Martinho, R.P.; Baptista, A.P.M. Characterization of laboratory and industrial CrN/CrCN/diamond-like carbon coatings. *Thin Solid Films* **2014**, *550*, 278–284. [CrossRef]

13. Rubshtein, A.P.; Vladimirov, A.B.; Korkh, Y.V.; Ponosov, Y.S.; Plotnikov, S.A. The composition, structure and surface properties of the titanium-carbon coatings prepared by PVD technique. *Surf. Coat. Technol.* **2017**, *309*, 680–686. [CrossRef]

14. Silva, F.J.G.; Martinho, R.P.; Alexandre, R.; Baptista, A.M. Wear resistance of TiAlSiN thin coatings. *J. Nanosci. Nanotechnol.* **2012**, *12*, 9094–9101. [CrossRef] [PubMed]

15. Sam, Z. *Thin Films and Coatings: Toughening and Toughness Characterization*, 1st ed.; CRC Press: Boca Raton, FL, USA, 2015; pp. 377–463.

16. Mubarak, A.M.A.; Hamzah, E.H.E.; Tofr, M.T.M. Review of Physical Vapour Deposition (PVD) Techniques for Hard Coating. *Jurnal Mekanikal* **2005**, *20*, 42–51.

17. Mattox, D.M. *Handbook of Physical Vapor Deposition (PVD) Processing*; William Andrew: Amsterdam, The Netherlands, 2010; p. 792.

18. Holmberg, K.; Matthews, A. *Coatings Tribology: Properties, Mechanisms, Techniques and Applications in Surface Engineering*, 2nd ed.; Elsevier: Amsterdam, The Netherlands, 2009; p. 576.

19. Tracton, A.A. *Coatings Technology Handbook*, 3rd ed.; CRC Press: Boca Raton, FL, USA, 2006.

20. Martin, P.M. *Handbook of Deposition Technologies for Films and Coatings*, 3rd ed.; Elsevier: Amsterdam, The Netherlands, 2010.

21. Mattox, D.M. *The Foundations of Vacuum Coating Technology*; Noyes Publications: Norwich, UK, 2003.

22. Silva, F.J.G.; Baptista, A.P.M.; Pereira, E.; Teixeira, V.; Fan, Q.H.; Fernandes, A.J.S.; Costa, F.M. Microwave plasma chemical vapour deposition diamond nucleation on ferrous substrates with Ti and Cr interlayers. *Diam. Relat. Mater.* **2002**, *11*, 1617–1622. [CrossRef]

23. Silva, F.J.G. Nanoindentation on Tribological Coatings. In *Applied Nanoindentation in Advanced Materials*; Tiwari, A., Natarajan, S., Eds.; Wiley: New York, NY, USA, 2017; pp. 111–133.

24. Silva, F.J.G.; Fernandes, A.J.S.; Costa, F.M.; Teixeira, V.; Baptista, A.P.M.; Pereira, E. Tribological behavior of CVD diamond films on steel substrates. *Wear* **2003**, *255*, 846–853. [CrossRef]

25. Damm, D.D.; Contin, A.; Barbieri, F.C.; Trava-Airoldi, V.J.; Barquete, D.M.; Corat, E.J. Interlayers applied to CVD diamond deposition on steel substrate: A review. *Coatings* **2017**, *7*, 141. [CrossRef]

26. Trucchi, D.M.; Bellucci, A.; Girolami, M.; Mastellone, M.; Orlando, S. Surface texturing of CVD diamond assisted by ultra short laser pulses. *Coating* **2017**, *7*, 185. [CrossRef]

27. Voottipruex, P.; Bergado, D.T.; Lam, L.G.; Hino, T. Back-analyses of flow parameters of PVD improved soft Bangkok clay with and without vacuum preloading from settlement data and numerical simulations. *Geotext. Geomembr.* **2014**, *42*, 457–467. [CrossRef]

28. Bobzin, K.; Brinkmann, R.; Mussenbrock, T.; Bagcivan, N.; Brugnara, R.; Schäfer, M.; Trieschmann, J. Continuum and kinetic simulations of the neutral gas flow in an industrial physical vapor deposition reactor. *Surf. Coat. Technol.* **2013**, *237*, 176–181. [CrossRef]

29. Kapopara, J.; Mengar, A.; Chauhan, K.; Rawal, S. CFD Analysis of Sputtered TiN Coating. *Mater. Today Proc.* **2017**, *4*, 9390–9393. [CrossRef]

30. Kapopara, J.M.; Mengar, A.R.; Chauhan, K.V.; Patel, N.P.; Rawal, S.K. Modelling and analysis of sputter deposited ZrN coating by CFD. In Proceedings of the IConAMMA 2016 International Conference on Advances in Materials and Manufacturing Applications, Bangalore, India, 14–16 July 2016; p. 012205.

31. Skordaris, G.; Bouzakis, K.; Kotsanis, T.; Charalampous, P.; Bouzakis, E.; Breidenstein, B.; Bergmann, B.; Denkena, B. Effect of PVD film's residual stresses on their mechanical properties, brittleness, adhesion and cutting performance of coated tools. *CIRP J. Manuf. Sci. Technol.* **2017**, *18*, 145–151. [CrossRef]

32. Skordaris, G.; Bouzakis, K.; Kotsanis, T.; Charalampous, P.; Bouzakis, E.; Lemmer, O.; Bolz, S. Film thickness effect on mechanical properties and milling performance of nano-structured multilayer PVD coated tools. *Surf. Coat. Technol.* **2016**, *307*, 452–460. [CrossRef]

33. Fox-Rabinovich, G.; Paiva, J.M.; Gershman, I.; Aramesh, M.; Cavelli, D.; Yamamoto, K.; Dosbaeva, G.; Veldhuis, S. Control of self-organized criticality through adaptivebehavior of nano-structured thin film coatings. *Entropy* **2016**, *18*, 290. [CrossRef]

34. Hoche, H.; Groß, S.; Oechsner, M. Development of new PVD coatings for magnesium alloys with improved corrosion properties. *Surf. Coat. Technol.* **2014**, *259*, 102–108. [CrossRef]

35. Korhonena, H.; Syväluoto, A.; Leskinen, J.T.T.; Lappalainen, R. Optically transparent and durable Al_2O_3 coatings for harsh environments by ultra short pulsed laser deposition. *Opt. Laser Technol.* **2018**, *98*, 373–384. [CrossRef]

36. Inspektor, A.; Salvador, P.A. Architecture of PVD coatings for metalcutting applications: A review. *Surf. Coat. Technol.* **2014**, *257*, 138–153. [CrossRef]

37. Momeni, S.; Tillmann, W. Investigation of these lf-healing sliding wear characteristics of NiTi-based PVD coatings on tool steel. *Wear* **2016**, *368*, 53–59. [CrossRef]

38. Michailidis, N. Variations in the cutting performance of PVD-coated tools in milling Ti6Al4V, explained through temperature-dependent coating properties. *Surf. Coat. Technol.* **2016**, *304*, 325–329. [CrossRef]

39. Gouveia, R.M.; Silva, F.J.G.; Reis, P.; Baptista, A.P.M. Machining duplex stainless steel: Comparative study regarding end mill coated tools. *Coatings* **2016**, *6*, 51. [CrossRef]

40. Martinho, R.P.; Silva, F.J.G.; Baptista, A.P.M. Cutting forces and wear analysis of Si3N4 diamond coated tools in high speed machining. *Vacuum* **2008**, *82*, 1415–1420. [CrossRef]

41. Nunes, V.; Silva, F.J.G.; Andrade, M.F.; Alexandre, R.; Baptista, A.P.M. Increasing the lifespan of high-pressure die cast molds subjected to severe wear. *Surf. Coat. Technol.* **2017**, *332*, 319–331. [CrossRef]

42. Fernandes, L.; Silva, F.J.G.; Andrade, M.F.; Alexandre, R.; Baptista, A.P.M.; Rodrigues, C. Increasing the stamping tools lifespan by using Mo and B4C PVD coatings. *Surf. Coat. Technol.* **2017**, *325*, 107–119. [CrossRef]

43. Vereschaka, A.; Kataeva, E.; Sitnikov, N.; Aksenenko, A.; Oganyan, G.; Sotova, C. Influence of Thickness of Multilayered Nano-Structured Coatings Ti-TiN-(TiCrAl)N and Zr-ZrN-(ZrCrNbAl)N on Tool Life of Metal Cutting Tools at Various Cutting Speeds. *Coatings* **2018**, *8*, 44. [CrossRef]

44. Baptista, A.; Silva, F.J.G.; Pinto, G.; Porteiro, J.; Míguez, J.; Fernandes, L. On the physical vapour deposition (PVD) process evolution. *Procedia Manuf.* **2018**, in press.

45. Pinto, G.; Silva, F.J.G.; Baptista, A.; Porteiro, J.; Míguez, J.; Fernandes, L. A critical review on the numerical simulation related to physical vapour deposition. *Procedia Manuf.* **2018**, in press.

46. Maity, S. Optimization of processing parameters of in-situ polymerization of pyrrole on woollen textile to improve its thermal conductivity. *Prog. Org. Coat.* **2017**, *107*, 48–53. [CrossRef]

47. Kim, M.; Kim, S.; Kim, T.; Lee, D.K.; Seo, B.K.; Lim, C.S. Mechanical and thermal properties of epoxy composites containing zirconium oxide impregnated halloysite nanotubes. *Coatings* **2017**, *7*, 231. [CrossRef]

48. Hu, N.; Khan, M.; Wang, Y.; Song, X.; Lin, C.; Chang, C.; Zeng, Y. Effect of Microstructure on the Thermal Conductivity of Plasma Sprayed Y_2O_3 Stabilized Zirconia (8% YSZ). *Coatings* **2017**, *7*, 198. [CrossRef]

49. Silva, F.J.G.; Casais, R.C.B.; Martinho, R.P.A.; Baptista, P.M. Mechanical and tribological characterization of TiB2 thin films. *J. Nanosci. Nanotechnol.* **2012**, *12*, 9187–9194. [CrossRef] [PubMed]

50. Silva, F.J.G.; Martinho, R.P.; Andrade, M.; Baptista, A.P.M.; Alexandre, R. Improving the wear resistance of moulds for the injection of glass fibre–reinforced plastics using PVD coatings: A comparative study. *Coatings* **2017**, *7*, 28. [CrossRef]

51. Abdullah, M.Z.B.; Ahmad, M.A.; Abdullah, A.N.; Othman, M.H.; Hussain, P.; Zainuddin, A. Metal release of multilayer coatings by physical vapour deposition (PVD). *Procedia Eng.* **2016**, *148*, 254–260. [CrossRef]

52. Imbeni, V.; Martini, C.; Lanzoni, E.; Poli, G.; Hutchings, I.M. Tribological behaviour of multi-layered PVD nitride coatings. *Wear* **2001**, *251*, 997–1002. [CrossRef]

53. Wasa, K.; Kitabatake, M.; Adachi, H. *Thin Film Materials Technology: Sputtering of Compound Materials*; William Andrew: Amsterdam, The Netherlands, 2004.

54. Decher, G.; Schlenoff, J.B. *Multilayer Thin Films: Sequential Assembly of Nanocomposite Materials*; Wiley: Weinheim, Germany, 2003.

55. Silva, F.J.G. Estudo da Estrutura e Comportamento Tribológico de Revestimentos Duros e Ultra-Duros Executados no Vácuo. Master's Thesis, FEUP, Porto, Portugal, 2001. (In Portuguese)

56. Bass, R.B.; Lichtenberger, L.T.; Lichtenberger, A.W. Effects of Substrate Preparation on the Stress of Nb Thin Films. *IEEE Trans. Appl. Supercond.* **2003**, *13*, 3298–3300. [CrossRef]

57. Karabacak, T.; Senkevich, J.J.; Wang, G.C.; Lu, T.M. Stress reduction in sputter deposited films using nanostructured compliant layers by high working-gas pressures. *J. Vac. Sci. Technol. A* **2005**, *23*, 986–990. [CrossRef]

58. Estrada-Martínez, J.L.; Banda, J.A.M.; García, U.P.; Muñoz, J.R.; Muñoz, J.L.F.; Lira, M.M.; Ángel, O.Z. Set-up method on properties of $Ba_xSr_{1-x}TiO_3$ thin films deposited by RF-magnetron co-sputtering by projecting temperature and stoichiometric effect. *Preprints* **2017**, 2017100043. [CrossRef]

59. Plummer, J.D.; Deal, M.; Griffin, P.B. *Silicon VLSI Technology: Fundamentals, Practice, and Modeling*; Prentice Hall: Upper Saddle River, NJ, USA, 2000; p. 817.

60. Rother, B.; Ebersbach, G.; Gabriel, H.M. Substrate-rotation systems and productivity of industrial PVD processes. *Surf. Coat. Technol.* **1999**, *116–119*, 694–698. [CrossRef]

61. Martinho, R.P. Revestimentos PVD Mono e Multicamada Para Moldes Utilizados na Injecção de Plásticos Reforçados. Master's Thesis, FEUP, Porto, Portugal, 2009. (In Portuguese)

62. Steimann, P.A.; Hintermann, H.E. Adhesion of TiC and Ti(C, N) coatings on steel. *J. Vac. Sci. Technol. A* **1985**, *3*, 2394–2400. [CrossRef]

63. Barshilia, H.C.; Ananth, A.; Khan, J.; Srinivas, G. Ar + H_2 plasma etching for improved adhesion of PVD coatings on steel substrates. *Vacuum* **2012**, *86*, 1165–1173. [CrossRef]

64. Lente, H.V.; Til, J.I.V. Articulation of sustainability in the emerging field of nanocoatings. *J. Clean. Prod.* **2008**, *16*, 967–976. [CrossRef]

65. Gassner, M.; Figueiredo, M.R.; Schalk, N.; Franz, R.; Weiß, C.; Rudigier, H.; Holzschuh, H.; Bürgind, W.; Pohler, M.; Czettl, C.; Mitterer, C. Energy consumption and material fluxes in hard coating deposition processes. *Surf. Coat. Technol.* **2016**, *299*, 49–55. [CrossRef]

66. Klocke, F.; Döbbeler, B.; Binder, M.; Kramer, N.; Grüter, R.; Lung, D. Ecological evaluation of PVD and CVD coating systems in metal cutting processes. In Proceedings of the 11th Global Conference on Sustainable Manufacturing, Berlin, Germany, 23–25 September 2013.

67. Hogmark, S.; Jacobson, S.; Larsson, M. Design and evaluation of tribological coatings. *Wear* **2000**, *246*, 20–33. [CrossRef]

68. Jiang, X.; Yang, F.C.; Chenc, W.C.; Lee, J.W.; Chang, C.L. Effect of nitrogen-argon flow ratio on the microstructural and mechanical properties of AlSiN thin films prepared by high power impulse magnetron sputtering. *Surf. Coat. Technol.* **2017**, *320*, 138–145. [CrossRef]

69. Li, B.S.; Wang, T.G.; Ding, J.; Cai, Y.; Shi, J.; Zhang, X. Influence of N_2/Ar Flow Ratio on Microstructure and Properties of the AlCrSiN Coatings Deposited by High-Power Impulse Magnetron Sputtering. *Coatings* **2018**, *8*, 3. [CrossRef]

70. Alamia, J.; Bolz, S.; Sarakinos, K. High power pulsed magnetron sputtering: Fundamentals and applications. *J. Alloys Compd.* **2009**, *483*, 530–534. [CrossRef]

71. Rother, B.; Jehn, H.A.; Gabriel, H.M. Multilayer hard coatings by coordinated substrate rotation modes in industrial PVD deposition systems. *Surf. Coat. Technol.* **1996**, *86–87*, 207–211. [CrossRef]

72. Bobzin, K.; Brögelmann, T.; Kalscheuer, C.; Liang, T. High-rate deposition of thick (Cr,Al) ON coatings by high speed physical vapor deposition. *Surf. Coat. Technol.* **2017**, *322*, 152–162. [CrossRef]

73. Liu, M.J.; Zhang, M.; Zhang, Q.; Yang, G.J.; Li, C.X.; Li, C.J. Gaseous material capacity of open plasma jet in plasma spray-physical vapor deposition process. *Appl. Surf. Sci.* **2018**, *428*, 877–884. [CrossRef]

74. Kolzer Vacuum Coating Equipment. Available online: http://www.kolzer.com (accessed on 15 July 2018).

75. CemeCon—The Tool Coating. Available online: http://www.cemecon.de/en (accessed on 15 July 2018).

76. High Performance Coatings for Tools and Components Oerlikon Balzers. Available online: https://www.oerlikon.com (accessed on 15 July 2018).

77. Teer Coatings. Available online: http://www.teercoatings.co.uk/index.php?page=13 (accessed on 4 October 2018).

78. Panjana, M.; Cekada, M.; Panjan, P.; Zalar, A.; Peterman, T. Sputtering simulation of multilayer coatings in industrial PVD system with three-fold rotation. *Vacuum* **2008**, *82*, 158–161. [CrossRef]

79. Mitterer, C. PVD and CVD hard coatings. In *Comprehensive Hard Materials*; Sarin, V.K., Ed.; Elsevier: Amsterdam, The Netherlands, 2014; Volume 2, pp. 449–467. [CrossRef]

80. Linss, V. Comparison of the large-area reactive sputter processes of ZnO: Al and ITO using industrial size rotatable targets. *Surf. Coat. Technol.* **2016**, *290*, 43–57. [CrossRef]

81. Linss, V. Challenges in the industrial deposition of transparent conductive oxide materials by reactive magnetron sputtering from rotatable targets. *Thin Solid Films* **2017**, *634*, 149–154. [CrossRef]

82. Rizzo, A.; Valerini, D.; Capodieci, L.; Mirenghi, L.; Benedetto, F.D.; Protopapa, M.L. Reactive bipolar pulsed dual magnetron sputtering of ZrN films: The effect of duty cycle. *Appl. Surf. Sci.* **2018**, *427*, 994–1002. [CrossRef]

83. Wackelgard, E.; Hultmark, G. Industrially sputtered solar absorber surface. *Sol. Energy Mater. Sol. Cells* **1998**, *54*, 165–170. [CrossRef]

84. Rösemann, N.; Ortner, K.; Petersen, J.; Bäke, M.; Bräuer, G.; Rösler, J. Microstructure of gas flow sputtered thermal barrier coatings: Influence of bias voltage. *Surf. Coat. Technol.* **2017**, *332*, 22–29. [CrossRef]

85. Rösemann, N.; Ortner, K.; Petersen, J.; Schadow, T.; Bäker, M.; Bräuer, G.; Rösler, J. Influence of bias voltage and oxygen flow rate on morphology and crystallographic properties of gas flow sputtered zirconia coatings. *Surf. Coat. Technol.* **2015**, *276*, 668–676. [CrossRef]

86. Weirather, T.; Czettl, C.; Polcik, P.; Kathrein, M.; Mitterer, C. Industrial-scale sputter deposition of $Cr_{1-x}Al_xN$ coatings with $0.21 \leq x \leq 0.74$ from segmented targets. *Surf. Coat. Technol.* **2013**, *232*, 303–310. [CrossRef]

87. Bagcivan, N.; Bobzin, K.; Theiß, S. $(Cr_{1-x}Al_x)N$: A comparison of direct current, middle frequency pulsed and high power pulsed magnetron sputtering for injection molding components. *Thin Solid Films* **2013**, *528*, 180–186. [CrossRef]

88. Anders, A. A review comparing cathodic arcs and high power impulse magnetron sputtering (HiPIMS). *Surf. Coat. Technol.* **2014**, *257*, 308–325. [CrossRef]

89. Gudmundsson, J.T.; Brenning, N.; Lundin, D.; Helmersson, U. High power impulse magnetron sputtering discharge. *J. Vac. Sci. Technol. A* **2012**, *30*, 030801. [CrossRef]

90. Lin, J.L.; Sproul, W.D.; Moore, J.J.; Wu, Z.L.; Lee, S.; Chistyakov, R.; Abraham, B. Recent advances in modulated pulsed power magnetron sputtering for surface engineering. *JOM* **2011**, *63*, 48–58. [CrossRef]

91. Yazdi, M.; Lomello, F.; Wang, J.; Sanchette, F.; Dong, Z.; White, T.; Wouters, Y.; Schuster, F.; Billard, A. Properties of TiSiN coatings deposited by hybrid HiPIMS and pulsed-DC magnetron co-sputtering. *Vacuum* **2014**, *109*, 43–51. [CrossRef]

92. Alami, J.; Maric, Z.; Busch, H.; Klein, F.; Grabowy, U.; Kopnarski, M. Enhanced ionization sputtering: A concept for superior industrial coatings. *Surf. Coat. Technol.* **2014**, *255*, 43–51. [CrossRef]

93. Wu, Z.L.; Li, Y.G.; Wu, B.; Lei, M.K. Effect of microstructure on mechanical and tribological properties of TiAlSiN nanocomposite coatings deposited by modulated pulsed power magnetron sputtering. *Thin Solid Films* **2015**, *597*, 197–205. [CrossRef]

94. Oliveira, J.C.; Ferreira, F.; Anders, A.; Cavaleiro, A. Reduced atomic shadowing in HiPIMS: Role of the thermalized metal ions. *Appl. Surf. Sci.* **2018**, *433*, 934–944. [CrossRef]

95. Lin, J.; Wei, R. A comparative study of thick TiSiCN nanocomposite coatings deposited by dcMS and HiPIMS with and without PEMS assistance. *Surf. Coat. Technol.* **2018**, *338*, 84–95. [CrossRef]

96. Bandorf, R.; Sittinger, V.; Bräuer, G. High Power Impulse Magnetron Sputtering–HIPIMS. In *Comprehensive Materials Processing*, 1st ed.; Hashmi, S., Ed.; Elsevier: Amsterdam, The Netherlands, 2014; pp. 75–99. [CrossRef]

97. Li, C.; Tian, X.; Gong, C.; Xu, J. The improvement of high power impulse magnetron sputtering performance by an external unbalanced magnetic field. *Vacuum* **2016**, *133*, 98–104. [CrossRef]

98. Li, C.; Tian, X.; Gong, C.; Xu, J.; Liu, S. Synergistic enhancement effect between external electric and magnetic fields during high power impulse magnetron sputtering discharge. *Vacuum* **2017**, *143*, 119–128. [CrossRef]

99. Tiron, V.; Velicu, I.L.; Mihăilă, I.; Popa, G. Deposition rate enhancement in HiPIMS through the control of magnetic field and pulse configuration. *Surf. Coat. Technol.* **2018**, *337*, 484–491. [CrossRef]

100. Ganesan, R.; Akhavan, B.; Dong, X.; McKenzie, D.R.; Bilek, M.M.M. External magnetic field increases both plasma generation and deposition rate in HiPIMS. *Surf. Coat. Technol.* **2019**, *352*, 671–679. [CrossRef]

101. Wu, B.; Haehnlein, I.; Shchelkanov, I.; Lain, J.M.; Patel, D.; Uhlig, J.; Jurczyk, B.; Leng, Y.; Ruzic, D.N. Cu films prepared by bipolar pulsed high power impulse magnetron sputtering. *Vacuum* **2018**, *150*, 216–221. [CrossRef]

102. Bras, P.; Frisk, C.; Tempez, A.; Niemi, E.; Björkman, C.P.; Brasa, P.; Frisk, C.; Tempez, A.; Niemi, E.; Platzer-Björkman, C. Ga-grading and Solar Cell Capacitance Simulation of an industrial $Cu(In,Ga)Se_2$ solar cell produced by an in-line. *Vacuum* **2017**, *636*, 367–374. [CrossRef]

103. Tapia, E.; Iranzo, A.; Pino, F.; Rosa, F.; Salva, J. Methodology for thermal design of solar tubular reactors using CFD techniques. *Int. J. Hydrog. Energy* **2016**, *41*, 19525–19538. [CrossRef]

104. Abdel-Fattah, A.; Fateen, S.; Moustafa, T.; Fouad, M. Three-dimensional CFD simulation of industrial Claus reactors. *Chem. Eng. Res. Des.* **2016**, *112*, 78–87. [CrossRef]

105. Bouaouina, B.; Mastail, C.; Besnard, A.; Mareus, R.; Nita, F.; Michel, A.; Abadias, G. Nanocolumnar TiN thin film growth by oblique angle sputter-deposition: Experiments vs. simulations. *Mater. Des.* **2018**, *160*, 338–349. [CrossRef]

106. Sliwa, A.; Mikuła, J.; Gołombek, K.; Tanski, T.; Kwasny, W.; Bonek, M.; Brytan, Z. Prediction of the properties of PVD/CVD coatings with the use of FEM analysis. *Appl. Surf. Sci.* **2016**, *388*, 281–287. [CrossRef]

107. Venkateswara Raoa, R. The Significant Application of FEM to Evaluate the Mechanical Properties of Thin Films. *Procedia Mater. Sci.* **2014**, *6*, 1260–1265. [CrossRef]

108. Bouzakis, K.; Skordaris, G.; Klocke, F.; Bouzakis, E. A FEM-based analytical–experimental method for determining strength properties gradation in coatings after micro-blasting. *Surf. Coat. Technol.* **2009**, *203*, 2946–2953. [CrossRef]

109. Prades, L.; Dorado, A.; Climent, J.; Guimerà, X.; Chiva, S.; Gamisans, X. CFD modeling of a fixed-bed biofilm reactor coupling hydrodynamics and biokinetics. *Chem. Eng. J.* **2017**, *313*, 680–692. [CrossRef]

110. Phuan, Y.; Ismail, H.; Garcia-Segura, S.; Chong, M. Design and CFD modelling of the anodic chamber of a continuous PhotoFuelCell reactor for water treatment. *Process Saf. Environ. Prot.* **2017**, *111*, 449–461. [CrossRef]

111. Adebiyi, D.; Popoola, A.; Botef, I. Experimental Verification of Statistically Optimized Parameters for Low-Pressure Cold Spray Coating of Titanium. *Metals* **2016**, *6*, 135. [CrossRef]

112. Silva, A.; Monteiro, C.; Souza, V.; Ferreira, A.; Jaimes, R.; Fontoura, D.; Nunhez, J. Fluid dynamics and reaction assessment of diesel oil hydrotreating reactors via CFD. *Fuel Process. Technol.* **2017**, *166*, 17–29. [CrossRef]

113. Sen, S.; Lake, M.; Kroppen, N.; Farber, P.; Wilden, J.; Schaaf, Ç. Self-propagating exothermic reaction analysis in Ti/Al reactive films using experiments and computational fluid dynamics simulation. *Appl. Surf. Sci.* **2017**, *396*, 1490–1498. [CrossRef]

114. Depla, D.; Leroy, W.P. Magnetron sputter deposition as visualized by Monte Carlo modeling. *Thin Solid Films* **2012**, *520*, 6337–6354. [CrossRef]

115. Trieschmann, J. Neutral gas simulation on the influence of rotating spokes on gas rare faction in high-power impulse magnetron sputtering. *Contrib. Plasma Phys.* **2017**, *58*, 394–403. [CrossRef]

Article

Combination of Electrodeposition and Transfer Processes for Flexible Thin-Film Thermoelectric Generators

Hiroki Yamamuro, Naoki Hatsuta, Makoto Wachi, Yoshihiro Takei and Masayuki Takashiri *

Department of Materials Science, Tokai University, Hiratsuka 259-1292, Japan; dakyrasit103@gmail.com (H.Y.);
www.cp_877-fayu.ett-n0want@ezweb.ne.jp (N.H.); km47416a@ja2.so-net.ne.jp (M.W.);
bboy_downy.c2oenvy_us11kb.feel@ezweb.ne.jp (Y.T.)
* Correspondence: takashiri@tokai-u.jp; Tel.: +81-463-58-1211

Received: 19 October 2017; Accepted: 21 December 2017; Published: 3 January 2018

Abstract: To reduce consumption for ambient assisted living (AAL) applications, we propose the design and fabrication of flexible thin-film thermoelectric generators at a low manufacturing cost. The generators were fabricated using a combination of electrodeposition and transfer processes. N-type Bi_2Te_3 films and p-type Sb_2Te_3 films were formed on a stainless-steel substrate employing potentiostatic electrodeposition using a nitric acid-based bath, followed by a transfer process. Three types of flexible thin-film thermoelectric generators were fabricated. The open circuit voltage (V_{oc}) and maximum output power (P_{max}) were measured by applying a temperature difference between the ends of the generator. The thin-film generators obtained using thermoplastic sheets with epoxy resin exhibited a V_{oc} that was tens of millivolts. In particular, the contact resistance of the thin-film generator decreased when silver paste was inserted at the junctions between the n- and p-type films. The most flexible thin-film generator fabricated in this study exhibited a P_{max} of 10.4 nW at a temperature difference of 60 K. The current performance of the generators was too low, but we innovated a combination process to prepare them. It is expected to increase the performance by further decreasing the micro-cracks and contact resistance in the generators.

Keywords: electrodeposition; transfer process; flexible thin films; thermoelectric generators

---

1. Introduction

Environmental energy harvesting has recently emerged as a viable technique to supplement battery supplies in energy-constrained embedded systems. The technologies used for energy harvesting have been studied in terms of ambient heat, airflow, and vibration, among others [1–3]. In particular, heat energy exists in all places and can be transferred to electric energy using thermoelectric generators [4–6]. The produced electric energy is used by low-consumption electronics for ambient assisted living (AAL) applications such as biosensors [7,8]. However, employing thermoelectric generators for AAL applications requires that miniaturized generators be produced at a low manufacturing cost while still maintaining thermoelectric performance as high as that achieved with large-scale generators.

Flexible thermoelectric generators fabricated using thin-film technology might satisfy the aforementioned requirements [9–11]. To date, many thin-film thermoelectric devices have been fabricated using various film deposition methods such as flash evaporation [12–14], sputtering [15–17], and electrodeposition [18,19]. In addition, thin-film thermoelectric materials possess favorable features not exhibited by bulk materials. The presence of nanostructured materials, including superlattices [20–22], nanocrystals [23–25], nanoporous structures [26–28], and inducing stresses [29–31], enhances thermoelectric performance. Thermoelectric performance is defined as the figure of merit, $ZT = S^2\sigma T/\kappa$, where S is the Seebeck coefficient, σ is electrical conductivity, T is the absolute temperature, and κ is thermal conductivity.

Among thin-film deposition methods, electrodeposition is most favorable for reducing the manufacturing cost since high deposition rates can be achieved and since the use of a vacuum system and a large power supply is not necessary [32–34]. However, a drawback of electrodeposition is that thin films can only be deposited on conductive substrates, making it difficult to fabricate working thermoelectric generators because the generated electric current mainly passes through conductive substrates. Therefore, to produce thin-film thermoelectric generators at a lower manufacturing cost, a combined method employing electrodeposition and transfer processes is the most feasible approach [35].

In this study, three types of flexible thin-film thermoelectric generators were fabricated using the combination method with different adhesive insulating sheets and varying the connecting approaches between the films. The n- and p-type thin films were obtained by potentiostatic electrodeposition nitric acid based bathes, followed by the transfer process. We used bismuth telluride (Bi_2Te_3) and antimony telluride (Sb_2Te_3) as the n- and p-type thermoelectric materials, respectively, because these materials exhibit a higher figure of merit near room temperature (RT) and similar thermal expansion rates, leading to the manufacture of devices with a high durability [36]. The in-plane thermoelectric properties of the n- and p-type thin films were evaluated at RT. The open circuit voltage (V_{oc}) and maximum output power (P_{max}) were measured by applying a temperature difference between the ends of the generator.

2. Experimental Section

Prior to the fabrication of flexible thin-film thermoelectric generators, we prepared n-type Bi_2Te_3 and p-type Sb_2Te_3 thin films using electrodeposition and measured their thermoelectric properties. The Bi_2Te_3 and Sb_2Te_3 thin films were prepared by potentiostatic electrodeposition using a standard three-electrode cell. The basic setup used for the electrodeposition of both thin films has been described in our previous reports [19,37,38]. A stainless-steel substrate with a thickness of 80 μm was chosen as the working electrode (electrode area: 1.5 cm^2) because of its excellent corrosion resistance. A platinum-coated titanium mesh on a titanium plate was used as the counter electrode (electrode area: 1.5 cm^2). An Ag/AgCl (saturated KCl) electrode was used as the reference electrode. Prior to use, the working and counter electrodes were degreased with sodium hydroxide and hydrochloric acid solutions and washed with deionized water. A potentiostat/galvanostat (Hokuto Denko, Tokyo, Japan, HA-151B) was used to control the voltage at 0.1 V. The film deposition was implemented without stirring at RT. For fabricating the Bi_2Te_3 thin films, nitric acid (0.4 M; diluted with deionized water) containing 2.0 mM $Bi(NO_3)_3$ (99.9%, Kojundo Chemical Laboratory, Sakado, Japan) and 1.3 mM TeO_2 (99.9%, Kojundo Chemical Laboratory, Sakado, Japan) was used as the electrolyte solution. The initial pH was approximately 1.0. For preparing the Sb_2Te_3 thin films, nitric acid (0.4 M; diluted with deionized water) containing 1.3 mM SbF_3 (99.9%, Kojundo Chemical Laboratory, Sakado, Japan) and 1.6 mM TeO_2 (99.9%, Kojundo Chemical Laboratory, Sakado, Japan) was used as the electrolyte solution. The initial pH was approximately 1.0.

After completing electrodeposition, to avoid any complications arising from electrical conduction occurring through the stainless-steel substrate, the film was fixed on a glass plate using epoxy resin, and the thin film was subsequently removed from the stainless-steel substrate. The in-plane Seebeck coefficient, S, of the thin films was measured at RT with an accuracy of ±10%. The Seebeck coefficient was measured by connecting one side of the film to a heater and the other side to a heat sink kept at RT, with a temperature difference of <4 K between both the sides. The in-plane electrical conductivity, σ, was measured at RT using the four-point probe method (RT-70V, NAPSON, Tokyo, Japan) with an accuracy of ±10%. Both the Seebeck coefficient and electrical conductivity were measured thrice at different portions of each sample to extract the average values. Finally, the in-plane power factor, $S^2\sigma$, which is one of the indices for evaluating the thermoelectric performance, was calculated from the measured Seebeck coefficient and electrical conductivity. The method of power measurement of the flexible thin-film thermoelectric generators is described in Section 3.3. Surface morphologies of the antimony telluride thin films were examined using a scanning electron microscope (SEM; S-4800, Hitachi and JSM-6301F, JEOL, Tokyo, Japan). The fabrication process of the flexible thin-film thermoelectric generators is described in Section 3.2.

3. Results and Discussion

3.1. Thermoelectric and Structual Properties of n-Type Bi_2Te_3 and p-Type Sb_2Te_3 Thin Films

Based on our previous study, the conditions of electrodeposition of n-type Bi_2Te_3 and p-type Sb_2Te_3 films were determined [39,40]. Table 1 presents the in-plane thermoelectric properties, Seebeck coefficient, electrical conductivity, and power factor of the thin films. The n-type Bi_2Te_3 thin film exhibited a Seebeck coefficient of -63.0 μV/K, an electrical conductivity of 79.0 S/cm, and a power factor of 0.3 μW/(cm·K^2). On the other hand, the p-type Sb_2Te_3 thin film exhibited a Seebeck coefficient of 172 μV/K, an electrical conductivity of 3.2 S/cm, and a power factor of 0.1 μW/(cm·K^2). The thermoelectric properties of both films were lower than that of the corresponding bulk materials [41]. This is because the films exhibited small grains, as shown in SEM images (Figure 1). In our previous reports, we performed the bending test on sputtered Bi_2Te_3 and Sb_2Te_3 thin films [16,17]. We found that the surface morphologies of the sputtered films were changed by the bending conditions, but the thermoelectric properties were not greatly changed. Therefore, we considered that the electrodeposited Bi_2Te_3 and Sb_2Te_3 thin films in this study would show similar trends. To further increase thermoelectric properties in the future, we plan to carry out not only additional treatments such as thermal annealing [42–44] but also investigate the effects of thermal annealing, i.e., the diffusion between the thermoelectric thin films and the stainless-steel substrate, and the cause of micro-cracks in the films during the transfer of thermoelectric thin films to the flexible sheets. After fabricating the flexible thin-film generators, we calculated the electrical resistance of the generators based on the thermoelectric properties of the n- and p-type thin films.

Table 1. In-plane thermoelectric properties of the n- and p-type thin films.

Column Title	S (μV/K)	σ (S/cm)	Power Factor, $S^2\sigma$ (μW/(cm·K^2))
n-type film	-62.8	79	0.3
p-type film	172.3	3.2	0.1

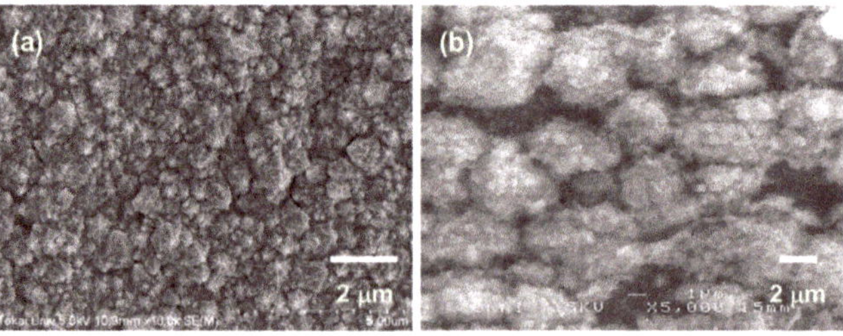

Figure 1. SEM images of the surface morphology and grain structure of (**a**) Bi_2Te_3 and (**b**) Sb_2Te_3 thin films.

3.2. Fabrication of Flexible Thin-Film Thermoelectric Generators

After determining the electrodeposition conditions of n-type Bi_2Te_3 and p-type Sb_2Te_3 thin films, we fabricated the flexible thin-film thermoelectric generators. Figure 2 shows the schematic process flow diagram employed for the fabrication of three thermoelectric generators. For the Type 1 generator, we prepared stainless-steel substrates covered with protection tape, which were partially clipped to achieve rectangle-shaped patterns. The position and shape of the substrate used for the fabrication of the p-type film correspond to those used for the fabrication of the n-type film. The n- and p-type films were electrodeposited on the substrates with thicknesses of 3.1 and 4.1 μm, respectively. The electrodeposition conditions employed for the fabrication of the n- and p-type thin films for the generators were the same as those presented in Section 3.1. After removing the protection tapes from the substrates, only the rectangle-shaped n- and p-type films were left on each stainless steel substrate. Next, the substrate and an adhesive polyimide tape (Kapton® film 650S#25, TERAOKA, Tokyo, Japan) were pasted together and rubbed hard on the tape surface by a finger. When the adhesive polyimide tape was removed from the substrate, the rectangle-shaped films were transferred onto the tape. These rectangle-shaped films were connected together, by electrically connecting the n- and p-type rectangle-shaped films in series. Noted that n–p contacts are not good for the thermoelectric effect because they likely generate a nonlinear resistance effect. To simplify the process flow, we did not insert metals between the n–p contacts. Finally, the generator fabrication was completed by inserting the metal electrodes.

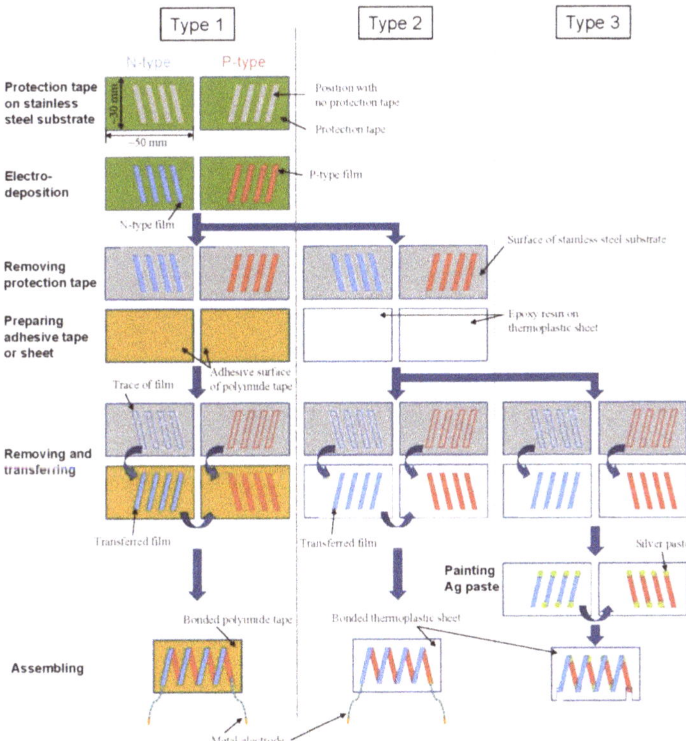

Figure 2. Schematic process flow diagram of the three types of flexible thin-film thermoelectric generators.

For fabricating the Type 2 generator, the electrodeposition process employed for the Type 1 generator was used. After electrodeposition, the protection tapes were removed from the substrates, and the n- and p-types rectangle-shaped films were then transferred onto a thin epoxy resin (CA-147, CEMEDINE, Tokyo, Japan) partially pasted on thermoplastic sheets (Laminating Film, JOINTEX, Tokyo, Japan). We used epoxy resin instead of the adhesive polyimide tape in the transfer process because the adhesion strength between the films and the epoxy resin is stronger than that between the films and the adhesive polyimide tape.

The n- and p-type rectangle-shaped films were connected together when two of the thermoplastic sheets were heat-sealed with a laminating machine while the metal electrodes were inserted.

We fabricated the Type 3 generator by tightly connecting the n- and p-type rectangular films. For fabricating the Type 3 generator, the same process employing the electrodeposition and the transfer process as that used for the Type 2 generator was employed. After the transfer process, silver pastes were painted on both ends of the n- and p-type rectangle-shaped films. The n- and p-type films were connected together, and the silver paste was then air-dried. Two of the thermoplastic sheets were heat-sealed using a laminating machine.

Figure 3 shows the photographs of three types of flexible thin-film thermoelectric generators. The Type 1 generator (Figure 3a) was composed of four pairs of n- and p-type films with a length of 30 mm and a width of 3 mm. The generator fabricated with the thin films was 40 mm wide, 30 mm high, and 60 μm thick. As shown in Figure 3b, the configuration of the Type 2 generator was mostly the same as that of Type 1 generator except for the use of thermoplastic sheets and the thickness of 0.6 mm. The thickness of the Type 2 generator was 10 times higher than that of the Type 1 generator, but the Type 2 generator also exhibited flexibility. As shown in Figure 3c, the Type 3 generator was composed of two pairs of n- and p-type films, and each rectangle-shaped film was 25 mm long and 4.5 mm wide, different from films prepared in the Type 1 and Type 2 generators. This is because we tried several times to fabricate the Type 3 generator with the same layout as that of the Type 2 generator. However, those trials ended in failure because some films broke during the process. Therefore, in order to firstly complete the generators using the combination processes with a well conducting current, we reduced the number of films and the film width was extended. The Type 3 generator was 35 mm wide and 25 mm high. The thickness of the Type 3 generator was 0.6 mm, which is the same as that of the Type 2 generator. In the Type 3 generator, the metal electrodes were not inserted, but both ends of the rectangle-shaped films were exposed to air by removing the corresponding part of the thermoplastic sheets.

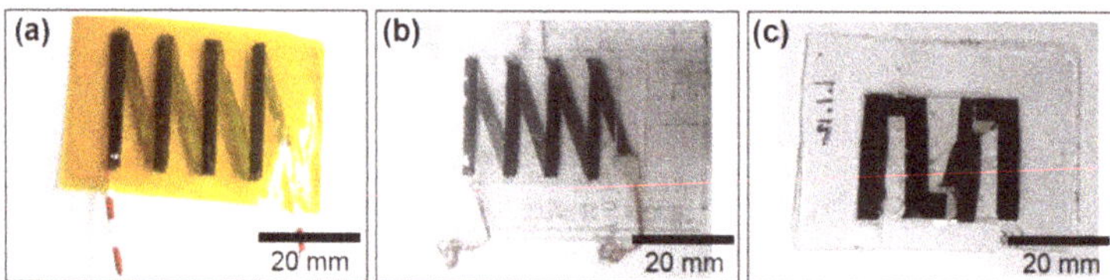

Figure 3. Photographs of the three types of flexible thin-film thermoelectric generators. (**a**) Type 1; (**b**) Type 2; (**c**) Type 3.

3.3. Performance of Flexible Thin-Film Thermoelectric Generators

We initially measured the total resistance (R_{total}) of the flexible thin-film thermoelectric generators, as listed in Table 2. The R_{total} value of the Type 1 generator was not measured because the rectangle-shaped thin films were partially broken during the transfer process. When the thin films were removed from the substrate, adhesive polyimide tape was extremely bent with the stretching state. As a result, micro-cracks appeared in the thin films, leading to partial breaking.

Table 2. Total electrical resistance of the flexible thin-film generators.

Type of Generator	R_{total} (kΩ)
Type 1	N/A
Type 2	0
Type 3	12

The Type 2 generator exhibited an R_{total} of 60 kΩ. Based on the measured electrical conductivity of n- and p-type thin films as listed in Table 1 and the film sizes as described in Section 3.2, the resistance for only the films (R_{film}) in the Type 2 generator was expected to be 31 kΩ. As a result, the R_{total} value in the Type 2 generator was approximately two times higher than the R_{film} value in the Type 2 generator, indicating that the Type 2 generator exhibited a relatively high contact resistance (R_{cont}) of 29 kΩ at the junctions between the n- and p-type thin films. On the other hand, the measured R_{total} in the Type 3 generator was 12 kΩ, and R_{film} was expected to be 8.6 kΩ. As a result, R_{cont} was estimated to be 3.4 kΩ, significantly lower than that obtained with the Type 2 generator because the junctions between n- and p-type films in the Type 3 generator were tightly connected using the silver paste, and the number of junctions was low.

To measure the performance of the flexible thin-film thermoelectric generators, a temperature difference was applied between the ends of the generators. The temperatures on both sides were monitored using the thermocouples (K-type) attached to the generator. In this study, we put the temperature difference at a maximum of 60 K, and it was difficult to further increase the temperature difference because the temperature of the cold side increased due to the thermal conduction from the hot side. When a temperature difference was generated, the open circuit voltage (V_{oc}) was measured using a digital multi-meter (TakedaRiken, Tokyo, Japan, TR6841). The V_{oc} of the flexible thin-film thermoelectric generators as a function of temperature difference is shown in Figure 4. We here confirmed that the relationship between the V_{oc} and temperature difference obtained reproducibility and reliability for repetitive operations. The results obtained for the Type 1 generator are not presented in this figure because V_{oc} was not measured owing to the partial breaking of the films. The V_{oc} value of the Type 2 and Type 3 generators increased as the temperature difference was increased. At a temperature difference of 60 K, the V_{oc} values of the Type 2 and Type 3 generators were 9.4 and 22.4 mV, respectively. The V_{oc} value of the Type 3 generator was 2.4 times higher than that of the Type 2 generator even though the number of the n–p pairs in the Type 3 generator was smaller than that in the Type 2 generator. This is because the higher resistance, which includes the film resistance and contact resistance, of the Type 2 generator caused a larger voltage drop as compared to that in the Type 3 generator. We conclude that the contact resistance is responsible for the voltage drop, based on our preliminarily experiment. In the preliminarily experiment, we prepared the Type 2 generator with the same layout as the Type 3 generator but without Ag paste connection and measured the output voltage. As a result, the voltage of the generator prepared in the preliminarily experiment decreased compared to those of the Type 2 and Type 3 generators.

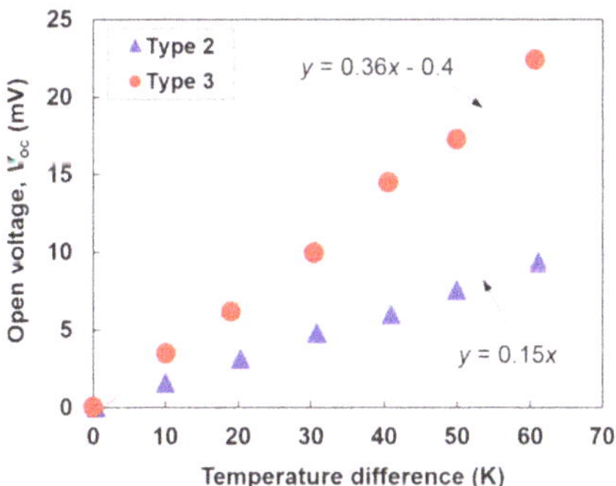

Figure 4. Open circuit voltage (V_{oc}) of the flexible thin-film thermoelectric generators as a function of temperature difference.

The maximum output power (P_{max}) of the flexible thin-film thermoelectric generators as a function of the temperature difference is shown in Figure 5. The P_{max} is expressed as $P_{max} = V_{oc}^2/4R_{total}$, and the accuracy of measurement was approximately ±10%. In the Type 2 generator, P_{max} was estimated to be 0.4 nW at a temperature difference of 60 K. On the other hand, the P_{max} value of the Type 3 generator drastically improved. P_{max} increased exponentially with an increase in the temperature difference. At a temperature difference of 60 K, the P_{max} of the Type 3 generator was 10.4 nW, which is 26 times higher than that of the Type 2 generator. Therefore, the performance of flexible thin-film thermoelectric generators can be improved to reduce the contact resistance, but the performance was still low compared to that of the generators fabricated using dry processing [4,6,45] and was not sufficient to operate potential applications such as CMOS image sensors [46], so there is room for improvement. However, we demonstrated the successful fabrication of flexible thin-film generators, which were obtained by the combination of electrodeposition and transfer processes, and this technology suggests the possible benefits of fabricating flexible thin-film generators at a low manufacturing cost.

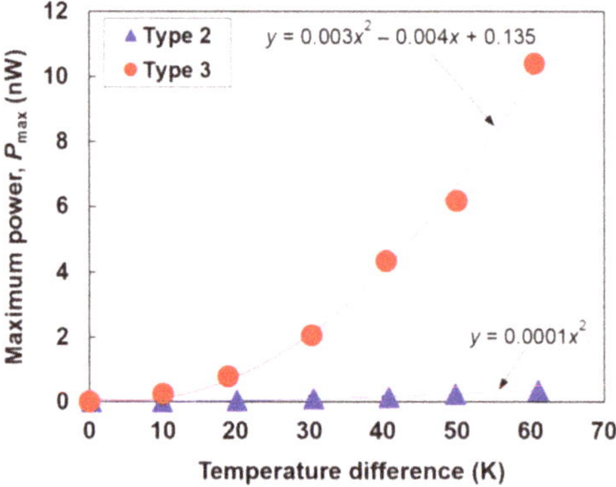

Figure 5. Maximum power (P_{max}) of the flexible thin-film thermoelectric generators as a function of temperature difference.

4. Conclusions

We fabricated three types of flexible thin-film thermoelectric generators, which consisted of n-type Bi_2Te_3 films and p-type Sb_2Te_3 films, using a combination of electrodeposition and transfer processes. Electrodeposition was performed on a stainless-steel substrate using potentiostatic electrodeposition with nitric acid based baths. The transfer process was performed using adhesive polyimide tapes or thermoplastic sheets with epoxy resin. The open circuit voltage (V_{oc}) and maximum output power (P_{max}) were measured by applying a temperature difference between the ends of the generator. The best generator obtained in this study exhibited a P_{max} of 10.4 nW at a temperature difference of 60 K. The performance of the thin-film generators improved, with a decrease in the number of micro-cracks in the thin films and on achieving a tight connection between the n- and p-type films using silver pastes. The current performance of the generators was too low, but we innovated the combination process to prepare the thin-film generators.

Acknowledgments: This study was partly supported by JSPS KAKENHI Grant Number 16K06752. The authors wish to thank Michael Faudree, Mitsuaki Okuhata, Momoko Takahashi, Takumi Makioka, Masaki Yamaguchi, Yuki Kimura, Akihiro Kobayashi, Kosuke Takano, Ken Hagiwara, and Akito Kawahira at Tokai University for experimental support.

Author Contributions: Hiroki Yamamuro and Masayuki Takashiri conceived and designed the experiments; Naoki Hatsuta, Makoto Wachi and Yoshihiro Takei performed the experiments; Naoki Hatsuta, Makoto Wachi and Yoshihiro Takei analyzed the data; Hiroki Yamamuro contributed reagents/materials/analysis tools; Hiroki Yamamuro and Masayuki Takashiri wrote the paper.

Conflicts of Interest: The authors declare no conflict of interest.

References

1. Beeby, S.P.; Tudor, M.J.; White, N.M. Energy harvesting vibration sources for microsystems applications. *Meas. Sci. Technol.* **2006**, *17*, R175–R195. [CrossRef]
2. Paradiso, J.A.; Starner, T. Energy scavenging for mobile and wireless electronics. *IEEE Pervasive Comput.* **2005**, *4*, 18–27. [CrossRef]
3. Kulah, H.; Najafi, K. Energy Scavenging from low-frequency vibrations by using frequency up-conversion for wireless sensor applications. *IEEE Sens. J.* **2008**, *8*, 261–268. [CrossRef]
4. Takayama, K.; Takashiri, M. Multi-layered-stack thermoelectric generators using p-type Sb_2Te_3 and n-type Bi_2Te_3 thin films by radio-frequency magnetron sputtering. *Vacuum* **2017**, *144*, 164–171. [CrossRef]
5. Kurosaki, J.; Yamamoto, A.; Tanaka, S.; Cannon, J.; Miyazaki, K.; Tsukamoto, H. Fabrication and evaluation of a thermoelectric microdevice on a free-standing substrate. *J. Electron. Mater.* **2009**, *38*, 1326–1330. [CrossRef]
6. Mizoshiri, M.; Mikami, M.; Ozaki, K. p-Type Sb_2Te_3 and n-Type Bi_2Te_3 films for thermoelectric modules deposited by thermally assisted sputtering method. *Jpn. J. Appl. Phys.* **2013**, *52*, 06GL07. [CrossRef]
7. Chen, X.; Xu, S.; Yao, N.; Shi, Y. 1.6 V nanogenerator for mechanical energy harvesting using PZT nanofibers. *Nano Lett.* **2010**, *10*, 2133–2137. [CrossRef] [PubMed]
8. Zhang, H.; Yang, Y.; Hou, T.C.; Su, Y.; Hu, C.; Wang, Z.L. Triboelectric nanogenerator built inside clothes for self-powered glucose biosensors. *Nano Energy* **2013**, *2*, 1019–1024. [CrossRef]
9. Francioso, L.; Pascali, C.D.; Farella, I.; Martucci, C.; Creti, P.; Siciliano, P.; Perrone, A. Flexible thermoelectric generator for ambient assisted living wearable biometric sensors. *J. Power Sources* **2011**, *196*, 3239–3243. [CrossRef]
10. Fan, P.; Zheng, Z.; Li, Y.; Lin, Q.; Luo, J.; Liang, G.; Cai, X.; Zhang, D.; Ye, F. Low-cost flexible thin-film thermoelectric generator on zinc based thermoelectric materials. *Appl. Phys. Lett.* **2015**, *106*, 073901. [CrossRef]
11. Zhu, W.; Deng, Y.; Cao, L. Light-concentrated solar generator and sensor based on flexible thin-film thermoelectric device. *Nano Energy* **2017**, *34*, 463–471. [CrossRef]
12. Zhao, D.; Min, Z. Enhanced thermoelectric performance of Ga-added $Bi_{0.5}Sb_{1.5}Te_3$ films by flash evaporation. *Intermetallics* **2012**, *31*, 321–324. [CrossRef]
13. Takashiri, M.; Tanaka, S.; Miyazaki, K. Determination of the origin of crystal orientation for nanocrystalline bismuth telluride-based thin films prepared by use of the flash evaporation method. *J. Electron. Mater.* **2014**, *43*, 1881–1889. [CrossRef]
14. Takashiri, M.; Imai, K.; Uyama, M.; Hagino, H.; Tanaka, S.; Miyazaki, K.; Nishi, Y. Comparison of crystal growth and thermoelectric properties of n-type Bi-Se-Te and p-type Bi-Sb-Te nanocrystalline thin films: Effects of homogeneous irradiation with an electron beam. *J. Appl. Phys.* **2014**, *115*, 214311. [CrossRef]
15. Böttner, H.; Nurnus, J.; Gavrikov, A.; Kühner, G.; Jägle, M.; Kunzel, C.; Eberhard, D.; Plescher, G.; Schubert, A.; Schlereth, K.H. New thermoelectric components using microsystem technologies. *J. Microelectromech. Syst.* **2004**, *13*, 414–420. [CrossRef]
16. Kusagaya, K.; Takashiri, M. Investigation of the effects of compressive and tensile strain on n-type bismuth telluride and p-type antimony telluride nanocrystalline thin films for use in flexible thermoelectric generators. *J. Alloy. Compd.* **2015**, *653*, 480–485. [CrossRef]
17. Kusagaya, K.; Hagino, H.; Tanaka, S.; Miyazaki, K.; Takashiri, M. Structural and thermoelectric properties of nanocrystalline bismuth telluride thin films under compressive and tensile strain. *J. Electron. Mater.* **2015**, *44*, 1632–1636. [CrossRef]
18. Snyder, G.J.; Lim, J.R.; Huang, C.; Fleurial, J. Thermoelectric microdevice fabricated by a MEMS-like electrochemical process. *Nat. Mater.* **2003**, *2*, 528–531. [CrossRef] [PubMed]
19. Matsuoka, K.; Okuhata, M.; Takashiri, M. Dual-bath electrodeposition of n-type Bi-Te/Bi-Se multilayer thin films. *J. Alloy. Compd.* **2015**, *649*, 721–725. [CrossRef]
20. Venkatasubramanian, R.; Siivola, E.; Colpitts, T.; O'Quinn, B. Thin-film thermoelectric devices with high room-temperature figures of merit. *Nature* **2001**, *413*, 597–602. [CrossRef] [PubMed]
21. Harman, T.C.; Taylor, P.J.; Walsh, M.P.; LaForge, B.E. Quantum dot superlattice thermoelectric materials and devices. *Science* **2002**, *297*, 2229–2232. [CrossRef] [PubMed]
22. Zhang, Y.; Chen, Y.; Gong, C.; Yang, J.; Qian, R.; Wang, Y. Optimization of superlattice thermoelectric materials and microcoolers. *J. Microelectromech. Syst.* **2007**, *16*, 1113–1119. [CrossRef]

23. Poudel, B.; Hao, Q.; Ma, Y.; Lan, Y.; Minnich, A.; Yu, B.; Yan, X.; Wang, D.; Muto, A.; Vashaee, D.; et al. High-thermoelectric performance of nanostructured bismuth antimony telluride bulk alloys. *Science* **2008**, *320*, 634–638. [CrossRef] [PubMed]

24. Yamauchi, K.; Takashiri, M. Highly oriented crystal growth of nanocrystalline bismuth telluride thin films with anisotropic thermoelectric properties using two-step treatment. *J. Alloy. Compd.* **2017**, *698*, 977–983. [CrossRef]

25. Kudo, S.; Hagino, H.; Tanaka, S.; Miyazaki, K.; Takashiri, M. Determining the thermal conductivities of nanocrystalline bismuth telluride thin films using the differential 3ω method while accounting for thermal contact resistance. *J. Electron. Mater.* **2015**, *44*, 2021–2025. [CrossRef]

26. Lee, J.; Galli, G.A.; Grossman, J.C. Nanoporous Si as an efficient thermoelectric material. *Nano Lett.* **2008**, *8*, 3750–3754. [CrossRef] [PubMed]

27. Kashiwagi, M.; Hirata, S.; Harada, K.; Zheng, Y.; Miyazaki, K.; Yahiro, M.; Adachi, C. Enhanced figure of merit of a porous thin film of bismuth antimony telluride. *Appl. Phys. Lett.* **2011**, *98*, 023114. [CrossRef]

28. Takashiri, M.; Tanaka, S.; Hagino, H.; Miyazaki, K. Combined effect of nanoscale grain size and porosity on lattice thermal conductivity of bismuth-telluride-based bulk alloys. *J. Appl. Phys.* **2012**, *112*, 084315. [CrossRef]

29. Biswas, K.; He, J.; Zhang, Q.; Wang, G.; Uher, C.; Dravid, V.P.; Kanatzidis, M.G. Strained endotaxial nanostructures with high thermoelectric figure of merit. *Nat. Chem.* **2011**, *3*, 160–166. [CrossRef] [PubMed]

30. Inamoto, T.; Takashiri, M. Experimental and first-principles study of the electronic transport properties of strained Bi_2Te_3 thin films on a flexible substrate. *J. Appl. Phys.* **2016**, *120*, 125105. [CrossRef]

31. Inamoto, T.; Morikawa, S.; Takashiri, M. Combined infrared spectroscopy and first-principles calculation analysis of electronic transport properties in nanocrystalline Bi_2Te_3 thin films with controlled strain. *J. Alloys Compd.* **2017**, *702*, 229–235. [CrossRef]

32. Frantz, C.; Stein, N.; Zhang, Y.; Bouzy, E.; Picht, O.; Toimil-Molares, M.E.; Boulanger, C. Electrodeposition of bismuth telluride nanowires with controlled composition in polycarbonate membranes. *Electrochim. Acta* **2012**, *69*, 30–37. [CrossRef]

33. Xiao, F.; Hangarter, C.; Yoo, B.; Rheem, Y.; Lee, K.H.; Myung, N.V. Recent progress in electrodeposition of thermoelectric thin films and nanostructures. *Electrochim. Acta* **2008**, *53*, 8103–8117. [CrossRef]

34. Manzano, C.V.; Abad, B.; Rojo, M.M.; Borca-Tasciuc, T.; Gonzalez, M.M. Anisotropic effects on the thermoelectric properties of highly oriented electrodeposited Bi_2Te_3 films. *Sci. Rep.* **2016**, *6*, 19129. [CrossRef] [PubMed]

35. Ye, Y.; Mao, Y.; Wang, F.; Lu, H.; Qu, L.; Dai, L. Solvent-free functionalization and transfer of aligned carbon nanotubes with vapor-deposited polymer nanocoatings. *J. Mater. Chem.* **2011**, *21*, 837–842. [CrossRef]

36. Bell, L.E. Cooling, heating, generating power, and recovering waste heat with thermoelectric systems. *Science* **2008**, *321*, 1457–1461. [CrossRef] [PubMed]

37. Matsuoka, K.; Okuhata, M.; Hatsuta, N.; Takashiri, M. Effect of composition on the properties of bismuth telluride thin films produced by galvanostatic electrodeposition. *Trans. Mater. Res. Soc. Jpn.* **2015**, *40*, 383–387. [CrossRef]

38. Okuhata, M.; Takemori, D.; Takashiri, M. Effect of pulse frequency on structural and thermoelectric properties of bismuth telluride thin films by electrodeposition. *ECS Trans.* **2017**, *75*, 133–141. [CrossRef]

39. Takemori, D.; Okuhata, M.; Takashiri, M. Thermoelectric properties of electrodeposited bismuth telluride thin films by thermal annealing and homogeneous electron beam irradiation. *ECS Trans.* **2017**, *75*, 123–131. [CrossRef]

40. Hatsuta, N.; Takemori, D.; Takashiri, M. Effect of thermal annealing on the structural and thermoelectric properties of electrodeposited antimony telluride thin films. *J. Alloy. Compd.* **2016**, *685*, 147–152. [CrossRef]

41. Scherrer, H.; Scherrer, S. Thermoelectric Materials. In *CRC Handbook of Thermoelectrics*; Rowe, D.M., Ed.; CRC Press: New York, NY, USA, 1995; pp. 211–237.

42. Takashiri, M.; Kurita, K.; Hagino, H.; Tanaka, S.; Miyazaki, K. Enhanced thermoelectric properties of phase-separating bismuth selenium telluride thin films via a two-step method. *J. Appl. Phys.* **2015**, *118*, 065301. [CrossRef]

43. Takashiri, M.; Asai, Y.; Yamauchi, K. Structural, optical, and transport properties of nanocrystalline bismuth telluride thin films treated with homogeneous electron beam irradiation and thermal annealing. *Nanotechnology* **2016**, *27*, 335703. [CrossRef] [PubMed]

44. Kudo, S.; Tanaka, S.; Miyazaki, K.; Nishi, Y.; Takashiri, M. Anisotropic analysis of nanocrystalline bismuth telluride thin films treated by homogeneous electron beam irradiation. *Mater. Trans.* **2017**, *58*, 513–519. [CrossRef]
45. Fan, P.; Zheng, Z.; Cai, Z.; Chen, T.; Liu, P.; Cai, X.; Zhang, D.; Liang, G.; Luo, J. The high performance of a thin-film thermoelectric generator with heat flow running parallel to film surface. *Appl. Phys. Lett.* **2013**, *102*, 033904. [CrossRef]
46. Hanson, S.; Seok, M.; Lin, Y.S.; Foo, Z.Y.; Kim, D.; Lee, Y.; Liu, N.; Ylvester, S.D.; Blaauw, D. A Low-voltage processor for sensing applications with picowatt standby mode. *IEEE J. Solid-State Circuits* **2010**, *45*, 759–767. [CrossRef]

Article

Defect-Free Large-Area (25 cm^2) Light Absorbing Perovskite Thin Films Made by Spray Coating

Mehran Habibi, Amin Rahimzadeh, Inas Bennouna and Morteza Eslamian *

University of Michigan-Shanghai Jiao Tong University Joint Institute, Shanghai 200240, China;
mhabibi82@sjtu.edu.cn (M.H.); amin.rahimzadeh@sjtu.edu.cn (A.R.); inas.bennouna@etu.univ-nantes.fr (I.B.)
* Correspondence: Morteza.Eslamian@sjtu.edu.cn or Morteza.Eslamian@gmail.com; Tel.: +86-213-420-7249; Fax: +86-213-420-6525

Academic Editor: Alessandro Lavacchi
Received: 20 January 2017; Accepted: 9 March 2017; Published: 12 March 2017

Abstract: In this work, we report on reproducible fabrication of defect-free large-area mixed halide perovskite ($CH_3NH_3PbI_{3-x}Cl_x$) thin films by scalable spray coating with the area of 25 cm^2. This is essential for the commercialization of the perovskite solar cell technology. Using an automated spray coater, the film thickness and roughness were optimized by controlling the solution concentration and substrate temperature. For the first time, the surface tension, contact angle, and viscosity of mixed halide perovskite dissolved in dimethylformamide (DMF) are reported as a function of the solution concentration. A low perovskite solution concentration of 10% was selected as an acceptable value to avoid crystallization dewetting. The determined optimum substrate temperature of 150 °C, followed by annealing at 100 °C render the highest perovskite precursor conversion, as well as the highest possible droplet spreading, desired to achieve a continuous thin film. The number of spray passes was also tuned to achieve a fully-covered film, for the condition of the spray nozzle used in this work. This work demonstrates that applying the optimum substrate temperature decreases the standard deviation of the film thickness and roughness, leading to an increase in the quality and reproducibility of the large-area spray-on films. The optimum perovskite solution concentration and the substrate temperature are universally applicable to other spray coating systems.

Keywords: mixed halide perovskite; large area perovskite; spray coating; perovskite solution physical properties; perovskite film optimization

---

1. Introduction

Within the past few years, a tremendous effort has been made to increase the power conversion efficiency (PCE) of perovskite solar cells (PSCs). In spite of achieving remarkable PCEs in the research labs, as high as 22.1% [1], two main obstacles still hinder the development of this technology: The device instability and the lack of knowledge and experience for large scale and large area device fabrication. Development of commercial methods for the fabrication of large area PSCs is one of the prerequisites for their commercialization, and it is as important as stabilizing the PSC performance. It is generally expected that, by increasing the film surface area, the defect and pinhole density would increase; therefore, research on the development of large area solar cells is essential. An ideal perovskite film must have a fully-covered monocrystalline structure with high uniformity and low roughness. Obtaining such ideal films is challenging if not impossible, due to the special behavior of the halide perovskite materials, which is the tendency to crystallize in a polycrystalline structure upon deposition, making the resulting thin films prone to dewetting due to crystallization (crystallization dewetting), and, therefore, the emergence of pinholes [2]. Perovskite crystal growth in all directions, including the direction normal to the film, tends to shrink and disintegrate the film, resulting in a decrease in the film coverage and an increase in the roughness. An ideal perovskite film must have a thickness within the range suitable for charge generation and transfer, as well. Thickness of the mixed halide perovskite films should be limited to 1 µm or so, dictated by the maximum diffusion length of the generated excitons in the perovskite structure [3]. Therefore, controlling the detrimental effect of the crystallization dewetting to achieve a fully-covered film, which also has desirable thickness and low roughness is quite challenging, especially when the film is deposited by a scalable method.

Solution-processed deposition of a thin film of perovskites may be performed using various casting methods, such as spin coating, dip-coating, doctor blading, spray coating, inkjet printing, screen printing, drop casting, slot-die coating, etc. Some of the aforementioned techniques, such as spin coating, in spite of providing precise controllability on the film morphology (thickness, coverage and roughness), are generally limited to batch processes and/or thin films with small effective surface areas, making them unsuitable for real-world applications. In the lab-scale and mostly using spin coating, various treatments are usually applied on the small-area perovskite films to reduce the roughness and increase the coverage and homogeneity and improve the crystalline structure. These methods include but are not limited to solvent engineering [4–6], manipulating the stoichiometry of the perovskite precursors (e.g., ratio of PbI_2/MAI solutions, where MAI stands for methylammonium iodide) [7,8], introducing additives to the perovskite solution (e.g., water, 1,8-diiodooctane(DIO)) [9,10], and controlling the annealing temperature and time [11]. However, preparation of large area perovskite films (>1 cm^2) with uniform characteristics across the film is harder to accomplish, compared with the films with small areas ($\leq$0.1 cm^2). Enlarging the perovskite surface area causes a decline in the cell performance. Figure 1 compares the PCE of several PSCs made under identical conditions, but using various deposition methods and effective surface areas. The figure confirms a systematic drop in the PCE after enlarging the active area of similar cells, or as a result of module fabrication by connecting various small-area cells in series, in order to increase the effective area [7,12–14].

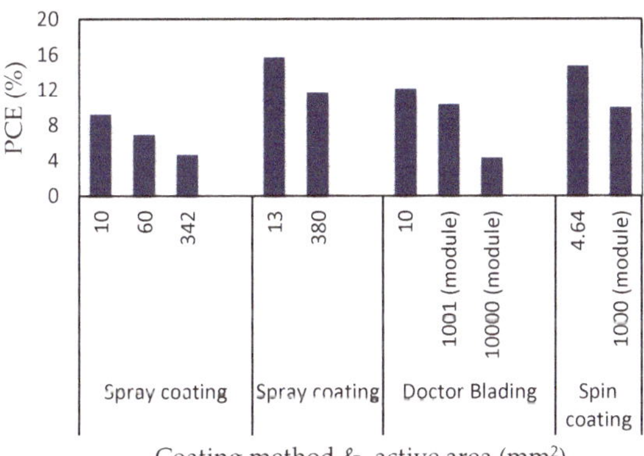

Figure 1. Degradation in the PCE of the PSCs made by various deposition techniques after enlarging the active area of individual cells, or by fabricating modules through connecting several small-area cells in series (data were taken from Refs. [7,12–14]). The precursor solutions associated with the mentioned coating processes are MAPbI3, MA(IxBr1−x)3, PbI2, and MAPbI3, respectivley, and from left to right. In the doctor blading case, the perovskite layer was obtained by dip coating of the blade-coated PbI2 film in the methylamonium iodide solution [12].

Following the aforementioned argument regarding the need for the fabrication of large-area solar cells using scalable methods, in this work, spray coating is used to produce large-area perovskite thin films with favorable morphological and light absorbing characteristics. This work focuses on the optimization of the perovskite light harvester layer only, and the fabrication of the entire device with large area is postponed to future works. Several advantages of spray coating compared to the other large scale casting techniques including the touch-free, low-cost and fast process, the possibility for deposition on flexible or rough substrates, and its capability for producing ultrathin films, assure its high potential for the roll-to-roll fabrication of solar cells on a large scale [7,15]. Spray coating may be also combined with shadow masks for pattern printing. Recently, it has been reported that spray-on films show better thermal stability compared to spun-on films [16]. Huang et al. [16] demonstrated that the prolonged annealing time required after spin coating may adversely affect the stability of spun-on films, whereas the spray-on perovskite films show high thermal stability, which originates from better crystallinity of spray-on perovskite films. In their study, they also observed better optoelectronic characteristics in spray-on perovskite films compared to spun-on counterparts, due to their higher carrier lifetime and better charge transfer capability. Despite the advantages of spray coating, fabrication of fully-covered and homogeneous thin (perovskite) films with desired low thickness and roughness is challenging. This is because the phenomenon of the liquid atomization and spraying is a random and stochastic process, which works based on transient impact of numerous droplets of different sizes across the wetted area. The droplets may first form a stable or unstable thin liquid film and then dry to form a thin solid film, or each individual droplet or patch of several merged droplets may dry to form a thin solid film. These uncertainties may cause unpredicted characteristics in the ensuing thin solid films [17]. The photovoltaic characteristics of a solar cell, such as its open-circuit voltage (V_{OC}) and fill factor (FF), are directly influenced by the quality of the film. Voids and pinholes in the perovskite films caused by a poor spray coating process may make short circuit pathways between the above and underneath layers of the perovskite, which may result in a decrease in the device shunt resistance and degradation of the device performance. Therefore, some pre-treatments, post-treatments, or additional processes have been suggested to achieve a desired spray-on thin film. For instance, Ramesh et al. [7] used a simple airbrush pen to spray $CH_3NH_3PbI_3$ perovskite precursor solution and tuned the ratio of the MAI to PbI_2 precursor solutions, spray flow rate, substrate temperature, and annealing temperature to achieve a desired film. In another work, Chandrasekhar et al. used electrostatic spray coating to spray the MAI solution onto a pre-cast PbI_2 film, where a more uniform and dense perovskite film with lager crystals was obtained, compared to that of the conventional spray coating [18]. Heo and coworkers [19] synthesized $CH_3NH_3PbI_{3-x}Cl_x$ powder and then dissolved it in a mixture of DMF (dimethylformamide) and GBL (g-butyrolactone). To adjust the perovskite crystal size in the spray-on film, they controlled the evaporation rate of DMF by changing the ratio GBL to DMF. Abdollahi Nejand et al. [20] sprayed a concentrated solution of $CH_3NH_3PbI_{3-x}Cl_x$, which caused the creation of columnar film of perovskite. Then, the film was exposed to low-pressure vapor of DMF to partially dissolve the crystals; the weakened film was then compacted by a cold-roll press. Through this method, the film coverage and the device performance were improved. Concurrent spraying of perovskite precursors using two spray nozzles is another suggested technique to control the film composition and achieve a pinhole-free film of perovskite [14]. In the literature, spraying of the MAI solution over a pre-cast PbI_2 film in a sequential deposition has been reported, as well [8,18,21–23]. Zabihi et al. sprayed perovskite precursor solutions sequentially, using two spray nozzles, on an ultrasonically vibrating substrate to form a mixed halide perovskite film [22]. The substrate vibration resulted in improved mixing and uniform deposition of the perovskite layer. In a recent work, we also used spray coating in a perovskite solar cell to fabricate a uniform PbI_2 layer and then converted it to single-halide $MAPbI_3$ perovskite film via pulsed-spray coating and drop casting of the MAI solution atop the PbI_2 layer [23].

Although spray deposition of a uniform and pinhole-free film of mixed halide perovskites (e.g., $CH_3NH_3PbI_{3-x}Cl_x$) in a PSC with planar structure is challenging, the mixed halide perovskites are more advantageous over single halide perovskites, e.g., $CH_3NH_3PbI_3$, due to larger charge carrier diffusion lengths (near 10 times), which results in a higher charge collection efficiency [19]. Therefore, in this work, single-step spray deposition of the mixed halide perovskite $CH_3NH_3PbI_{3-x}Cl_x$ is adopted. Then, the morphological and optoelectronic characteristics of the fabricated perovskite films with a square area of 25 cm^2 are optimized by systematically tuning the important process parameters, i.e., the solution concentration, substrate temperature, and the number of spray passes, while other parameters are pre-optimized and kept constant during the experiments. The best concentration of the solution of $CH_3NH_3PbI_{3-x}Cl_x$ dissolved in DMF is determined based on measuring the physical characteristics of the perovskite solution and also considering the detrimental effect of the crystallization dewetting that occurs at high solution concentrations [2]. The surface tension, contact angle and viscosity of the perovskite precursor solution are the main physical properties that govern the droplet impact and the coating process, and therefore are measured and reported in this work. The second important parameter, which is controlled to achieve a fully-covered film, is the substrate temperature. To begin with, a reasonable range of high temperatures is chosen due to the better spreading of droplets on high substrate temperatures (deduced by the measured contact angles versus temperature), and then the best substrate temperature is found based on the conversion of precursors to perovskite. Finally, the number of spray passes is tuned to achieve a fully-covered film with the lowest roughness and desired thickness. Standard deviation of the roughness and thickness data obtained from the spray-on films fabricated on a hotplate are lower than their counterparts sprayed on substrates kept at the ambient temperature. This fact reveals the strong feature of the high substrate temperature to increase the reproducibility of the spray-on films. It is important to distinguish between the substrate temperature in spray coating and the long-duration annealing temperature performed after the deposition process. In this work, all deposited samples were annealed on a hotplate at 100 °C for two hours.

2. Experimental

2.1. Materials and Methods

Lead chloride ($PbCl_2$, 98.5%), and N,Ndimethylformamide (DMF, 99.8%) were supplied by Sigma-Aldrich, St. Louis, MO, USA. Methylammonium iodide (MAI, 99.5%) was purchased from Xi'an Polymer Light Technology Corp. (Xi'an, China). MAI and $PbCl_2$ were mixed in 3:1 molar ratio and then dissolved in DMF in various concentrations (5% to 50% weight ratio: Ratio of the solid precursors mass to the total mass of the solvent and solid precursors in the solution). The $MAPbI_{3-x}Cl_x$ solution was heated to 60 °C and stirred on a magnetic stirrer overnight, and then cooled at room temperature. Small (1×1 cm^2) and large (6×6 cm^2) fluorine-doped tin oxide (FTO)-coated glass substrates with an average roughness of near 14.5 nm were cleaned by a mixture of deionized water and soap, acetone and isopropyl alcohol in an ultrasonic bath, sequentially. Then, the substrates were exposed to UV-Ozone irradiation for 15 min. Spray coating was performed by replacing the spray nozzle of an automatic spray coating system (Holmarc, Opto-Mechatronics Pvt. Ltd., Model HO-TH-04, Kochi, India) with a low flow rate commercial airbrush pen. This was done because the liquid container of the original ultrasonic nozzle of the Holmarc machine is excessively large, resulting in the wastage of the perovskite solution. The installed airbrush generates a fine mist, is easy to control, and is suitable for the fabrication of perovskite films. The speed and position of the spray nozzle in the x–y plane were controlled by software to move the nozzle in a continuous and raster movement to simulate an industrial spray coating process. The perovskite solution was atomized by the pressurized air at constant pressure and air flow rate. Parameters of the spray coating process are summarized in Table 1. Pictures of the spray coating machine and the spray nozzle are shown in Figure 2.

Table 1. Spray coating parameters for the fabrication of perovskite films.

Spray Parameters	Value
Nozzle to substrate distance (cm)	12
Nozzle speed (mm/s)	150
Flow rate (µL/s)	15
Air pressure (bar)	3.5
Number of passes	40, 70, 100
Substrate temperature (°C)	25–200

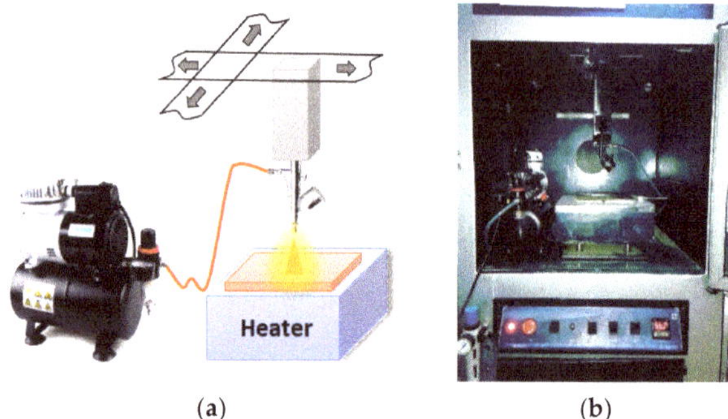

Figure 2. (**a**) schematic of the spray coating process and the aparatus. (**b**) picture of the Holmarc spray coating system, which accomodates the spray nozzle and the hotplate shown in (**a**).

The relative humidity of Shanghai during the spray deposition of the perovskite films was in the range of 35% to 50% in different days. Although the humidity has a detrimental effect on the perovskite quality, it mainly affects the perovskite film during and after annealing. Therefore, to minimize the effect of the humidity, the spray-on films, made in the ambient conditions in few minutes, were transferred to the glovebox after deposition and initial drying on the high temperature substrate, for complete drying and annealing on a hotplate at 100 °C, for two hours. In some of the tests (for determination of the crystallization dewetting), the samples were formed on small-size FTO-coated glass (1×1 cm^2) by spin coating at 2500 rpm for 20 s, and then the samples were annealed based on the same procedure mentioned above for the spray-on films. Perovskite films made on small area FTO-coated glass were encapsulated with poly (methyl methacrylate) (PMMA) to enhance the perovskite stability against humidity during the X-ray diffraction (XRD) and absorption tests.

2.2. Film and Device Characterization

Optical microscopy images of perovskite films were taken by a confocal laser scanning microscope (CLSM 700, Carl Zeiss AG, Oberkochen, Germany) and scanning electron microscopy (SEM) images were obtained using a Hitachi microscope, Model S-3400N, Tokyo, Japan. Average thickness and roughness of the films were measured by a stylus profilometer (KLA-Tencor P7, Milpitas, CA, USA). Roughness values were measured and averaged along six randomly chosen lines of 100 μm long, on two similar samples, in each case. Thickness values were also measured by the same instrument, in four randomly-chosen spots near the edge of the films with respect to the uncovered areas of the FTO-coated glass, and were averaged.

The conversion of the MAI and PbCl$_2$ precursors to mixed halide perovskite and the absorption of the perovskite films were evaluated by X-ray diffraction (XRD, model D5005, Bruker, Billerica, MA, USA) and UV-Visible absorption spectrophotometry (EV300, Thermo Fisher Scientific, Waltham, MA, USA), respectively. Physical properties of the perovskite solutions were measured, as well. The solution surface tension and contact angle were measured, using Theta Lite Optical Tensiometer (Biolin Scientific AB, Gothenburg, Sweden), and the viscosity was measured using a digital rotary viscometer (model NDJ-8S, Jiangsu Zhengji Instruments Co., Ltd., Jintan, China) in various concentrations, defined as the weight fraction of the solute (perovskite precursors) in the solution. A digital infrared camera (FLIR C2, Tallinn, Estonia) was used to confirm the substrate temperature, which is generated and maintained by an electric heater.

3. Results and Discussion

Knowledge of the concentration-dependent physical properties of the perovskite precursor solution, such as the contact angle, surface tension, and viscosity, would lead to better understanding of the droplet impact behavior during spray deposition, and also would help proper selection of the solution concentration. This is because the droplet impact dynamics is governed by the Ohnesorge number, a dimensionless group based on the liquid properties, as well as Reynolds and Weber numbers, which include the effect of the droplet momentum upon impact. Assuming constant impingement velocity and substrate texture and surface energy, concentration of the perovskite precursor solution and the substrate temperature control the spreading behavior of the perovskite droplets after impinging on the substrate, and therefore control the morphology of the resulting perovskite film. Physical properties of the precursor solution may also affect the thin film characteristics prepared by other casting methods. Various solution concentrations (9.5–33 wt.%) [20,24,25] have been used by others as the starting concentration for the fabrication of perovskite films, but, to the best of our knowledge, this has not been done systematically. Therefore, here we measured the contact angle, surface tension and viscosity of the mixed halide perovskite $MAPbI_{3-x}Cl_x$ dissolved in DMF solvent in a wide range of concentrations (5–50 wt.%). Table 2 lists the average measured properties along with the standard deviation of the measurements.

Table 2. Some of the physical properties of mixed halide perovskite solution ($MAPbI_{3-x}Cl_x$ in DMF) at various concentrations (wt.% of solute in the solution). MAI and $PbCl_2$ powders were mixed in 3:1 molar ratio and then dissolved in DMF to achieve various concentrations.

Concentration (wt.%)	Viscosity (mPa.s)	Surface Tension (mN/m)	Contact Angle (°)
0	0.887 ± 0.02	37.29 ± 0.02	12.39 ± 0.04
5	1.0 ± 0.0	36.81 ± 0.02	16.35 ± 0.24
10	1.0 ± 0.0	35.18 ± 0.02	20.79 ± 0.10
20	1.59 ± 0.016	34.39 ± 0.06	24.00 ± 0.15
30	2.0 ± 0.0	31.44 ± 0.03	29.26 ± 0.11
40	2.0 ± 0.0	29.80 ± 0.03	30.40 ± 0.15
50	6.72 ± 0.135	28.87 ± 0.02	31.65 ± 0.21

Some studies, e.g., Mullins and Sekerka [26], suggest that some ionic solutions may act like a surfactant, in that, their surface tension may decrease with the solution concentration, and, in fact, our data of perovskite solutions at different concentrations comply with the reported observations for other similar solutions. Figure 3a shows a decrease in the surface tension of the perovskite solution with the solution concentration. On the other hand, examination of the perovskite solution droplets on glass substrates shows that the contact angle increases with concentration (Figure 3b). Young's equation, i.e., $\gamma\cos\theta = \gamma_{sg} - \gamma_{ls}$, relates the binary surface tensions with the equilibrium contact angle on a solid surface. In Young's equation, γ is the liquid–air surface tension or simply surface tension, γ_{sg} is the solid-gas surface tension, γ_{ls} is the liquid–solid surface tension, and θ is the equilibrium contact angle. Our results show that, as the solution concentration increases, γ decreases on one hand and contact angle increases on the other hand ($\cos\theta$ decreases), thus the solid–liquid surface tension, γ_{ls} must increase, since the solid–gas surface tension γ_{sg} remains constant for the same glass substrate next to the same surrounding gas (air). Some numerical works by Sear [27] and theoretical works by Djikaev and Ruckenstein [28] have shown the substrate effect on crystal nucleation. For instance, crystal nucleation is greater in the vicinity of the gas–liquid–solid triple line, since the concentration is the highest at the triple line, due to the coffee stain or coffee ring effect. Hence, to suppress the inhomogeneous crystal nucleation due to the coffee-ring effect, increasing the spreading of the solution droplets, as well as increasing the droplet drying rate, could be effective. In this work, this has been achieved by using treated FTO-coated glass substrates to increase wettability, as well as by choosing a low solution concentration (e.g., 10%), to achieve a low contact angle.

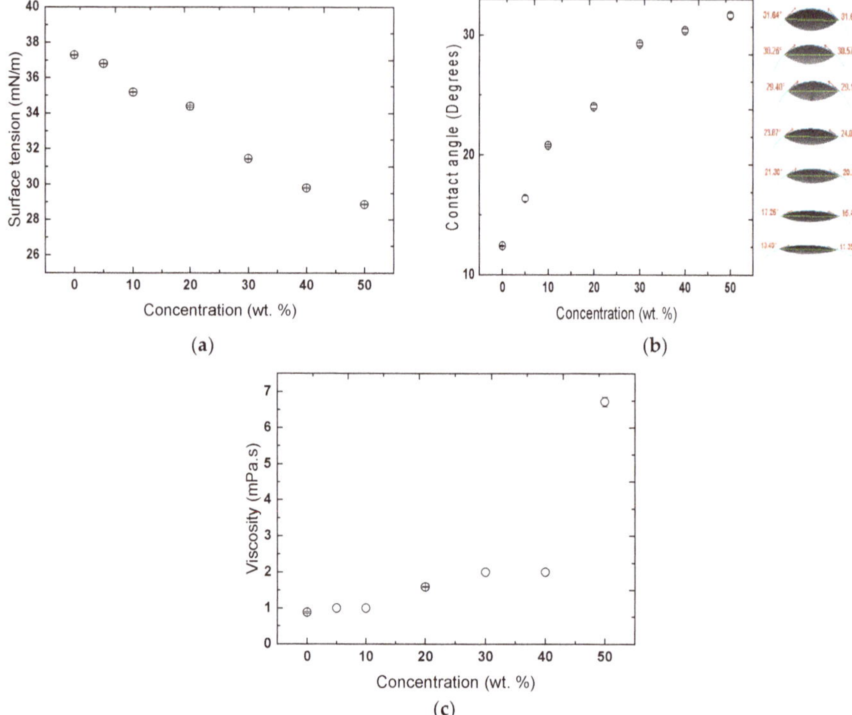

Figure 3. Variation of the physical properties of the mixed halide perovskite solution (MAPbI$_{3-x}$Cl$_x$ in DMF) versus solution concentration in wt.% of solute in solution. (**a**) liquid–air surface tension; (**b**) contact angle, and (**c**) viscosity. Note: The data and the associated errors are listed in Table 1.

Viscosity is another physical property of the liquid solution that affects the liquid film spreading and wetting, and is believed to increase nonlinearly as the solvent evaporates and the liquid film partially dries [29]. In droplet-based coating methods, viscosity, through the Ohnesorge number, affects droplet impact dynamics and spreading. As shown in Figure 3c, viscosity increases somewhat linearly at concentrations up to about 40 wt.%. Then, it sharply increases as the concentration increases to 50 wt.%, which corroborates the nonlinear behavior of the viscosity with the concentration. This is because, at high concentrations, the solution starts to show non-Newtonian behavior.

High concentration of the perovskite solution may cause the formation of larger crystals in the film and therefore surface dewetting, i.e., crystallization dewetting [2], which obviously leads to the formation of rough films with pinholes (Figure 4). All films in Figure 4 are spun under the same conditions, while the concentration changes from 10% to 30% and 50% from left to right, respectively. Since the data from Figure 3 clearly show that the solution droplets with lower concentration have smaller contact angles, and therefore higher surface wettability, and Figure 4 shows that smaller detrimental crystallization dewetting occurs when the perovskite thin films are made using a solution concentration of 10 wt.%, this concentration is selected as the favorable concentration for spray coating of the mixed halide perovskite MAPbI$_{3-x}$Cl$_x$ dissolved in DMF. At this low concentration, the crystallization dewetting is the minimum and spreading is the maximum, both favoring the formation of a continuous and defect-free film.

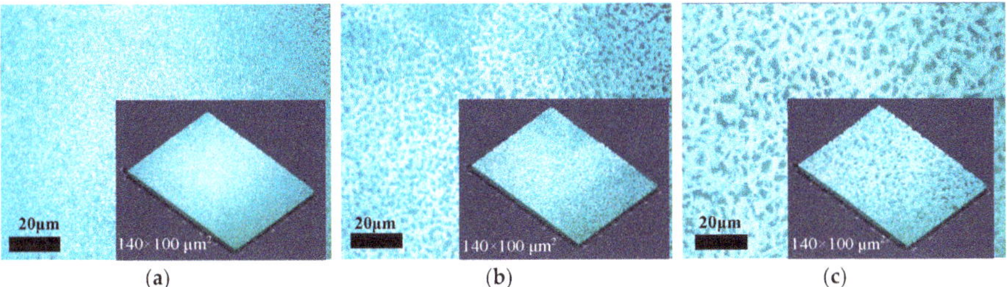

Figure 4. Optical images of spun-on perovskite films at different concentrations. The solution concentrations are (**a**) 10 wt.%; (**b**) 30 wt.%; and (**c**) 50 wt.%. This figure shows that high concentrations result in the formation of large crystals, and therefore, occurrence of crystallization dewetting.

Having determined the ideal solution concentration, now we will find the optimum substrate temperature. Spray deposition of perovskite solution on a low temperature substrate prolongs the drying time, which may adversely affect the film coverage [24]. This is due to the prolonged crystal growth at low temperatures and therefore the occurrence of crystallization dewetting. Figure 5 shows the range of the substrate temperatures (during the coating process) that have been used in recent published papers [3,7,8,13,16,18–22,24,25,30–37], in order to fabricate the perovskite layer of PSCs, mostly by a scalable technique. These techniques include spray coating, doctor blading, slot-die coating, roller coating, inkjet printing, and drop casting. In most of the methods, temperature of the substrate was raised to facilitate the coating process. However, in most of those works, a systematic procedure was not followed. Hence, in this work, the best substrate temperature for rapid drying and in situ heat treatment during spray deposition is selected based on two criteria; first, a range of temperatures are selected to achieve the best droplet spreading upon impingement on a hot substrate. To this end, the contact angles of 10 wt.% perovskite solution droplets were evaluated right after droplet release on glass substrates kept at various temperatures (25 °C to 200 °C with 25 °C intervals) (Figure 6a). The heated substrate increases the evaporation rate, which is beneficial for arresting the crystal growth and suppressing the crystallization dewetting. In addition, the data of Figure 6a show that a higher substrate temperature improves the film coverage by decreasing the contact angle, because of better droplet spreading and surface wettability. Therefore, based on the contact angle measurements, relatively high substrate temperatures seem to be more effective for a better coverage. It is noted that recent research on the thermal and thermodynamic stability of perovskites proves that high annealing temperatures for a long time results in the decomposition of the perovskite structure [38,39]; however, here in this work, the films are exposed to a high substrate temperature for a short period of time during the spray deposition, which lasts only for several seconds. It is also noted that excessive substrate temperatures above the Leidenfrost temperature must be avoided due to the formation of a vapor film over the substrate, which would interrupt the droplet spreading and the coating process.

In the second step, we further narrow down the choices of the substrate temperature by evaluating the conversion of the perovskite precursors to perovskite. We define the conversion ratio using the XRD patterns of the perovskite films, as the ratio of the PbI_2 peak intensity at 12.7° to the perovskite peak intensity at 14.2°. PbI_2 is an impurity in the perovskite film and has to be eliminated or minimized. Figure 6b illustrates the XRD patterns of four samples fabricated on substrates kept at 25, 100, 150 and 200 °C. It is noted that these are the substrate temperatures, and all samples were annealed at 100 °C after deposition. We observe that the intensity of the PbI_2 peak changes with the substrate temperature. Figure 6c shows the conversion ratio obtained at various substrate temperatures. It is observed that the maximum conversion occurs near 150 °C, which is near the boiling point of DMF, i.e., 153 °C. Low conversion ratio at temperatures lower than 150 °C is ascribed to initial incomplete conversion of precursors, whereas, at higher temperatures than the boiling point of DMF, it may be attributed to rapid evaporation of the solvent and the lack of enough solvent for a complete conversion. This observation is in agreement with the report of Mallajosyula and coworkers that considered substrate temperature of 145 °C during doctor blading of MAPbI$_3$ [37]. Therefore, according to the contact angle measurements of perovskite droplets at various substrate temperatures and also the conversion ratio data, 150 °C is chosen as the optimum substrate temperature, since, at this temperature, spreading of perovskite solution droplets and also conversion ratio of perovskite attain their maximum values.

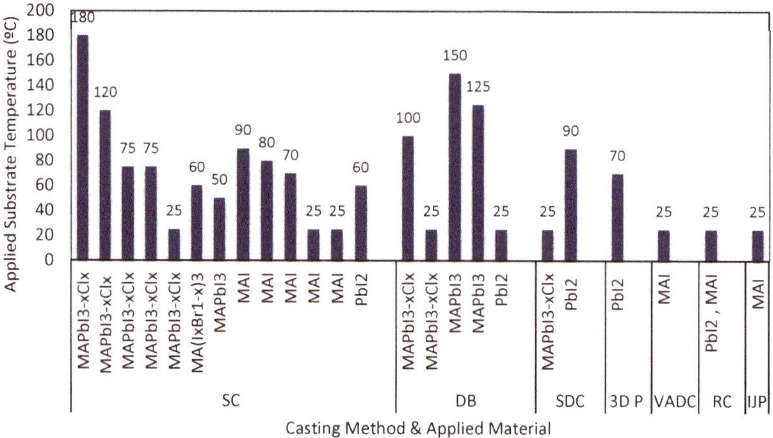

Figure 5. Employed substrte temeratures during the fabrication of perovsite films by various (large scale) techniques. (data taken from Refs. [3,7,8,13,16,18–22,24,25,30–37]). Abreviations: SC: Spray coating; DB: Doctor blading; SDC: Slot-die coating; 3D P: 3D printing based on SDC and N$_2$ gas-blowing; VADC: Vibration-assisted drop casting; RC: Roller coating; IJP: Inkjet printing.

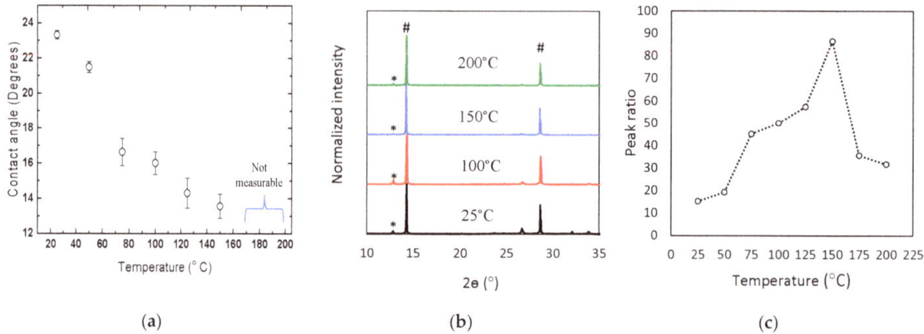

Figure 6. (**a**) contact angle of the perovskite solution droplets (MAPbI$_{3-x}$Cl$_x$ in DMF) versus substrate temperature; (**b**) XRD patterns of MAPbI$_{3-x}$Cl$_x$ perovskite films, spray deposited on FTO-coated glass substrates at various substrate temperatures (# and * represent the perovskite and PbI$_2$ peaks, respectively); (**c**) the quantified conversion ratios at various substrate temperatures. At temperatures higher than the boiling point of DMF (>153 °C), measurement of the contact angle was found to be not accurate, due to rapid solvent evaporation. It is noted that, after deposition, all perovskite films were annealed at 100 °C for 2 h.

The third and last parameter to optimize is the number of spray passes for the used spray nozzle and spray conditions (Table 1). The number of spray passes affects the film coverage, thickness, and roughness. In general, photons with long wavelengths are better absorbed in thicker films, which could translate to a higher current density in the active layer [25]; however, enhanced absorption occurs at the cost of increased charge recombination and the loss of the current density [25]. The limitation of charge diffusion length in perovskite films also confines the admissible span of thicknesses. In $MAPbI_{3-x}Cl_x$ mixed halide perovskite, the charge diffusion length has been measured to be around 1000 nm, while, in $MAPbI_3$ single halide perovskite, it is near 100 nm [3]. Therefore, there is a specific range of thicknesses in which the highest current from a perovskite film could be extracted. In this work, adjusting the thickness was performed by using multiple spray passes. A single spray pass is the travelling of the spray nozzle arm over the substrate with a specific nozzle velocity. Multiple-pass spraying is comprised of several consecutive single spray passes back and forth in a predesigned raster pattern to improve the coverage and achieve a desired thickness. Figure 7 shows the measured thickness and roughness values of the perovskite films, sprayed using 40, 70, and 100 passes, on the substrates kept at the ambient temperature. Coverage measurements revealed that, at the used spray flow rate and spray nozzle speed (Table 1), spraying using 40 and 70 passes was not enough to cover the film completely. Raising the passes to 100 resulted in full coverage of the substrate. Obviously, the film thickness increases by raising the number of spray passes; however, the 40-pass-spray-on film showed the highest roughness, which is due to the excessive number of pinholes. However, in the two other cases, increasing the number of passes from 70 to 100, resulted in an increase in the roughness again (Figure 7b). This is in agreement with the result of other reports, which declared that the roughness of spray-on perovskite films increases by increasing the thickness [25]. It is noted that in less-crystalline or amorphous films, such as polymeric films, the correlation between the number of spray passes and roughness may be somewhat different from what was observed here for perovskite films, simply because, in polymeric films, the phenomenon of crystallization dewetting and the formation of large grains is absent or it is insignificant [40].

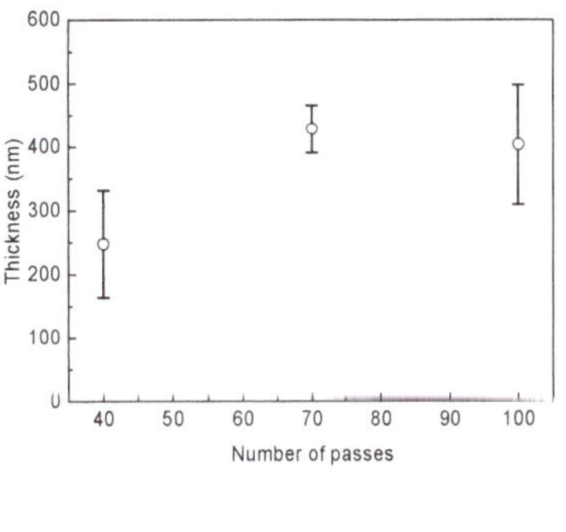

(a)

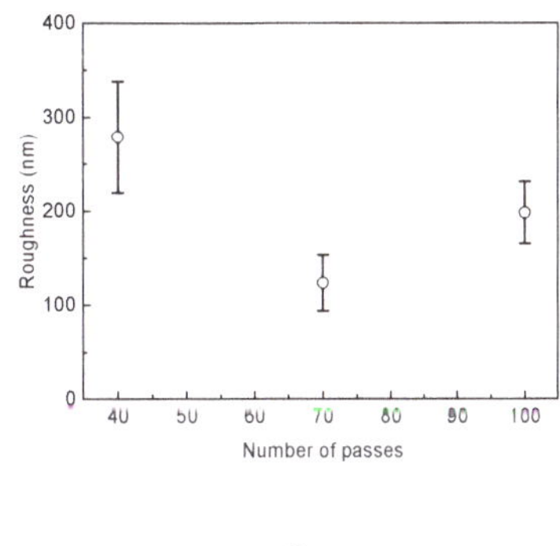

(b)

Figure 7. Measured (**a**) thickness and (**b**) roughness of spray-on perovskite films deposited on the substrates kept at the ambient temperature, at various number of spray passes.

To fabricate the best functional perovskite film, i.e., lowest roughness and proper thickness with negligible PbI_2 impurity, the spraying process was repeated using different passes (40, 70 and 100) on hotplates at the optimum temperature of 150 °C, which yields the best conversion ratio (Figure 8). Again, the full coverage is achieved at 100 passes, while the thickness and roughness decreased in all cases, compared with the samples fabricated on the substrates kept at the ambient temperature.

Figure 9 shows the SEM images of spray-on perovskite films deposited on the hotplates kept at 150 °C, while the number of passes varies. The average coverage, denoted as C on the images, obtained from the SEM images, reveals that a fully-covered film was obtained after 100 passes. The coverage was obtained by image processing using the ImageJ software (version 1.51j, National Institutes of Health (NIH), Bethesda, MD, USA).

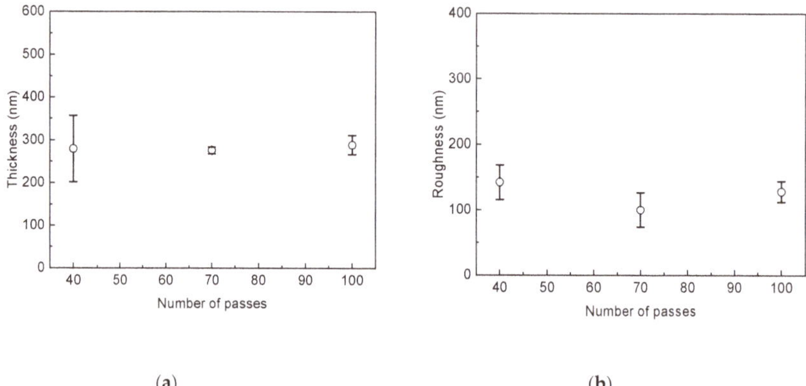

(a) (b)

Figure 8. Measured (**a**) thickness and (**b**) roughness of spray-on perovskite films deposited on the substrates kept at 150 °C, at various number of spray passes.

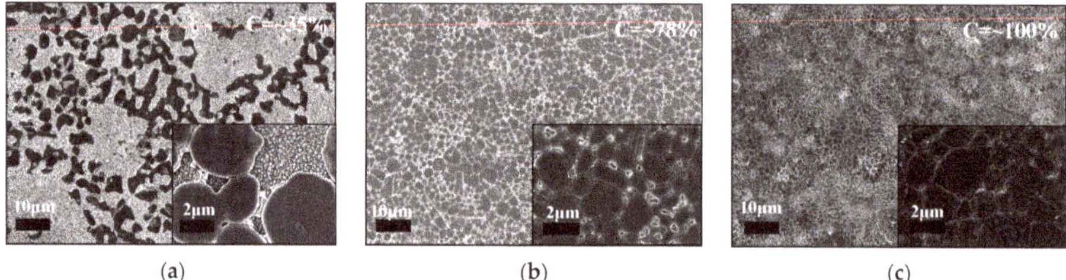

(a) (b) (c)

Figure 9. SEM images of spray-on perovskite films deposited on FTO-coated glass substrates at substrate temperatures of 150 °C. The number of spray passes are (**a**) 40; (**b**) 70; and (**c**) 100. Letter C refers to the percentage of coverage of perovskite layer on the substrates.

Interestingly, as depicted in Figure 10, the application of the high temperature substrate results in a decrease in the standard deviation of roughness and thickness data of spray-on films with respect to their counterparts fabricated at the ambient temperature. This is an important factor to increase the device reproducibility of large-area perovskite films. According to Figure 8a, the average thickness values of all three spray-on samples is less than 300 nm, which is acceptable for mixed halide perovskite films to perform well in a PSC [24]. In Figure 10, the spray-on films made using 70 passes show the minimum standard deviation in their thicknesses. The same films have the lowest roughness as well (Figure 8b). However, these films suffer from inadequate coverage (78%). Thus, the film made using 100 passes on the hotplate kept at 150 °C is considered as the best film.

The perovskite films, sprayed using 100 passes on the substrates kept at 25 and 150 °C, are compared using optical and SEM images shown in Figure 11a,b. The absorbance spectra of the same films are also shown in Figure 11c. According to the thickness values given in Figures 7a and 8a, the film sprayed using 100 passes on the substrate kept at the ambient temperature is thicker than the film sprayed on the hotplate at 150 °C, while its absorbance shown in Figure 11c is lower in a wide range of wavelengths. This may have originated from two sources. Firstly, high roughness in spray-on films may decrease the light absorbance [41]. Based on the roughness data of Figures 7b and 8b, the former film is rougher, i.e., its local variation of thickness is higher than the latter film. Inset optical images in Figure 11a,b also verify that the film fabricated on the hotplate is smoother than the film sprayed on the substrate kept at the ambient temperature, thus it can absorb the light more efficiently. Secondly, according to the conversion ratio data shown in Figure 6c, there is more

PbI$_2$ impurity in the perovskite film fabricated at 25 °C, with respect to the film deposited on the hotplate kept at 150 °C. Therefore, the film deposited on the hotplate is more functional and thus capable of better photon absorption, considering that the bandgap of the perovskite films is suitable for the absorption of photons in the visible range. Figure 11d is a picture of the fabricated film (5 × 5 cm^2), made using the optimum conditions of 100 spray passes deposited on the hotplate kept at 150 °C, using a perovskite solution concentration of 10 wt.%. The raster movement of the spray nozzle is also illustrated schematically on this picture. This spray-on film is semitransparent and thus has the capability to be used as a second absorber layer to make tandem perovskite–copper indium gallium diselenide (CIGS) or perovskite–silicon solar cells.

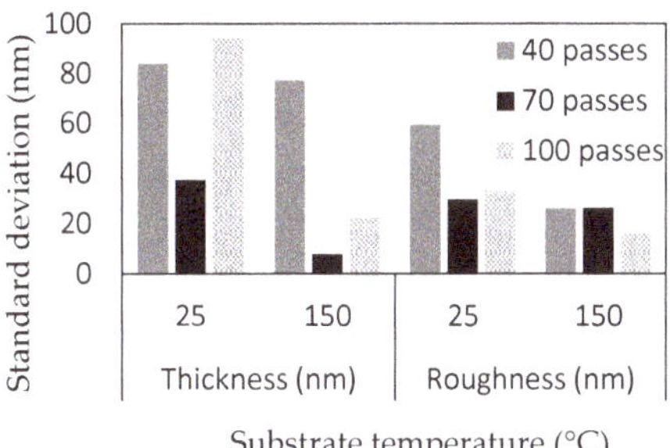

Figure 10. Standard deviation of the roughness and thickness data of perovskite films fabricated by various spray passes on the substrates kept at 25 and 150 °C. High substrate temperature improves reproducibility of the data.

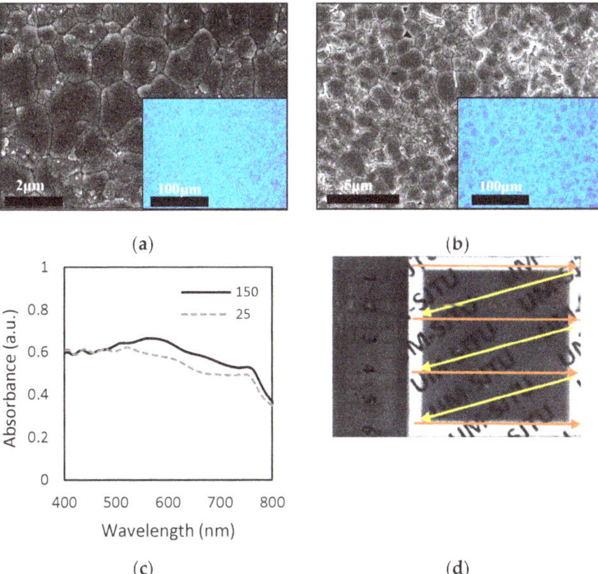

Figure 11. SEM and optical (inset) images of spray-on perovskite films made at substrate temperature of (**a**) 25 °C and (**b**) 150 °C. (**c**) absorbance spectra of perovskite films sprayed on FTO-coated glass kept at 150 °C; (**d**) the spray-on film deposited on the hotplate kept at 150 °C and the deposition pattern based on a zigzag movement of the spray nozzle over the substrate. The films were made at optimum condition of 100 spray passes using a 10 wt.% solution concentration.

4. Conclusions

A continuous, uniform, and pinhole-free perovskite thin film was fabricated by spray coating with an effective squared area of 25 cm^2. Optimum concentration of the mixed halide perovskite precursor (10 wt.% of MAPbI$_{3-x}$Cl$_x$ in DMF) and substrate temperature (150 °C) were determined as two essential parameters that strongly affect the morphology of the spray-on film. The optimum concentration and substrate temperature may be generalized to other spray systems. Then, for the spray nozzle used in this study, for the given spray flow rate and nozzle speed, the number of spray passes was optimized (100 spray passes). The combination of the best solution concentration, substrate temperature, and the number of spray passes produced a fully functional perovskite film with high surface area suitable for the commercialization of the technology. We also measured the contact angle, surface tension, and viscosity of the mixed halide perovskite solution in different concentrations, in order to find the optimum range of concentration. In order to choose the best substrate temperature, at first, we measured the contact angle of perovskite droplets on the hotplates at various temperatures. It was observed that a higher substrate temperature leads to a higher surface wettability (lower contact angle). At 150 °C, which is around the boiling point of DMF, we also observed the highest conversion ratio (efficient conversion of perovskite precursors to perovskite), which is a measure of the percentage of PbI$_2$ impurity in the perovskite film. Spraying of 10 wt.% mixed halide perovskite using 100 passes on a hotplate kept at 150 °C resulted in a fully-covered film with the lowest thickness and roughness, compared to the films made on the substrates kept at room temperature. Standard deviation of the thickness and roughness of the films made on hot substrates also decreased, which is an indication of increased reproducibility of the process. Although the fabricated large-area perovskite films demonstrated superior morphological and optoelectronic functionality, it is noted that their photovoltaic performance would be determined when used in a complete device. The fabrication of such large area devices is challenging and entails the optimization of all layers, which was beyond the scope of this work.

Acknowledgments: Research funding from the Shanghai Municipal Education Commission in the framework of the Oriental scholar and distinguished professor designation and funding from the National Natural Science Foundation of China (NSFC) is acknowledged.

Author Contributions: Mehran Habibi and Morteza Eslamian conceived the work; Mehran Habibi designed and performed the experiments regarding the optimization of perovskite layers, conducted characterization tests, and discussed the results. Amin Rahimzadeh performed the tests concerning the physical properties of perovskite solutions and interpreted the results. Inas Bennouna assisted with the experiments and characterizations; and Mehran Habibi, Amin Rahimzadeh and Morteza Eslamian wrote the paper. All authors read and approved the paper.

Conflicts of Interest: The authors declare no conflict of interest.

References

1. Efficiency chart. Available online: http://www.nrel.gov/pv/assets/images/efficiency_chart.jpg (accessed on 10 March 2017).
2. Habibi, M.; Rahimzadeh, A.; Eslamian, M. On dewetting of thin films due to crystallization (crystallization dewetting). *Eur. Phys. J. E Soft Matter.* **2016**, *39*, 30. [CrossRef] [PubMed]
3. Habibi, M.; Zabihi, F.; Ahmadian-Yazdi, M.R.; Eslamian, M. Progress in emerging solution-processed thin film solar cells–Part II: Perovskite solar cells. *Renew. Sustain. Energy Rev.* **2016**, *62*, 1012–1031. [CrossRef]
4. Cai, B.; Zhang, W.-H.; Qiu, J. Solvent engineering of spin-coating solutions for planar-structured high-efficiency perovskite solar cells. *Chin. J. Catal.* **2015**, *36*, 1183–1190. [CrossRef]
5. Jeon, N.J.; Noh, J.H.; Kim, Y.C.; Yang, W.S.; Ryu, S.; Seok, S.I. Solvent engineering for high-performance inorganic-organic hybrid perovskite solar cells. *Nat. Mater.* **2014**, *13*, 897–903. [CrossRef] [PubMed]
6. Ahmadian-Yazdi, M.; Zabihi, F.; Habibi, M.; Eslamian, M. Effects of Process Parameters on the Characteristics of Mixed-Halide Perovskite Solar Cells Fabricated by One-Step and Two-Step Sequential Coating. *Nanoscale Res. Lett.* **2016**, *11*, 408. [CrossRef] [PubMed]

7. Ramesh, M.; Boopathi, K.M.; Huang, T.Y.; Huang, Y.C.; Tsao, C.S.; Chu, C.W. Using an airbrush pen for layer-by-layer growth of continuous perovskite thin films for hybrid solar cells. *ACS Appl. Mater. Interfaces* **2015**, *7*, 2359–2366. [CrossRef] [PubMed]

8. Mohammadian, N.; Alizadeh, A.H.; Moshaii, A.; Gharibzadeh, S.; Alizadeh, A.; Mohammadpour, R.; Fathi, D. A two-step spin-spray deposition processing route for production of halide perovskite solar cell. *Thin Solid Films* **2016**, *616*, 754–759. [CrossRef]

9. Gong, X.; Li, M.; Shi, X.-B.; Ma, H.; Wang, Z.-K.; Liao, L.-S. Controllable Perovskite Crystallization by Water Additive for High-Performance Solar Cells. *Adv. Funct. Mater.* **2015**, *25*, 6671–6678. [CrossRef]

10. Liang, P.W.; Liao, C.Y.; Chueh, C.C.; Zuo, F.; Williams, S.T.; Xin, X.K.; Lin, J.; Jen, A.K. Additive enhanced crystallization of solution-processed perovskite for highly efficient planar-heterojunction solar cells. *Adv. Mater.* **2014**, *26*, 3748–3754. [CrossRef] [PubMed]

11. Lau, C.F.J.; Deng, X.; Ma, Q.; Zheng, J.; Yun, J.S.; Green, M.A.; Huang, S.; Ho-Baillie, A.W.Y. CsPbIBr$_2$ Perovskite Solar Cell by Spray-Assisted Deposition. *ACS Energy Lett.* **2016**, *1*, 573–577. [CrossRef]

12. Razza, S.; Di Giacomo, F.; Matteocci, F.; Cinà, L.; Palma, A.L.; Casaluci, S.; Cameron, P.; D'Epifanio, A.; Licoccia, S.; Reale, A.; et al. Perovskite solar cells and large area modules (100 cm^2) based on an air flow-assisted PbI$_2$ blade coating deposition process. *J. Power Sources* **2015**, *277*, 286–291. [CrossRef]

13. Yeo, J.-S.; Lee, C.-H.; Jang, D.; Lee, S.; Jo, S.M.; Joh, H.-I.; Kim, D.-Y. Reduced graphene oxide-assisted crystallization of perovskite via solution-process for efficient and stable planar solar cells with module-scales. *Nano Energy* **2016**, *30*, 667–676. [CrossRef]

14. Tait, J.G.; Manghooli, S.; Qiu, W.; Rakocevic, L.; Kootstra, L.; Jaysankar, M.; Masse de la Huerta, C.A.; Paetzold, U.W.; Gehlhaar, R.; Cheyns, D.; et al. Rapid composition screening for perovskite photovoltaics via concurrently pumped ultrasonic spray coating. *J. Mater. Chem. A* **2016**, *4*, 3792–3797. [CrossRef]

15. Markus, H.; Henrik, F.D.; Krebs, F.C. Development of Lab-to-Fab Production Equipment Across Several Length Scales for Printed Energy Technologies, Including Solar Cells. *Energy Technol.* **2015**, *3*, 293–304.

16. Huang, H.; Shi, J.; Zhu, L.; Li, D.; Luo, Y.; Meng, Q. Two-step ultrasonic spray deposition of CH$_3$NH$_3$PbI$_3$ for efficient and large-area perovskite solar cell. *Nano Energy* **2016**, *27*, 352–358. [CrossRef]

17. Eslamian, M. Spray-on thin film PV solar cells: Advances, potentials and challenges. *Coatings* **2014**, *4*, 60–84. [CrossRef]

18. Chandrasekhar, P.S.; Kumar, N.; Swami, S.K.; Dutta, V.; Komarala, V.K. Fabrication of perovskite films using an electrostatic assisted spray technique: the effect of the electric field on morphology, crystallinity and solar cell performance. *Nanoscale* **2016**, *8*, 6792–6800. [CrossRef] [PubMed]

19. Heo, J.H.; Lee, M.H.; Jang, M.H.; Im, S.H. Highly efficient CH3NH3PbI$_{3-x}$Cl$_x$ mixed halide perovskite solar cells prepared by re-dissolution and crystal grain growth via spray coating. *J. Mater. Chem. A* **2016**, *4*, 17636–17642. [CrossRef]

20. Abdollahi, N.B.; Gharibzadeh, S.; Ahmadi, V.; Shahverdi, H.R. New Scalable Cold-Roll Pressing for Post-treatment of Perovskite Microstructure in Perovskite Solar Cells. *J. Phys. Chem. C* **2016**, *120*, 2520–2528. [CrossRef]

21. Jung, Y.S.; Hwang, K.; Scholes, F.H.; Watkins, S.E.; Kim, D.Y.; Vak, D. Differentially pumped spray deposition as a rapid screening tool for organic and perovskite solar cells. *Sci. Rep.* **2016**, *6*, 20357. [CrossRef] [PubMed]

22. Zabihi, F.; Ahmadian-Yazdi, M.R.; Eslamian, M. Fundamental Study on the Fabrication of Inverted Planar Perovskite Solar Cells Using Two-Step Sequential Substrate Vibration-Assisted Spray Coating (2S-SVASC). *Nanoscale Res. Lett.* **2016**, *11*, 71. [CrossRef] [PubMed]

23. Habibi, M.; Ahmadian-Yazdi, M.R.; Eslamian, M. Optimization of spray coating for the fabrication of planar perovskite solar cells. *At. Sprays* **2017**. under review.

24. Barrows, A.T.; Pearson, A.J.; Kwak, C.K.; Dunbar, A.D.F.; Buckley, A.R.; Lidzey, D.G. Efficient planar heterojunction mixed-halide perovskite solar cells deposited via spray-deposition. *Energy Environ. Sci.* **2014**, *7*, 2944. [CrossRef]

25. Das, S.; Yang, B.; Gu, G.; Joshi, P.C.; Ivanov, I.N.; Rouleau, C.M.; Aytug, T.; Geohegan, D.B.; Xiao, K. High-Performance Flexible Perovskite Solar Cells by Using a Combination of Ultrasonic Spray-Coating and Low Thermal Budget Photonic Curing. *ACS Photonics* **2015**, *2*, 680–686. [CrossRef]

26. Mullins, W.W.; Sekerka, R.F. Morphological Stability of a Particle Growing by Diffusion or Heat Flow. *J. Appl. Phys.* **1963**, *34*, 323–329. [CrossRef]

27. Sear, R.P. Nucleation at contact lines where fluid–fluid interfaces meet solid surfaces. *J. Phys. Condens. Matter.* **2007**, *19*, 466106. [CrossRef]

28. Djikaev, Y.S.; Ruckenstein, E. Thermodynamics of Heterogeneous Crystal Nucleation in Contact and Immersion Modes. *J. Phys. Chem. A* **2008**, *112*, 11677–11687. [CrossRef] [PubMed]

29. Rahimzadeh, A.; Eslamian, M. Stability of thin liquid films subjected to ultrasonic vibration and characteristics of the resulting thin solid films. *Chem. Eng. Sci.* **2017**, *158*, 587–598. [CrossRef]

30. Deng, Y.; Peng, E.; Shao, Y.; Xiao, Z.; Dong, Q.; Huang, J. Scalable fabrication of efficient organolead trihalide perovskite solar cells with doctor-bladed active layers. *Energy Environ. Sci.* **2015**, *8*, 1544–1550. [CrossRef]

31. Wei, Z.; Chen, H.; Yan, K.; Yang, S. Inkjet printing and instant chemical transformation of a CH3NH3PbI3/nanocarbon electrode and interface for planar perovskite solar cells. *Angew. Chem.* **2014**, *53*, 13239–13243. [CrossRef] [PubMed]

32. Hwang, K.; Jung, Y.S.; Heo, Y.J.; Scholes, F.H.; Watkins, S.E.; Subbiah, J.; Jones, D.J.; Kim, D.Y.; Vak, D. Toward large scale roll-to-roll production of fully printed perovskite solar cells. *Adv. Mater.* **2015**, *27*, 1241–1247. [CrossRef] [PubMed]

33. Schmidt, T.M.; Larsen-Olsen, T.T.; Carlé, J.E.; Angmo, D.; Krebs, F.C. Upscaling of Perovskite Solar Cells: Fully Ambient Roll Processing of Flexible Perovskite Solar Cells with Printed Back Electrodes. *Adv. Energy Mater.* **2015**, *5*, 69. [CrossRef]

34. Kim, J.H.; Williams, S.T.; Cho, N.; Chueh, C.-C.; Jen, A.K.Y. Enhanced Environmental Stability of Planar Heterojunction Perovskite Solar Cells Based on Blade-Coating. *Adv. Energy Mater.* **2015**, *5*, 1401229. [CrossRef]

35. Back, H.; Kim, J.; Kim, G.; Kyun, K.T.; Kang, H.; Kong, J.; Lee, H.S.; Lee, K. Interfacial modification of hole transport layers for efficient large-area perovskite solar cells achieved via blade-coating. *Sol. Energy Mater. Sol. Cells* **2016**, *144*, 309–315. [CrossRef]

36. Park, S.-M.; Noh, Y.-J.; Jin, S.-H.; Na, S.-I. Efficient planar heterojunction perovskite solar cells fabricated via roller-coating. *Solar Energy Mater. Solar Cells* **2016**, *155*, 14–19. [CrossRef]

37. Mallajosyula, A.T.; Fernando, K.; Bhatt, S.; Singh, A.; Alphenaar, B.W.; Blancon, J.-C.; Nie, W.; Gupta, G.; Mohite, A.D. Large-area hysteresis-free perovskite solar cells via temperature controlled doctor blading under ambient environment. *Appl. Mater. Today* **2016**, *3*, 96–102. [CrossRef]

38. Dualeh, A.; Gao, P.; Seok, S.I.; Nazeeruddin, M.K.; Grätzel, M. Thermal behavior of methylammonium lead-trihalide perovskite photovoltaic light harvesters. *Chem. Mater.* **2014**, *26*, 6160–6164. [CrossRef]

39. Brunetti, B.; Cavallo, C.; Ciccioli, A.; Gigli, G.; Latini, A. On the thermal and thermodynamic (in)stability of methylammonium lead halide perovskites. *Sci. Rep.* **2016**, *6*, 31896. [CrossRef] [PubMed]

40. Xie, Y.; Gao, S.; Eslamian, M. Fundamental study on the effect of spray parameters on characteristics of P3HT:PCBM active layers made by spray coating. *Coatings* **2015**, *5*, 488–510. [CrossRef]

41. Wengeler, L.; Schmitt, M.; Peters, K.; Scharfer, P.; Schabel, W. Comparison of large scale coating techniques for organic and hybrid films in polymer based solar cells. *Chem. Eng. Process. Process Intensif.* **2013**, *68*, 38–44. [CrossRef]

Article

Antireflection Coatings for Strongly Curved Glass Lenses by Atomic Layer Deposition

Kristin Pfeiffer [1,*], Ulrike Schulz [2], Andreas Tünnermann [1,2] and Adriana Szeghalmi [1,2,*]

[1]   Institute of Applied Physics, Abbe Center of Photonics, Friedrich Schiller University Jena, Albert-Einstein-Str. 15, 07745 Jena, Germany; Andreas.Tuennermann@iof.fraunhofer.de

[2]   Fraunhofer Institute for Applied Optics and Precision Engineering, Albert-Einstein-Str. 7, 07745 Jena, Germany; Ulrike.Schulz@iof.fraunhofer.de

*   Correspondence: kristin.pfeiffer@uni-jena.de (K.P.); a.szeghalmi@uni-jena.de (A.S.); Tel.: +49-3641-947859 (A.S.)

Received: 12 June 2017; Accepted: 3 August 2017; Published: 9 August 2017

Abstract: Antireflection (AR) coatings are indispensable in numerous optical applications and are increasingly demanded on highly curved optical components. In this work, optical thin films of SiO_2, Al_2O_3, TiO_2 and Ta_2O_5 were prepared by atomic layer deposition (ALD), which is based on self-limiting surface reactions leading to a uniform film thickness on arbitrarily shaped surfaces. $Al_2O_3/TiO_2/SiO_2$ and $Al_2O_3/Ta_2O_5/SiO_2$ AR coatings were successfully applied in the 400–750 nm and 400–700 nm spectral range, respectively. Less than 0.6% reflectance with an average of 0.3% has been measured on a fused silica hemispherical (half-ball) lens with 4 mm diameter along the entire lens surface at $0°$ angle of incidence. The reflectance on a large B270 aspherical lens with height of 25 mm and diameter of 50 mm decreased to less than 1% with an average reflectance < 0.3%. The results demonstrate that ALD is a promising technology for deposition of uniform optical layers on strongly curved lenses without complex in situ thickness monitoring.

Keywords: atomic layer deposition; optical coatings; antireflection; strongly curved surface

---

1. Introduction

Most optical systems contain a large number of lenses or other optical elements. Reflections at each interface reduce the intensity of the transmitted light and thus the overall efficiency of the systems. Reflection losses can be greatly reduced by applying antireflection (AR) coatings to the optical surfaces [1–4]. In addition, AR coatings attenuate the effect of ghost images that are caused by multiple reflection of light from lens surfaces. Optical interference coatings that are typically thin film multilayers of high-refractive and low-refractive index materials demand precise thickness control of each layer. Commonly, thin films applied in precision optics are produced by physical vapor deposition (PVD) [5,6]. Due to the line-of-sight nature of PVD, the surface of a convex lens that is normal to the deposition flux receives a higher amount of material than the edges of the lens. As indicated in Figure 1a, significant thickness gradients might occur on highly curved lenses. Consequently, the required film thickness might not be met over the entire surface of the lens, leading to a distortion of the resulting transmittance spectra. To achieve a better uniformity on curved substrates, complex technical modifications are necessary when using PVD methods, that includes f.e. the constant rotating and tilting of the lens during the deposition, with or without the usage of complicated shadowing masks [7–10]. Antireflection nanostructures are another approach to reduce reflection losses at curved surfaces [11–13]. Nevertheless, for outer lenses of optical systems multilayer AR coatings are preferably used due to their better cleanability and mechanical stability.

Atomic layer deposition (ALD) is an alternative and promising technology for uniform multilayer optical coatings [14–18]. We have previously shown a broadband AR coating on flat high refractive index glasses using SiO_2/HfO_2 multilayers [19]. Atomic layer deposition is also being considered for more complex interference coatings such as dichroic mirrors and narrow bandpass filters [16,20,21]. Atomic layer deposition is a modified form of chemical vapor deposition, where the precursors are sequentially exposed to the surface until saturation is reached [22]. Precursor pulses are separated by inert gas purging; as a result, no gas-phase reactions can take place and the chemical reactions are limited to the surface, see Figure 1b. A typical ALD cycle for the deposition of metal oxides contains four steps: precursor pulse, inert gas purge, oxidizing pulse and inert gas purge. In the case of precursors with low chemical reactivity, often a hold step is introduced after the precursor pulse. Hence, the precursor is trapped in the reactor to entirely react with the surface active groups. Due to this cyclic surface-controlled growth, ALD inherently offers precise thickness control, good thickness uniformity and high reproducibility. It is well known for its conformal film growth on complex nanostructures with high aspect ratios [23,24]. In this work, the capability of ALD for deposition of antireflection coatings on highly curved lenses has been analyzed.

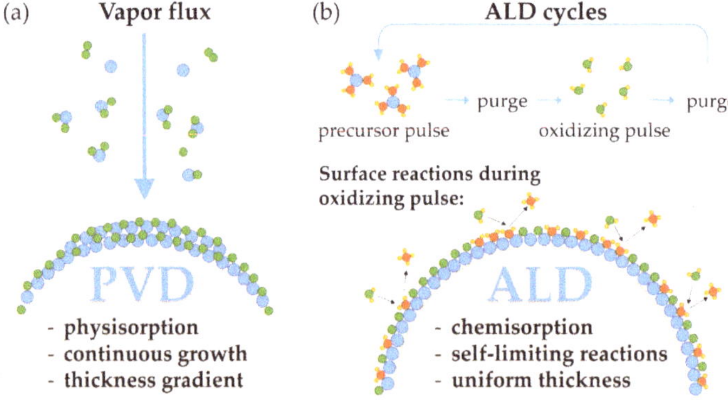

Figure 1. Illustration of (**a**) physical vapor deposition (PVD) deposition and (**b**) atomic layer deposition (ALD) on a hemispherical lens.

This paper first discusses single-layer properties and thickness uniformity of the SiO_2, Al_2O_3, TiO_2, and Ta_2O_5 coatings, then an AR design and its adjustment to the ALD coating is presented on flat glass substrates. Finally, ALD antireflection coatings are demonstrated on curved lenses, firstly on a half-ball lens and secondly on an asphere.

2. Materials and Methods

ALD-deposited SiO_2, Al_2O_3, Ta_2O_5 and TiO_2 thin films were used for the antireflection coatings. Depositions were carried out in an Oxford Instruments (Bristol, United Kingdom) OpAL™ ALD reactor and a Picosun Oy (Espoo, Finland) Sunale™ R200 ALD reactor with a showerhead setup for single-wafer processing. In the OpAL tool, thin films have been grown by plasma-enhanced ALD (PEALD) at substrate temperatures of 100 °C. In the Sunale tool, thermal ALD processes were performed at 300 °C.

All metal oxide films were grown from commercially available precursors. The low-index material SiO_2 was deposited using tris[dimethylamino]silane (3DMAS). Trimethylaluminium (TMA) was applied to deposit the mid-refractive index material Al_2O_3. The high-index materials TiO_2 and Ta_2O_5 were deposited from titanium(IV)isopropoxide (TTIP) and tantalum(V)ethoxide (Ta(OEt)$_5$), respectively. Process parameters are summarized in Table 1.

Table 1. Process parameter for depositing SiO_2, Al_2O_3, TiO_2 and Ta_2O_5 ALD thin films.

Material	Precursor, Source Temperature, Delivery Method	Oxidizing Agent	ALD Tool	ALD Cycle [Pulse/Purge/Gas Stabilization/Oxidizing Pulse/Purge] (in s)
SiO_2	3DMAS, 30 °C, vapor draw	O_2-plasma	OpAL	[0.4 + 4 (hold)/–/5/3/4]
Al_2O_3	TMA, 20 °C, vapor draw	O_2-plasma	OpAL	[0.04/3.5/2.5/5/3.5]
Al_2O_3	TMA, 20 °C, vapor draw	H_2O_2	Sunale	[0.1/4.0/–/0.2/4.0]
TiO_2	TTIP, 50 °C, bubbling	O_2-plasma	OpAL	[1.5/7.0/3.0/6.0/4.0]
Ta_2O_5	Ta(OEt)$_5$, 185 °C, pressure boost	H_2O_2	Sunale	[1.6/6.0/–/2.0/10]

The growth rates and the optical properties of the ALD thin films are determined from single-layer experiments on flat substrates. The growth rates (growth per cycles, GPC) were determined on Si samples by measuring the film thickness with a J.A. Woollam Co. (Lincoln, NV, USA) M-2000® spectroscopic ellipsometer. A Sentech Instruments GmbH (Berlin, Germany) SE850 spectroscopic ellipsometer was used for uniformity mapping of the film thickness on an 8 inch (200 mm) silicon wafer over 180 mm central area on the wafer.

Refractive indices were determined by spectrophotometry of 200 to 300 nm thin films coated on fused silica samples. The reflectance and transmittance spectra were measured with a PerkinElmer, Inc. (Waltham, MA, USA) Lambda 950 spectrophotometer equipped with a home-build accessory for absolute reflectance measurements [25].

For demonstration purposes, antireflection (AR) coatings were applied to a half-ball lens with a diameter of 4 mm and to an aspheric lens with a diameter of 50 mm and a center thickness of 25 mm. An Olympus K. K. USPM-RU-W NIR micro-spectrophotometer (Tokio, Japan) was used to measure the reflectance from a minute spot on different positions of the lens, whereas the lens is placed on a tilt stage and tilted to angles up to 60°. The tilted lens is then moved in the *x*-, *y*- and *z*-direction so that the light from a fixed source is focused on the lens surface and the light rays are perpendicular to the surface (AOI = 0°).

3. Results and Discussion

3.1. Characterization of ALD Thin Films

ALD processes for dielectric thin films have frequently been reported, whereas Al_2O_3 is the most investigated ALD material [26,27]. Al_2O_3 has been applied in ALD antireflection coatings in combination with TiO_2 [17,20] or Ta_2O_5 [28]. Next to this, SiO_2 is a very important low-index material that we recently applied in ALD optical coatings [19,21,28,29]. The properties of the single-layer films resulting from the ALD processes used in this work are summarized in Table 2. The listed GPC values have been used to calculate the necessary ALD cycles to reach the thicknesses of each layer of the following AR coatings.

Table 2. Growth rate on silicon substrates (growth per cycles (GPC) in Å/cycle), refractive index and thickness non-uniformity over a 200 mm area of deposited ALD thin films. The corresponding deposition temperatures are given in brackets.

Material/Properties	SiO_2 (100 °C)	Al_2O_3 (100 °C)	Al_2O_3 (300 °C)	Ta_2O_5 (300 °C)	TiO_2 (100 °C)
Tool	OpAL	OpAL	Sunale	Sunale	OpAL
GPC on Si	1.20	1.21	0.89	0.49	0.29
n @ 550 nm	1.46	1.62	1.66	2.21	2.44
NU% [1]	±1.5%	±1.5%	±2.1%	±4.0%	±2.0%

[1] thickness non-uniformity, defined as NU% = $(d_{max} - d_{min})/2d_{average} \times 100$.

Growth rates and refractive index of SiO_2 thin films are similar to films grown from other commercially available precursors, as BDEAS, BTBAS and AP-LTO®330 [30,31]. Alumina ALD thin films show a lower refractive index at lower deposition temperature [32] owing to a lower density at lower deposition temperatures [26]. The lower GPC of Al_2O_3 at higher deposition temperatures is attributed to less OH groups on the surface. Determined growth rates of Ta_2O_5 are comparable to growth rates reported for Ta_2O_5 thin films deposited using H_2O and $Ta(OEt)_5$, $Ta(NEt_2)_3$ or $Ta(NEt)(NEt_2)_3$ [33–35]. The reported GPC for PEALD TiO_2 using TTIP in the range of 0.3–0.6 Å/cycle are relatively low, whereas thin films grown from TDMAT, Ti-Prime or Ti-Star have slightly higher growth rates than films grown from TTIP [36–38].

Very good lateral film thickness uniformity in the reactor is a prerequisite to ensure a uniform coating on a lens surface. However, non-uniformity in ALD processes is not explicitly analyzed in most research articles. Most tool providers guarantee a standard deviation of the ALD coatings of ca. 1%–3% depending on the material and process conditions. Noteworthy, the upscaling of ALD processes in batch reactors with similar non-uniformity distribution on larger-area batches is feasible [16]. The ALD coatings deposited in the OpAL research tool have thickness non-uniformities (NU%), defined as $(d_{max} - d_{min})/2d_{average} \times 100$, of about ±1.5% ($Al_2O_3$, SiO_2) and ±2.0% (TiO_2). The processes in the Sunale R200 ALD reactor result in a thickness non-uniformity of about ±2.1% for Al_2O_3 and ±4.0% for Ta_2O_5, see Table 2. Elers et al. [39] discussed the sources of non-uniformities in ALD processes including overlapping precursor pulses due to short purge times, death pockets, etc., but also non-uniform gas and temperature distributions in the reactor chamber.

Figure 2a shows the surface mapping of a 200 mm wafer after thermal Al_2O_3 ALD process using 1156 cycles (TMA + H_2O). The alumina film thickness on the wafer in this thermal process does not show a statistical random distribution, but a specific and well-reproducible lower film thickness on the right side of the reactor chamber than on the left side. Interestingly, the precursor and purge gas inlet is on the side where lower film thickness is measured indicating that the precursor dose is sufficient. There might be a temperature gradient on the wafer due to the gases entering the reactor on the right side or the purge time and gas flow might be not sufficient due to inadequate inert gas distribution. In PEALD processes, rather concentric thickness contour lines have been observed (not shown here), whereby the maximum position can be adjusted by the flow rates of the precursor and purge gas. We have demonstrated the possibility to improve the film thickness uniformities by rotating the substrate. Figure 2b depicts a thickness mapping of a wafer where the thermal Al_2O_3 ALD process was stopped after 500 cycles, the wafer manually rotated by 180° and the process continued for another 500 cycles. The wafer rotation could significantly improve the thickness non-uniformity from 2.4% to 0.6%, calculated from 392 mapping points on a 180 mm wafer area.

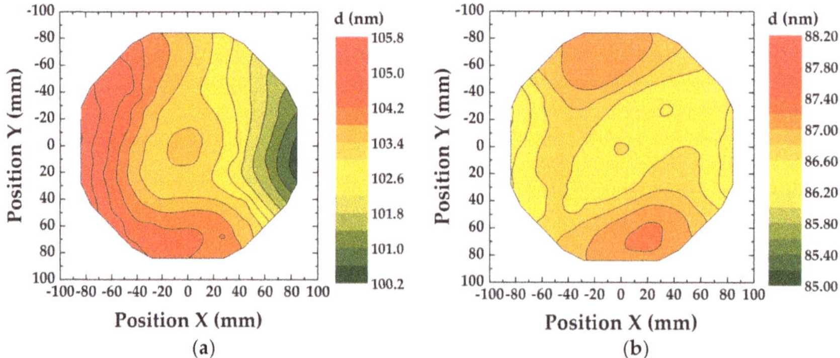

Figure 2. Thickness uniformity mapping of Al_2O_3 (thermal ALD) on a 200 mm wafer measured after (**a**) 1156 ALD cycles without rotation, (**b**) 1000 ALD cycles with manual sample rotation by 180° after 500 ALD cycles.

3.2. Antireflection Coatings on Plane Glass Substrates

An AR design consisting of seven layers has been calculated using the thin film software OptiLayer (version 11.65e, OptiLayer GmbH, Garching, Germany) to reduce the residual reflectance of a fused silica substrate from approximately 3.5% to less than 0.5% in the visible spectral range from 400 to 750 nm. Silicon dioxide was chosen as final layer, as its low refractive index will significantly improve the performance of the AR coating. ALD oxide films are typically amorphous, especially when deposited at low temperatures [40]. However, TiO_2 ALD thin films tend to crystallize at moderate deposition temperatures. The growth of crystallites leads to high surface roughness and, as a result, strong scattering of light. The surface roughness significantly increases for film thicknesses greater than about 40 nm [41]. The crystallization can be inhibited by inserting a thin Al_2O_3 interlayer [42]. In the first design AR-D1 (Table 3) this interlayer was not included into the design, whereas experimentally, the thick 63.9 nm TiO_2 has been split in two thinner TiO_2 layers by introducing a 1.5 nm thin Al_2O_3 interlayer to inhibit the crystallization.

The AR coating was first tested on a plane substrate. By applying the AR-D1 coating to a fused silica glass sample, the reflectance could be reduced to an average reflectance of 0.3% in the visible spectral range from 400 to 750 nm, see Figure 3a. Comparing the reflectance spectra, the AR-D1 coating shows a deviation from the AR-D1 design. It was found that the misfit between design and coating has two origins. First, the 1.5 nm thin Al_2O_3 layer needs to be taken into account when designing the AR coating. This presumption is based on the good agreement of the measured spectrum to calculated expected one that includes the interlayer, see Figure 3b. Thin ALD layers are well known to be very dense and pinhole-free and are intensively investigated for barrier coating [43]. Therefore, the reflections at the interfaces of this ultra-thin layer must be considered in the optical design.

Table 3. Designed layer thickness and necessary ALD cycles of AR coating on fused silica.

Material	AR-D1					AR-D2	
	Experimental			Recalculation		Experimental	
	Design (nm)	Coating (nm)	ALD Cycles	Actual Thickness (nm)	Actual GPC (Å/cycle)	Design and Coating (nm)	ALD Cycles
Al_2O_3	75.1	75.1	621	75.4	1.21	76.8	635
TiO_2	16.1	16.1	556	16.1	0.29	16.1	555
Al_2O_3	20.5	20.5	170	19.9	1.17	21.5	184
TiO_2	63.9	33.3	1150	32.5	0.28	37.5	1293
Al_2O_3	–	1.5	12	1.4	1.17	1.5	13
TiO_2	–	30.6	1054	31.0	0.29	24.3	837
Al_2O_3	13.2	13.2	109	12.8	1.17	14.1	120
TiO_2	25.00	25.00	862	25.3	0.29	24.2	834
SiO_2	92.3	92.3	769	90.2	1.17	92.7	792

Second, a recalculation of the actual thicknesses from the measured spectra using the Film Wizard™ software (version 8.5.0, Scientific Computing International, Carlsbad, CA, USA) showed that Al_2O_3 layers on TiO_2 are thinner as expected. The GPC on the underlying TiO_2 is only 1.17 Å/cycle instead of 1.21 Å/cycle on Si or fused silica. Also, SiO_2 thin films have a lower GPC on the underlying TiO_2 films of only 1.17 Å/cycle instead of the expected 1.20 Å/cycle. Altered growth rates on different substrates have been repeatedly observed and are possibly a reason of different OH group concentrations or irregular OH group distributions on the underlying surface [17].

The film thickness deviation has been 0.4 and 0.6 nm for the alumina layers and approximately 2 nm for silica. This deviation in film thicknesses results in slight deviation of the measured curve (coating AR-D1) and the corrected design curve in Figure 3b. Note that no in situ control of the film thicknesses has been applied during the ALD process. In situ monitoring might be necessary for more complex AR coatings or interference coatings such as narrow bandpass filters or dichroic mirrors [20,21].

A second AR coating AR-D2 was designed including the Al_2O_3 interlayer. Furthermore, the adapted GPC values were used for calculating the necessary ALD cycles of Al_2O_3 and SiO_2 layers on TiO_2, see Table 2. By applying these two corrections, the reflectance of the design and the coating are in an excellent agreement for a sample that was placed in the center of the substrate table, see Figure 4a.

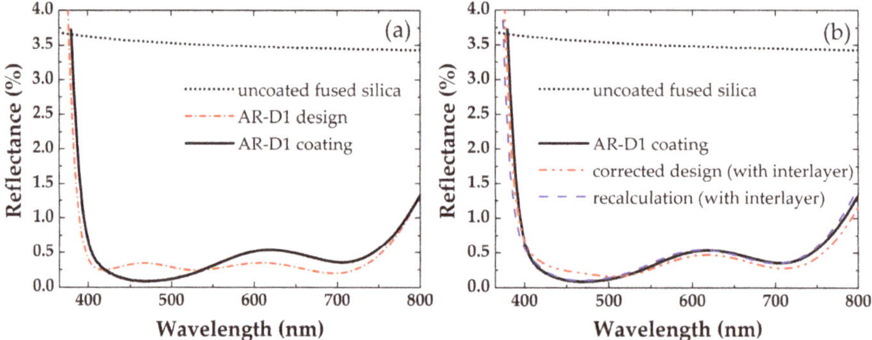

Figure 3. Reflectance spectra of AR-D1 on fused silica reference glass substrate: (**a**) Design and coating; (**b**) Corrected design (with interlayer) and recalculation from measured spectra (taking the Al_2O_3 interlayer into account).

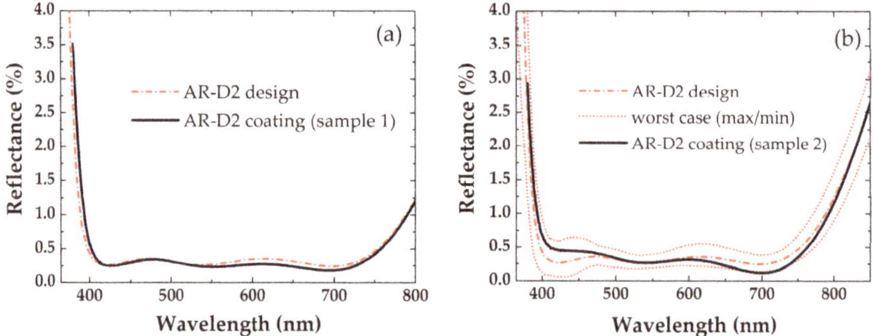

Figure 4. Measured and expected reflectance spectra of AR-D2 on fused silica reference glass substrate positioned: (**a**) at the center of the substrate table; (**b**) at approximately 75 mm from the center table.

As the thickness non-uniformity was expected to be the main origin of errors, a worst-case analysis was performed, whereas the maximum allowed thickness deviation was specified as the expected NU% of each material, see Table 2. The area between the dotted lines in Figure 4b indicates the worst-case error corridor of the calculated maximum possible deviations from the theoretical reflectance spectra. To estimate the influence of the NU experimentally, next to the fused silica substrate (sample 1) that was placed in the middle of the substrate table, a second substrate (sample 2) was positioned at approximately 75 mm from the center of the table during the deposition. The measured reflectance spectra of sample 2 lies within the worst-case error corridor, indicating that the small deviations to the AR design are most likely a consequence of the lateral film thickness non-uniformity on the substrate table.

3.3. Antireflection Coatings on a Half-Ball Lens

The antireflection coating AR-D2 was applied to a hemispherical lens. The refractive index of the lens was calculated from the measured reflectance spectra of the uncoated half-ball lens, which is slightly higher than the reflectance of the fused silica glass substrate, see Figure 5a. Due to the higher effective refractive index of the bare lens, the appearance of the expected and measured AR spectra on the lens differs from the spectra on the coated glass slab (compare Figures 4a and 5b). The measured spectra of the AR coating on the lens is in good agreement with the adapted AR design (Figure 5b).

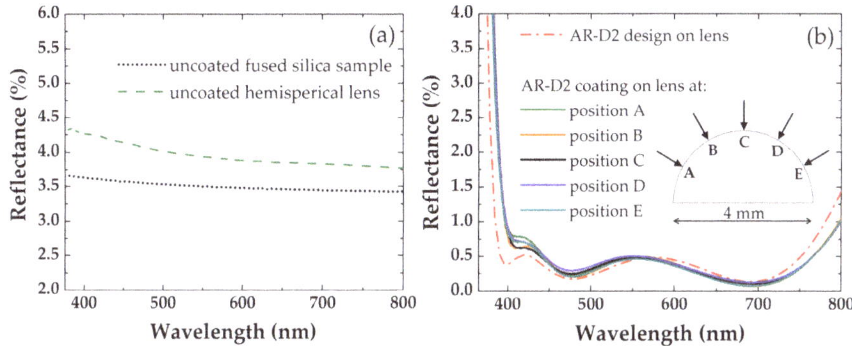

Figure 5. (**a**) Measured reflectance spectra of uncoated fused silica reference glass substrate and uncoated fused silica half-ball lens; (**b**) Measured reflectance spectra (AOI = 0°) of AR-coated fused silica half-ball lens at different positions of the lens.

It should be emphasized that the reflectance spectra are consistent at all positions on the lens. Hence, the AR coating was deposited uniformly on the hemispherical lens without any complex equipment to control the layer thickness.

An upright-positioned glass sample was used as reference sample for the edge of the glass plate since it is not possible to measure the reflectance at the very edge of the lens. As shown in Figure 6, the measured reflectance is in very good agreement with the design. The deposition occurs simultaneously on both sides of the glass sample and the measured spectra are identical on both sides of the substrates. The results show that the ALD-technology is not restricted to the radius of curvature,

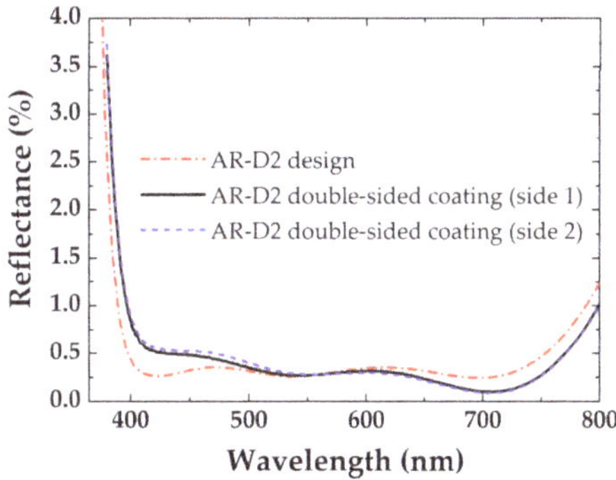

Figure 6. Measured and expected reflectance spectra (AOI = 0°) of AR-D2 deposited simultaneously on both sides of an upright-positioned flat fused silica glass substrate.

The AR performance of the coated lens depends on the position in the chamber due to the lateral thickness non-uniformity. Hence, it has been possible to obtain an excellent AR coating on a curved lens matching very well the design curve, see Figure 7.

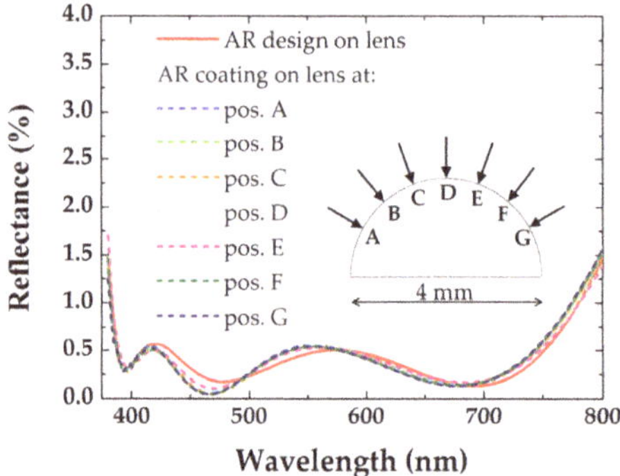

Figure 7. Measured and expected reflectance spectra (AOI = 0°) of AR-coated fused silica half-ball lens at different positions of the lens.

3.4. Antireflection Coating for an Aspheric Lens

To confirm that ALD AR-coatings can be also used to reduce reflection losses of larger lenses, a second antireflection coating was applied to a steeply curved aspheric lens with center height of 25 mm and a diameter of 50 mm. Ta_2O_5 was used as high-index material for the AR coating, since the grown Ta_2O_5 ALD thin films are amorphous and hence no additional Al_2O_3 interlayer is needed to inhibit crystal growth. TEM and SEM images of about 5 nm, 35 nm and 200 nm Ta_2O_5 thin films show an amorphous structure [33,44,45]. X-ray diffraction (XRD) measurements also confirmed the amorphous nature of 200 nm tanatala thin films grown from $Ta(OEt)_5$ at 300 °C. These spectra are not shown here.

The glass lens has a refractive index that is similar to that of B270. An AR-D3 coating (see Table 4) was designed to reduce the reflectance of a B270 substrate from approximately 4.0% to less than 0.5% in the visible spectral range from 400 nm to 700 nm. The first part of the coating design is based on the patented AR-hard® (Jena, Germany). A thin high-index layer is sandwiched by two thicker lower-index layers forming a symmetrical stack of three-quarter-wave optical thickness [46]. Silicon dioxide was chosen as final layer to attain a low residual reflectance. After completion of the Al_2O_3/Ta_2O_5 sequences in the Sunale R200 tool at a deposition temperature of 300 °C, the samples were unloaded to atmosphere and transferred to the OpAL tool for further processing of the top SiO_2 layer at 100 °C.

Table 4. Designed layer thickness and necessary ALD cycles of AR coating on B270.

Material	AR-D3	
	Thickness (nm)	ALD Cycles
Al_2O_3	101.6	1181
Ta_2O_5	11.2	208
Al_2O_3	186.9	2173
Ta_2O_5	35.0	714
Al_2O_3	21.8	253
Ta_2O_5	43.6	891
SiO_2	93.7	787

Figure 8 depicts the reflectance of the AR-D3 design and the AR-D3-coated lens. The reflectance spectra of the lens show a good match to the design. Minor deviations between design and the measured reflectance at the inclined surface of the lens (position A and E) may be attributed to a temperature gradient of the lens during deposition and to lateral thickness non-uniformity across the chamber.

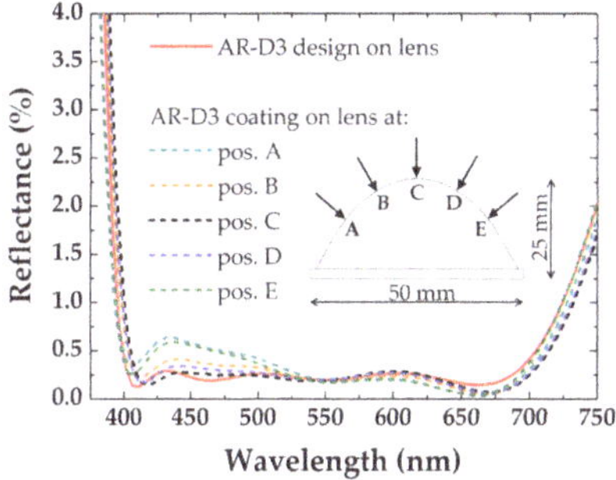

Figure 8. Measured and expected reflectance spectra (AOI = 0°) of the AR-coated steeply curved B270 aspherical lens at different positions of the lens.

4. Conclusions

Atomic layer deposition successfully applies to deposit antireflection coatings on strongly curved lenses. In particular, the average reflectance could be minimized to 0.3% for a fused silica half-ball lens with 4 mm diameter and a steeply curved B270 aspherical lens in the visible spectral range from 400 to 750 nm and 400 to 700 nm, respectively. Similar reflectance spectra across the entire lens surface at normal light incidence are a result of the very good conformality of ALD coatings. The good agreement between design and coatings confirms the precise thickness control of ALD thin films. Thickness monitoring was not necessary to reach the desired film thicknesses, but only the counting of ALD cycles. Moreover, it was demonstrated that the conformal deposition is not restricted to the radius of curvature of a lens, as an AR coating that was deposited simultaneously on both sides of a flat glass substrate showed identical spectra on both sides. Noteworthy, these antireflection coatings are demonstrated in two commercially available tools with significantly different configurations, indicating that ALD can become highly attractive for production purposes.

Further development of ALD coating equipment such as spatial ALD, atmospheric pressure ALD, and batch tools will increase the applicability of this technology for high volume applications. The slow deposition rate is considered as the main disadvantage of ALD. The long deposition times are generally the consequence of the required purge times between the precursor pulses. The possibility to perform double-sided coatings increases the throughput of this coating technology. Spatial ALD [47] is a promising approach to shorten the purge times, in that the substrate is moved to different precursor zones, hence precursor pulses are spatially separated and purge steps become dispensable. The use of batch coaters is another possibility to increase the throughput [16]. However, the lateral thickness uniformity needs to be improved for scale up to large-area. For a better uniformity, both the chamber design and the precursor chemistry needs to be improved. The development of precursors that are highly reactive and volatile, but at the same time thermally stable and non-corrosive, as well as the design of a tool, that comprises a uniform gas distribution, a homogeneous temperature and the absence of dead volumes remains a future challenge [39].

Although further research and developments are needed, ALD is a promising method to deposit optical thin films that can be prospectively applied for optical coatings on complex formed optical components due to the very good conformality of ALD coatings (convex and concave lenses, cylinders, ball lenses, etc.).

Acknowledgments: The research was supported by the Deutsche Forschungsgemeinschaft (DFG) (Emmy-Noether-Project SZ253/1-1) and the European Space Agency (ESA) (Contract No. 4000109161/13/NL/RA). This work was partially supported by the FhG Internal Programs under Grant No. Attract 066-601020. Kristin Pfeiffer thanks the Carl Zeiss Foundation for promoting her doctoral research studies. The authors gratefully acknowledge David Kästner for the micro-spectrophotometer measurements.

Author Contributions: Adriana Szeghalmi and Kristin Pfeiffer conceived and designed the experiments; Ulrike Schulz supported the design of the coatings; Kristin Pfeiffer performed the experiments, analyzed the data and wrote the paper. All authors critically revised the article.

Conflicts of Interest: The authors declare no conflict of interest.

References

1. Raut, H.K.; Ganesh, V.A.; Nair, A.S.; Ramakrishna, S. Anti-reflective coatings: A critical, in-depth review. *Energy Environ. Sci.* **2011**, *4*, 3779. [CrossRef]
2. Buskens, P.; Burghoorn, M.; Mourad, M.C.D.; Vroon, Z. Antireflective coatings for glass and transparent polymers. *Langmuir* **2016**, *32*, 6781–6793. [CrossRef] [PubMed]
3. Hedayati, K.M.; Elbahri, M. Antireflective coatings: Conventional stacking layers and ultrathin plasmonic metasurfaces, a mini-review. *Materials* **2016**, *9*, 497. [CrossRef] [PubMed]
4. Das, N.; Islam, S. Design and Analysis of nano-structured gratings for conversion efficiency improvement in GaAs solar cells. *Energies* **2016**, *9*, 690. [CrossRef]
5. Pulker, H.K. Optical coatings deposited by ion and plasma pvd processes. *Surf. Coat. Technol.* **1999**, *112*, 250–256. [CrossRef]
6. Thielsch, R.; Gatto, A.; Heber, J.; Kaiser, N. A comparative study of the UV optical and structural properties of SiO_2, Al_2O_3, and HfO_2 single layers deposited by reactive evaporation, ion-assisted deposition and plasma ion-assisted deposition. *Thin Solid Films* **2002**, *410*, 86–93. [CrossRef]
7. Liu, C.; Kong, M.; Chun, G.; Gao, W.; Li, B. Theoretical design of shadowing masks for uniform coatings on spherical substrates in planetary rotation systems. *Opt. Express* **2012**, *20*, 23790. [CrossRef] [PubMed]
8. Wang, Z.; West, P.R.; Meng, X.; Kinsey, N.; Shalaev, V.M.; Boltasseva, A. Angled physical vapor deposition techniques for non-conformal thin films and three-dimensional structures. *MRS Commun.* **2016**, *6*, 17–22. [CrossRef]
9. Gross, M.; Dligatch, S.; Chtanov, A.C. Optimization of coating uniformity in an ion beam sputtering system using a modified planetary rotation method. *Appl. Opt.* **2011**, *50*, C316–C320. [CrossRef] [PubMed]
10. Zhang, L.C.; Cai, X.K. Uniformity masks design method based on the shadow matrix for coating materials with different condensation characteristics. *Sci. World J.* **2013**, *10*, 160792. [CrossRef] [PubMed]
11. Diao, Z.; Kraus, M.; Brunner, R.; Dirks, J.H.; Spatz, J.P. Nanostructured stealth surfaces for visible and near-infrared light. *Nano. Lett.* **2016**, *16*, 6610–6616. [CrossRef] [PubMed]
12. Schulze, M.; Lehr, D.; Helgert, M.; Kley, E.B.; Tünnermann, A. Transmission enhanced optical lenses with self-organized antireflective subwavelength structures for the uv range. *Opt. Lett.* **2011**, *36*, 3924–3926. [CrossRef] [PubMed]
13. Taylor, C.D.; Busse, L.E.; Frantz, J.; Sanghera, J.S.; Aggarwal, I.D.; Poutous, M.K. Angle-of-incidence performance of random anti-reflection structures on curved surfaces. *Appl. Opt.* **2016**, *55*, 2203–2213. [CrossRef] [PubMed]
14. Riihelä, D.; Ritala, M.; Matero, R.; Leskelä, M. Introduction atomic layer epitaxy for the deposition of optical thin films. *Thin Solid Films* **1996**, *289*, 250–255. [CrossRef]
15. Kumagai, H.; Toyoda, K.; Kobayashi, K.; Obara, M.; Iimura, Y. Titanium oxide aluminum oxide multilayer reflectors for "water-window" wavelengths. *Appl. Phys. Lett.* **1997**, *70*, 2338–2340. [CrossRef]
16. Maula, J. Atomic layer deposition for industrial optical coatings. *Chin. Opt. Lett.* **2010**, *8*, 53–58. [CrossRef]
17. Li, Y.H.; Shen, W.D.; Zhang, Y.G.; Hao, X.; Fan, H.H.; Liu, X. Precise broad-band anti-refection coating fabricated by atomic layer deposition. *Opt. Commun.* **2013**, *292*, 31–35. [CrossRef]
18. Li, Y.H.; Shen, W.D.; Hao, X.; Lang, T.T.; Jin, S.Z.; Liu, X. Rugate notch filter fabricated by atomic layer deposition. *Appl. Opt.* **2014**, *53*, A270–A275. [CrossRef] [PubMed]
19. Pfeiffer, K.; Shestaeva, S.; Bingel, A.; Munzert, P.; Ghazaryan, L.; van Helvoirt, C.; Kessels, W.M.M.; Sanli, U.T.; Grévent, C.; Schütz, G.; et al. Comparative study of ALD SiO_2 thin films for optical applications. *Opt. Mater. Express* **2016**, *6*, 660–670. [CrossRef]

20. Szeghalmi, A.; Helgert, M.; Brunner, R.; Heyroth, F.; Gosele, U.; Knez, M. Atomic layer deposition of Al_2O_3 and TiO_2 multilayers for applications as bandpass filters and antireflection coatings. *Appl. Opt.* **2009**, *48*, 1727–1732. [CrossRef]

21. Shestaeva, S.; Bingel, A.; Munzert, P.; Ghazaryan, L.; Patzig, C.; Tünnermann, A.; Szeghalmi, A. Mechanical, structural, and optical properties of PEALD metallic oxides for optical applications. *Appl. Opt.* **2017**, *56*, C47–C59. [CrossRef] [PubMed]

22. Ritala, M.; Niinistö, J. Atomic layer deposition. In *Chemical Vapour Deposition: Precursors, Processes and Applications*; Jones, A.C., Hitchman, M.L., Eds.; Royal Society of Chemistry: Cambridge, UK, 2009; pp. 158–206.

23. Kariniemi, M.; Niinisto, J.; Vehkamaki, M.; Kemell, M.; Ritala, M.; Leskela, M.; Putkonen, M. Conformality of remote plasma-enhanced atomic layer deposition processes: An experimental study. *J. Vac. Sci. Technol. A* **2012**, *30*, 01A115. [CrossRef]

24. Siefke, T.; Kroker, S.; Pfeiffer, K.; Puffky, O.; Dietrich, K.; Franta, D.; Ohlidal, I.; Szeghalmi, A.; Kley, E.B.; Tünnermann, A. Materials pushing the application limits of wire grid polarizers further into the deep ultraviolet spectral range. *Adv. Opt. Mater.* **2016**, *4*, 1780–1786. [CrossRef]

25. Stenzel, O.; Wilbrandt, S.; Friedrich, K.; Kaiser, N. Realistische Modellierung der NIR/VIS/UV-optischen Konstanten dünner optischer Schichten im Rahmen des Oszillatormodells. *Vak. Forsch. Prax.* **2009**, *21*, 15–23. [CrossRef]

26. Groner, M.D.; Fabreguette, F.H.; Elam, J.W.; George, S.M. Low-temperature Al_2O_3 atomic layer deposition. *Chem. Mat.* **2004**, *16*, 639–645. [CrossRef]

27. Puurunen, R.L. Surface chemistry of atomic layer deposition: A case study for the trimethylaluminum/water process. *J. Appl. Phys.* **2005**, *97*, 121301. [CrossRef]

28. Pfeiffer, K.; Schulz, U.; Tünnermann, A.; Szeghalmi, A. $Ta_2O_5/Al_2O_3/SiO_2$—Antireflective coating for non-planar optical surfaces by atomic layer deposition. *Proc. SPIE* **2017**, *10115*, 1011513.

29. Ghazaryan, L.; Kley, E.B.; Tünnermann, A.; Szeghalmi, A. Nanoporous SiO_2 thin films made by atomic layer deposition and atomic etching. *Nanotechnology* **2016**, *27*, 255603. [CrossRef] [PubMed]

30. Dingemans, G.; van Helvoirt, C.A.A.; Pierreux, D.; Keuning, W.; Kessels, W.M.M. Plasma-assisted ALD for the conformal deposition of SiO_2: Process, material and electronic properties. *J. Electrochem. Soc.* **2012**, *159*, H277–H285. [CrossRef]

31. Putkonen, M.; Bosund, M.; Ylivaara, O.M.E.; Puurunen, R.L.; Kilpi, L.; Ronkainen, H.; Sintonen, S.; Ali, S.; Lipsanen, H.; Liu, X.W.; et al. Thermal and plasma enhanced atomic layer deposition of SiO_2 using commercial silicon precursors. *Thin Solid Films* **2014**, *558*, 93–98. [CrossRef]

32. Ylivaara, O.M.E.; Liu, X.W.; Kilpi, L.; Lyytinen, J.; Schneider, D.; Laitinen, M.; Julin, J.; Ali, S.; Sintonen, S.; Berdova, M., et al. Aluminum oxide from trimethylaluminum and water by atomic layer deposition: The temperature dependence of residual stress, elastic modulus, hardness and adhesion. *Thin Solid Films* **2011**, *552*, 124–135. [CrossRef]

33. Kukli, K.; Ritala, M.; Leskelä, M. Atomic layer epitaxy growth of tantalum oxide thin films from $Ta(OC_2H_5)_5$ and H_2O. *J. Electrochem. Soc.* **1995**, *142*, 1670–1675. [CrossRef]

34. Blanquart, T.; Longo, V.; Niinistö, J.; Heikkilä, M.; Kukli, K.; Ritala, M.; Leskelä, M. High-performance imido–amido precursor for the atomic layer deposition of Ta_2O_5. *Semicond. Sci. Technol.* **2012**, *27*, 074003. [CrossRef]

35. Hausmann, D.M.; de Rouffignac, P.; Smith, A.; Gordon, R.; Monsma, D. Highly conformal atomic layer deposition of tantalum oxide using alkylamide precursors. *Thin Solid Films* **2003**, *443*, 1–4. [CrossRef]

36. Zhao, C.; Hedhili, M.N.; Li, J.; Wang, Q.; Yang, Y.; Chen, L.; Li, L. Growth and characterization of titanium oxide by plasma enhanced atomic layer deposition. *Thin Solid Films* **2013**, *542*, 38–44. [CrossRef]

37. Potts, S.E.; Kessels, W.M.M. Energy-enhanced atomic layer deposition for more process and precursor versatility. *Coord. Chem. Rev.* **2013**, *257*, 3254–3270. [CrossRef]

38. Xie, Q.; Musschoot, J.; Deduytsche, D.; Vanmeirhaeghe, R.; Detavernier, C.; Van Den Berghe, S. Growth kinetics and crystallization behavior of TiO_2 films prepared by plasma enhanced atomic layer deposition. *J. Electrochem. Soc.* **2008**, *155*, H688–H692. [CrossRef]

39. Elers, K.E.; Blomberg, T.; Peussa, M.; Aitchison, B.; Haukka, S.; Marcus, S. Film uniformity in atomic layer deposition. *Chem. Vap. Depos.* **2006**, *12*, 13–24. [CrossRef]

40. Miikkulainen, V.; Leskela, M.; Ritala, M.; Puurunen, R.L. Crystallinity of inorganic films grown by atomic layer deposition: Overview and general trends. *J. Appl. Phys.* **2013**, *113*, 021301. [CrossRef]

41. Ratzsch, S.; Kley, E.B.; Tunnermann, A.; Szeghalmi, A. Influence of the oxygen plasma parameters on the atomic layer deposition of titanium dioxide. *Nanotechnology* **2015**, *26*, 024003. [CrossRef] [PubMed]

42. Ylivaara, O.M.E.; Kilpi, L.; Liu, X.W.; Sintonen, S.; Ali, S.; Laitinen, M.; Julin, J.; Haimi, E.; Sajavaara, T.; Lipsanen, H.; et al. Aluminum oxide/titanium dioxide nanolaminates grown by atomic layer deposition: Growth and mechanical properties. *J. Vac. Sci. Technol. A* **2017**, *35*, 01B105. [CrossRef]

43. Hoffmann, L.; Theirich, D.; Pack, S.; Kocak, F.; Schlamm, D.; Hasselmann, T.; Fahl, H.; Raupke, A.; Gargouri, H.; Riedl, T. Gas diffusion barriers prepared by spatial atmospheric pressure plasma enhanced ALD. *ACS Appl. Mater. Interfaces* **2017**, *9*, 4171–4176. [CrossRef] [PubMed]

44. Szeghalmi, A.; Senz, S.; Bretschneider, M.; Gösele, U.; Knez, M. All dielectric hard X-ray mirror by atomic layer deposition. *Appl. Phys. Lett.* **2009**, *94*, 133111. [CrossRef]

45. Mayer, M.; Grévent, C.; Szeghalmi, A.; Knez, M.; Weigand, M.; Rehbein, S.; Schneider, G.; Baretzky, B.; Schütz, G. Multilayer Fresnel zone plate for soft X-ray microscopy resolves sub-39 nm structures. *Ultramicrospcopy* **2011**, *111*, 1706–17011. [CrossRef] [PubMed]

46. Schulz, U.; Schallenberg, U.B.; Kaiser, N. Symmetrical periods in antireflective coatings for plastic optics. *Appl. Opt.* **2003**, *42*, 1346–1351. [CrossRef] [PubMed]

47. Poodt, P.; Cameron, D.C.; Dickey, E.; George, S.M.; Kuznetsov, V.; Parsons, G.N.; Roozeboom, F.; Sundaram, G.; Vermeer, A. Spatial atomic layer deposition: A route towards further industrialization of atomic layer deposition. *J. Vac. Sci. Technol. A* **2012**, *30*, 010802. [CrossRef]

Article

Continuous Tip Widening Technique for Roll-to-Roll Fabrication of Dry Adhesives

Sung Ho Lee [1], Hoon Yi [2], Cheol Woo Park [1], Hoon Eui Jeong [2,*] and Moonkyu Kwak [1,*]

[1]   School of Mechanical Engineering, Kyungpook National University, Daegu 41566, Korea;
     lee_sh@knu.ac.kr (S.H.L.); chwoopark@knu.ac.kr (C.W.P.)
[2]   Department of Mechanical Engineering, Ulsan National Institute of Science and Technology, Ulsan 34130,
     Korea; poemist@unist.ac.kr
*   Correspondence: hoonejeong@unist.ac.kr (H.E.J.); mkkwak@knu.ac.kr (M.K.);
     Tel.: +82-51-217-2339 (H.E.J.); +82-53-950-5573 (M.K.)

Received: 23 August 2018; Accepted: 28 September 2018; Published: 30 September 2018

Abstract: In this study, we reported continuous partial curing and tip-shaped modification methods for continuous production of dry adhesive with microscale mushroom-shaped structures. Typical fabrication methods of dry adhesive with mushroom-shaped structures are less productive due to the failure of large tips on pillar during demolding. To solve this problem, a typical pillar structure was fabricated through partial curing, and tip widening was realized through applying the proper pressure. Polyurethane acrylate was used in making the mushroom structure using two-step UV-assisted capillary force lithography (CFL). To make the mushroom structure, partial curing was performed on the micropillar, followed by tip widening. Dry adhesives with properties similar to those of typical mushroom-shaped dry adhesives were fabricated with reasonable adhesion force using the two-step UV-assisted CFL. This production technology was applied to the roll-to-roll process to improve productivity, thereby realizing continuous production without any defects. Such a technology is expected to be applied to various fields by achieving the productivity improvement of dry adhesives, which is essential for various applications.

Keywords: dry adhesive; biomimetics; continuous process; partial curing

---

1. Introduction

Microstructure-based dry adhesives, which were inspired by the feet of gecko lizards and beetles, have attracted attention in various applications due to their strong adhesion, repeatability, reversibility, and self-cleaning properties [1–13]. They are applied to a transfer or fixing device, such as a glass substrate and silicon wafer, and an attempt has been made to replace the existing electrostatic or vacuum chuck [8,14]. These dry adhesives are typically made from polymeric pillars of the size of a few hundred nanometers to tens of micrometers, and the tip of the pillar is shaped like a spatula or mushroom to maximize its adhesive properties [15–17]. Generally, dry adhesive production method is a molding technique, such as soft lithography; thus, manufacturing the aforementioned size structure is easy [4]. However, constructing a precisely designed tip shape on a master can be a complex process [4,8,18]. For example, in the case of producing mushroom-like microstructure dry adhesives on a master, the tip must be formed inside the silicon wafer to form a wide tip on the surface of the dry adhesive when replicated. This tip plays an important role in securing the adhesion; this role has been confirmed in studies that handle the comparison of adhesive strengths according to various tip shapes [15]. Based on the literature, a dry adhesive with a thin and wide area tip has a high adhesive strength, and a complicated process is required to produce a master for such a dry adhesive. An example of fabricating a master with a wide tip is to use the footing effect that occurs in the etch stop layer during deep RIE (Reactive Ion Etching) process [19]. If etching is performed to a thickness over the working depth simply by using the SOI (Silicon on insulator) wafer, then the etching ion does not proceed in the depth direction in the SiO_2 layer but is etched laterally. This process is called the footing effect, in which forming a wide tip on a master is relatively easy. However, this method has a disadvantage; it cannot precisely control the thickness or size of the tip. To solve this problem, the tip precisely fabricated on the wafer surface through the surface micromachining and the tip buried masters are manufactured by fusion bonding between the surface-machined wafers and bare wafer [5,8,18]. The fabricated master mold can be used to produce a dry adhesive after a simple release layer treatment; however, it is difficult to apply in a continuous production using a roll-to-roll process due to poor demolding (e.g., tip tearing depending on the polymer used) [20]. Particularly, in the roll-to-roll process, which can only use a flexible polymer mold, the breakage of the tip during demolding becomes evident. Therefore, the productivity should be improved through new methods. In the present work, we introduce an advanced UV molding technique called two-step UV-assisted capillary force lithography (CFL) to produce a mushroom-shaped dry adhesive without special demolding failure. This method can be applied to the roll-to-roll-based apparatus without any changes; thus, it is applied to the prototype roll-to-roll apparatus to test the mass production possibility. Moreover, the pull-off strength of the fabricated dry adhesive is measured. The measured adhesive strength of approximately 9.5 N/cm^2 is slightly lower than that of dry adhesives made by typical one-step molding [8]; nevertheless, this value is appropriate for application. In addition, higher adhesion can be achieved if the tip size can be widened in a future optimization process. The study of tip widening using partial curing has been experimentally proven in previous studies [21]. In this paper, we intend to implement this process continuously.

2. Materials and Methods

Polyurethane acrylate (PUA): PUA is an ultraviolet ray curable material composed of prepolymer with acrylate group, a photoinitiator, a crosslinker and a release agent. The PUA was dropped onto a master mold with pillar shaped microstructures, which were fabricated by surface micromachining. A PET (polyethylene terephthalate) film as supporting layer was covered on the dropped PUA, followed by UV exposure for 40 s (wavelength = 250–400 nm, dose = 100 mJ/cm^2). After UV exposure, the PET film was peeled off from the master with cured PUA structure.

Partially cured micropillar: Our fabrication method was based on the two-step process of capillary UV molding and additional tip shape modification. For the fabrication of PUA partially cured micropillars, the polydimethylsiloxane (PDMS) mold with micro holes was used with enough window for partial curing time. Then, the PUA resin was partially cured by UV exposure for 15–20 s (wavelength = 365 nm, intensity = 100 mW/cm^2).

Tip widening: The film was placed on the top of the partially cured pillar, and further curing was performed for 30 s while applying the appropriate pressure of 20–30 kPa (wavelength = 365 nm, intensity = 500 mW/cm^2). For a uniform pressure distribution, a thin PDMS block was placed as a buffer on top of the PET film prior to the application of pressure.

Measurement of pull-off force: We measured pull-off force and durability with custom built equipment [8]. To measure normal adhesion force between fabricated dry adhesive and target smooth substrate, the smooth substrate was made with a glass sample and was moved vertically with speed of 3 mm/s by a step motor connected crank. The device is equipped with a load cell capable of measuring the load in the z-axis direction, allowing measurement of the pull-off force including the preload. Pull-off forces were measured for various preloads, and marathon tests were conducted at the speed of 50 cycles per minute and 40 N/cm^2 preload.

3. Results and Discussion

Figure 1 shows the schematic of the two-step UV-assisted CFL and that of a continuous production apparatus. PDMS is a known porous material. In this study, PDMS was used to induce partial curing and fabricate hierarchical structures. In the case of the PDMS mold, the surface of the molded polymer film was exposed to air that was permeated through a porous PDMS mold. The trapped or permeated oxygen inhibited UV curing by scavenging radicals generated from the photoinitiator by UV [22]. Thus, the surface of the PUA resin in contact with the air remained tacky, whereas the resin beneath the surface cured completely. Some pre-dissolved oxygen existed in the liquid resin. This oxygen was rapidly consumed under UV exposure due to high mobility and reactivity of oxygen with a large number of initial radicals. Therefore, the formation of a tacky surface could be attributed to the diffusion of oxygen that was trapped in the mold cavity or permeated out of the mold. In a previous research on two-step CFL, micro/nano hierarchical structure was fabricated using a nano-structured mold after the microstructure fabrication [22]. A microstructure was fabricated, and a wide tip was then formed on the partially cured microstructure by pressurization using a flat substrate. The tip of the microstructure was widened by simply using a glass substrate, and the low adhesion between glass and PUA enabled the production of a mushroom structure without any surface treatment. After the process proof at the wafer level, this principle was used to realize continuous production using a roll-to-roll apparatus. The roll-to-roll apparatus was divided into microstructure and tip production sections. In the microstructure production section, a PDMS negative mold was attached to the roll, whereas the tip production section is composed of a simple quartz cylinder and a urethane roll. The partial curing phenomenon in the microstructure fabrication was theoretically predictable. On the basis of the literature, the oxygen saturation inside the polymer resin, which depends on the depth of the resin from the surface, can be expressed as [22]:

$$\frac{C_{O_2}(x)}{C_{O_2\_\text{surface}}} = \cosh\left(k/D_{O_2/\text{polymer}}\right)^{1/2} x - \tanh\left(k/D_{O_2/\text{polymer}}\right)^{1/2} L \cdot \sinh\left(k/D_{O_2/\text{polymer}}\right)^{1/2} x$$

where C_{O_2} is the oxygen concentration in the cavities, $D_{O_2/\text{polymer}}$ is the oxygen diffusivity coefficient in the polymer layer, $C_{O_2\_\text{surface}}$ is the surface concentration of oxygen, L is the depth of the cured polymer resin, x is distance from free-surface between air and resin. When the appropriate constants were substituted for this equation ($k = 1$, $D_{O_2/\text{polymer}} = 10^{-12}$ m$^2 \cdot$s^{-1}), the penetration of oxygen during the microstructure fabrication of PUA resin proceeded from the resin surface to approximately 4–5 µm. That is, the upper 4 µm of the microstructure remained in a state capable of subsequent patterning while still maintaining a non-cured tacky surface. The tip structure to be fabricated through this study was expected to express sufficiently by this partial curing process, in which the diameter was approximately 1 µm larger than the micropillar and the thickness was approximately 500 nm. To fabricate an appropriate partially cured micropillar, the intensity of UV LED at 365 nm wavelength was adjusted to 100 mW/cm^2 and exposed for a suitable time (15–20 s) to fabricate a microstructure whose surface remained tacky.

Thereafter, various pressures were applied to the tip widening. As a result, suitable pressure (25 kPa) was identified, and the mushroom structure was constructed. As shown in Figure 2, an extremely low pressure (10 kPa) did not sufficiently expand the tip, and an extremely high pressure (40 kPa) caused columnar deformation or collapse. At lower pressures, the not fully cured resin is irregularly wetted, clumped like contaminants, or wicked down the pillar to create an unusual pillar shape. (Figure 2a) To solve this problem, a structure with a wide tip is successfully fabricated by making perfect contact with high pressure. The column part is already cured and fixed to its own shape differently from the tacky surface of the tip part, but since it is not over-cured, it is possible to deform a little by strong pressure and exposure so that the upper part becomes a little wider column structure after the pressure applying process. The production speed is determined by the type of material, the time to produce partial curing, and the time required to derive the structural change through pressure. In this work, production speed was about 100 cm^2/min. We expect this result to be improved through future optimization studies.

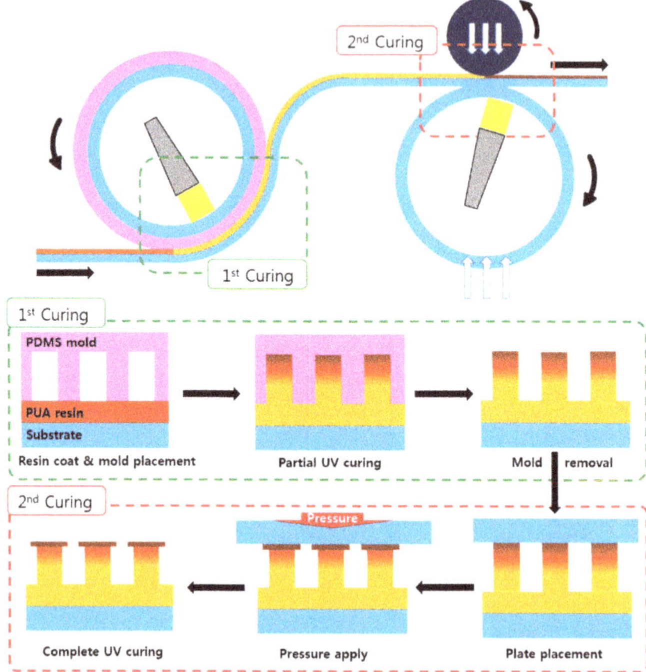

Figure 1. Schematic of the continuous fabrication process for mushroom-shaped dry adhesives and procedure for mushroom-shaped structures via two-step UV molding process.

Figure 2. SEM images of mushroom structures fabricated with various pressures: (**a**) 0 kPa, (**b**) 10 kPa, (**c**) 25 kPa.

After the appropriate pressurization, the mushroom structure was fabricated through second exposure, and the pull-off strength of the sample was measured by a typical adhesion measurement. The preload was fixed to 40 N, and the repeated adhesion test was performed using a load cell. We predicted the life of fabricated dry adhesive through the analysis of the tendency of decrease in adhesive force by using a marathon test equipment made from a simple crank system. The marathon test was conducted in a normal laboratory environment and not in a clean room; thus, various contaminants may be found.

Figure 3 shows the pull-off strength measurement results of the prepared dry adhesive samples. The measured adhesive strength was lower than that of the mushroom-shaped wide-tip microstructures ($\sim$20 N/cm^2) produced by conventional methods because forming a wide and thin tip necessary for strong adhesion was impossible, which may be solved by optimization of the subsequent process.

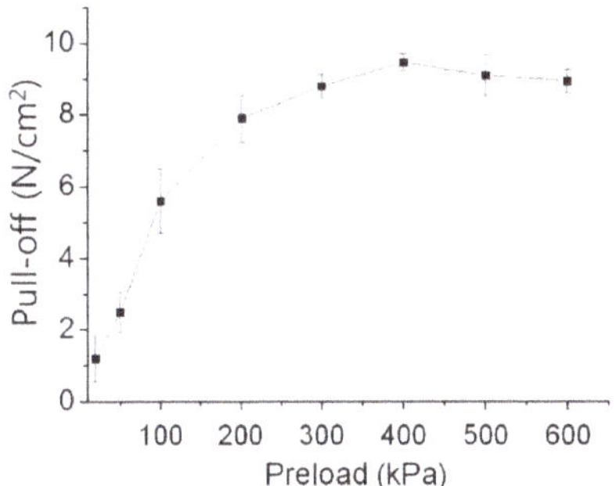

Figure 3. Measurement data of the pull-off strength of fabricated dry adhesives as a function of preload.

Selection of the resin, optimization of the primary curing time and UV intensity, the amount of pressure applied during the secondary curing, and the curing time should be considered for the production of a wide tip. Figure 4 shows the results of the marathon test for the durability of dry adhesives. The adhesives initially showed a maximum adhesive strength of 9.5 N/cm^2 and a 10% decrease in adhesion after repeated use of 5000 times. As shown in Figure 4, numerous contaminants were found on the dry adhesive surface observed after 5000 uses, and several defects were also found. Although not many, defects were present where the tip portion was torn out or the microstructures were matted together. Usually, the tip is torn rather than the column is broken. This can be seen as fatigue failure and breakage at the tip connection part where the greatest stress concentration occurs. After using the test for 5000 cycles, the surfaces of the dry adhesive and the substrate were cleaned using a commercial pressure sensitive adhesive tape. Subsequently, the recovered adhesive strength was 94% of the original adhesive strength. As shown in Figure 4, all of the contaminants, except for defects, were removed through the cleaning process, and some of the matting was recovered during the cleaning process using the adhesive tape, resulting in a slight increase in the adhesion. The subsequent 5000 additional adhesion tests resulted in the same level of adhesion of the contaminants and additional defects. However, the degradation tendency was considerably better than that of dry adhesives with conventional wide tips. Adhesive strength and lifetime are inversely related to each other; thus, the fabricated dry adhesives are likely to be fully utilized in applications where they have to be used for a long period of time with proper adhesive strength. The marathon test result shows reasonable adhesive strength of up to 10,000 times, and the test is expected to be used in various fields.

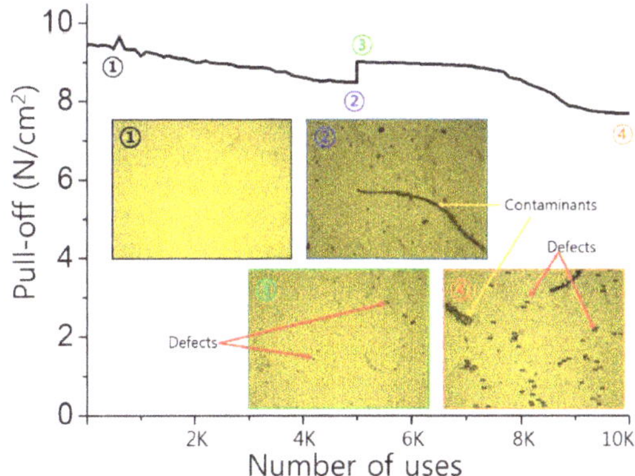

Figure 4. Durability test of the fabricated dry adhesives over 10,000 cycles of attachment and detachment. Inset microscopic images for dry adhesive surfaces at initial state, 5000 cycles before cleaning, 5000 cycles after cleaning, and 10,000 times usage.

4. Conclusions

We developed a continuous production technology of mushroom-shaped dry adhesives by utilizing the two-step UV-assisted CFL. The mushroom shape was produced by continuously producing the micropillars and by applying pressure to the partially cured micropillars, the tip portion was widened, and the fabricated dry adhesives exhibited an adhesive strength of 9.5 N/cm^2 on the glass substrate. In addition, the dry adhesives were excellent in terms of durability due to their shape characteristics and exhibited an adhesive strength of approximately 80% even in a marathon test of 10,000 times. The complicated mold-making process was necessary in the conventional continuous production technique. However, this process could be omitted to reduce the production cost and increase production efficiency. Furthermore, if the productivity is secured with the performance of a reasonable dry adhesive, then it can be applied to various application fields.

Author Contributions: Conceptualization, M.K.; Methodology, H.E.J. and M.K.; Experiments, H.Y. and S.H.L.; Writing-Original Draft Preparation, S.H.L. and M.K.; Writing-Review & Editing, M.K., H.E.J. and C.W.P.; Supervision, M.K. and H.E.J.; Funding Acquisition, M.K.

Funding: This work was supported by the National Research Foundation of Korea (2016R1A2B4007858) funded by the Korean Government (MSIP).

Acknowledgments: The authors thank Jeong Hyeon Lee for his technical support.

Conflicts of Interest: The authors declare no conflicts of interest.

References

1. Autumn, K. The gecko effect: Dynamic dry adhesive microstructures. *Am. Zool.* **2001**, *41*, 1382–1383.
2. Gorb, S.N.; Varenberg, M. Mushroom-shaped geometry of contact elements in biological adhesive systems. *J. Adhes. Sci. Technol.* **2007**, *21*, 1175–1183. [CrossRef]
3. Varenberg, M.; Pugno, N.M.; Gorb, S.N. Spatulate structures in biological fibrillar adhesion. *Soft Matter* **2010**, *6*, 3269–3272. [CrossRef]
4. Kwak, M.K.; Pang, C.; Jeong, H.E.; Kim, H.N.; Yoon, H.; Jung, H.S.; Suh, K.Y. Towards the next level of bioinspired dry adhesives: New designs and applications. *Adv. Funct. Mater.* **2011**, *21*, 3606–3616. [CrossRef]
5. Yi, H.; Hwang, I.; Lee, J.H.; Lee, D.; Lim, H.; Tahk, D.; Sung, M.; Bae, W.G.; Choi, S.J.; Kwak, M.K.; et al. Continuous and scalable fabrication of bioinspired dry adhesives via a roll-to-roll process with modulated ultraviolet-curable resin. *ACS Appl. Mater. Interfaces* **2014**, *6*, 14590–14599. [CrossRef] [PubMed]
6. Yi, H.; Kang, M.; Kwak, M.K.; Jeong, H.E. Simple and reliable fabrication of bioinspired mushroom-shaped micropillars with precisely controlled tip geometries. *ACS Appl. Mater. Interfaces* **2016**, *8*, 22671–22678. [CrossRef] [PubMed]
7. Hwang, I.; Kim, H.N.; Seong, M.; Lee, S.H.; Kang, M.; Yi, H.; Bae, W.G.; Kwak, M.K.; Jeong, H.E. Multifunctional smart skin adhesive patches for advanced health care. *Adv. Healthcare Mater.* **2018**, *7*, 1800275. [CrossRef] [PubMed]
8. Lee, S.H.; Kim, S.W.; Kang, B.S.; Chang, P.S.; Kwak, M.K. Scalable and continuous fabrication of bio-inspired dry adhesives with a thermosetting polymer. *Soft Matter* **2018**, *14*, 2586–2593. [CrossRef] [PubMed]
9. Hensel, R.; Moh, K.; Arzt, E. Engineering micropatterned dry adhesives: From contact theory to handling applications. *Adv. Funct. Mater.* **2018**, *28*, 1800865. [CrossRef]
10. Heepe, L.; Xue, L.; Gorb, S.N. *Bio-Inspired Structured Adhesives: Biological Prototypes, Fabrication, Tribological Properties, Contact Mechanics, and Novel Concepts*; Springer Nature: Heidelberg, Germany, 2017.
11. Zhao, T.; Yu, K.; Li, L.; Zhang, T.; Guan, Z.; Gao, N.; Yuan, P.; Li, S.; Yao, S.Q.; Xu, Q.H.; et al. Gold nanorod enhanced two-photon excitation fluorescence of photosensitizers for two-photon imaging and photodynamic therapy. *ACS Appl. Mater. Interfaces* **2014**, *6*, 2700–2708. [CrossRef] [PubMed]
12. Xue, L.; Kovalev, A.; Dening, K.; Eichler-Volf, A.; Eickmeier, H.; Haase, M.; Enke, D.; Steinhart, M.; Gorb, S.N. Reversible adhesion switching of porous fibrillar adhesive pads by humidity. *Nano Lett.* **2013**, *13*, 5541–5548. [CrossRef] [PubMed]

13. Xue, L.; Kovalev, A.; Thöle, F.; Rengarajan, G.T.; Steinhart, M.; Gorb, S.N. Tailoring normal adhesion of arrays of thermoplastic, spring-like polymer nanorods by shaping nanorod tips. *Langmuir* **2012**, *28*, 10781–10788. [CrossRef] [PubMed]

14. Jeong, H.E.; Lee, J.K.; Kim, H.N.; Moon, S.H.; Suh, K.Y. A nontransferring dry adhesive with hierarchical polymer nanohairs. *Proc. Natl. Acad. Sci. USA* **2009**, *106*, 5639–5644. [CrossRef] [PubMed]

15. del Campo, A.; Greiner, C.; Alvarez, I.; Arzt, E. Patterned surfaces with pillars with controlled 3D tip geometry mimicking bioattachment devices. *Adv. Mater.* **2007**, *19*, 1973–1977. [CrossRef]

16. Greiner, C.; Arzt, E.; del Campo, A. Hierarchical gecko-like adhesives. *Adv. Mater.* **2009**, *21*, 479–482. [CrossRef]

17. Boesel, L.F.; Greiner, C.; Arzt, E.; del Campo, A. Gecko-inspired surfaces: A path to strong and reversible dry adhesives. *Adv. Mater.* **2010**, *22*, 2125–2137. [CrossRef] [PubMed]

18. Kwak, M.K.; Jeong, H.E.; Suh, K.Y. Rational design and enhanced biocompatibility of a dry adhesive medical skin patch. *Adv. Mater.* **2011**, *23*, 3949–3953. [CrossRef] [PubMed]

19. Seok, S.; Lee, B.; Kim, J.; Kim, H.; Chun, K. A new compensation method for the footing effect in mems fabrication. *J. Micromech. Microeng.* **2005**, *15*, 1791. [CrossRef]

20. Sameoto, D.; Menon, C. A low-cost, high-yield fabrication method for producing optimized biomimetic dry adhesives. *J. Micromech. Microeng.* **2009**, *19*, 115002. [CrossRef]

21. Jeong, H.E.; Suh, K.Y. Precise tip shape transformation of nanopillars for enhanced dry adhesion strength. *Soft Matter* **2012**, *8*, 5375–5380. [CrossRef]

22. Jeong, H.E.; Kwak, R.; Khademhosseini, A.; Suh, K.Y. UV-assisted capillary force lithography for engineering biomimetic multiscale hierarchical structures: From lotus leaf to gecko foot hairs. *Nanoscale* **2009**, *1*, 331–338. [CrossRef] [PubMed]

Article

Preparation and Characterization of Bioplastics from Grass Pea Flour Cast in the Presence of Microbial Transglutaminase

C. Valeria L. Giosafatto [1], Asmaa Al-Asmar [1,2], Antonio D'Angelo [1], Valentina Roviello [3], Marilena Esposito [1] and Loredana Mariniello [1,*]

[1] Department of Chemical Sciences, University of Naples "Federico II", Complesso Universitario di Monte Sant'Angelo, Via Cinthia 4, 80126 Naples, Italy; giosafat@unina.it (C.V.L.G.); asmaa.alasmar@unina.it (A.A.-A.); antoniodangelo2601@gmail.com (A.D.); marilena.esposito2@unina.it (M.E.)

[2] Analysis, Poison control and Calibration Center (APCC), An-Najah National University, P.O. Box 7, Nablus, Palestine

[3] CeSMA, University of Naples "Federico II", 80126 Naples, Italy; valentina.roviello@unina.it

* Correspondence: loredana.mariniello@unina.it; Tel.: +39-081-2539470

Received: 22 October 2018; Accepted: 23 November 2018; Published: 28 November 2018

Abstract: The aim of this work was to prepare bioplastics, from renewable and biodegradable molecules, to be used as edible films. In particular, grass pea (*Lathyrus sativus* L.) flour was used as biopolymer source, the proteins of which were structurally modified by means of microbial transglutaminase, an enzyme able to catalyze isopeptide bonds between glutamines and lysines. We analyzed, by means of Zeta-potential, the flour suspension with the aim to determine which pH is more stable for the production of film-forming solutions. The bioplastics were produced by casting and they were characterized according to several technological properties. Optical analysis demonstrated that films cast in the presence of the microbial enzyme are more transparent compared to the untreated ones. Moreover, the visualization by scanning electron microscopy demonstrated that the enzyme-modified films possessed a more compact and homogeneous structure. Furthermore, the presence of microbial transglutaminase allowed to obtain film more mechanically resistant. Finally, digestion experiments under physiological conditions performed in order to obtain information useful for applying these novel biomaterials as carriers in the industrial field, indicated that the enzyme-treated coatings might allow the delivery of bioactive molecules in the gastro-intestinal tract.

Keywords: grass pea; bioplastics; mechanical properties; transglutaminase; Zeta-potential

1. Introduction

Nowadays life without plastics seems to be unimaginable because of their important role in our society and applications in almost all the areas of daily life, from packaging to food, medical and communication technology to cars. The majority of these plastics are based on very unsustainable fossil resources, causing pollution that affects the entire environment. According to Geyer et al. [1], 8300 million metric tons (Mt) as of virgin plastics have been produced to date and in 2015, approximately 6300 Mt of plastic waste had been generated, around 9% of which had been recycled, 12% was incinerated, and 79% was accumulated in landfills or in the natural environment. In order to reduce pollution from plastics, during the last few decades, researchers have been developing different technologies to produce new kind of biobased plastics and bioplastics that are similar or better than the traditional ones [2–4]. According to European Bioplastic [5], bioplastics are a large family of different materials that are either biobased and/or biodegradable. Among bioplastics, it is worthwhile to talk about edible films, that are important in the sector of food packaging and represent a potential new highly competitive market [6]. Edible films have received increasing attention mostly because of their advantages as components of food packaging over fossil-fuel materials [3,4]. An edible film is a preformed, thin layer, made of edible material, which can be placed either on or between food components, playing an important role on the conservation, distribution and marketing of foodstuff [7]. Some of its functions consist in protecting food products from mechanical damage, physical, chemical and microbiological activities [6,8,9]. The aim of this work was to prepare and characterize a new kind of hydrocolloid bioplastics, to be used as edible films, based on grass pea (*Lathyrus sativus* L.) flour, a legume from the family of *Fabaceae* [10,11]. Grass pea flour is very profitable because the legume is resistant to both abiotic (dryness, water stagnation and very poor and dry soils) and biotic (high capability to fix atmospheric nitrogen, high seeds and proteins yield) stresses [10]. The films were prepared by using grass pea flour either treated or not treated with microbial transglutaminase (mTGase, E.C. 2.3.2.13), an enzyme easily purified from the culture medium of *Streptoverticillium mobaraense* [12], able to catalyze the crosslinking of proteins via acyl transfer reactions between the γ-carboxamide group of glutamine residues and the ε-amino group of lysine residues, leading to the formation of inter-molecular and intra-molecular isopeptide bonds [13,14]. mTGase is Ca^{2+} independent, and it is active over a broad range of temperatures and pHs with an optimal activity at approximately 40 °C and pH of 7–7.5. These properties are important prerequisites for an application of an enzyme in the industrial sector. The film forming solutions prepared by using grass pea flour modified or not by mTGase have been characterized and the resulting bioplastics investigated according to their transparency, microstructure and mechanical properties. Moreover, digestibility studies carried out under physiological conditions were performed in order to apply such bioplastics in either food or pharmaceutical sector.

2. Materials and Methods

2.1. Materials

Grass pea seeds were bought in a local supermarket (Naples, Italy). Microbial transglutaminase (ACTIVA WM, Ajinomoto, Tokyo, Japan, specific activity 92 U/g) was purchased from Prodotti Gianni S.p.A. Milan, Italy. Glycerol, used as a plasticizer for the preparation of films, was purchased from Sigma (St. Louis, MO, USA). Acrylamide and Blue Brilliant Coomassie were purchased from Bio-Rad (Segrate, Milan, Italy). All other chemical reagents were purchased from the following companies: Amersham Pharmacia (Stockholm, Sweden), Merck (Rome, Italy), Roche (Grenzach-Wyhlen, Germany). The remaining chemicals and solvents used in this study were of analytical grade unless specified.

2.2. Grass Pea Flour Characterization

2.2.1. Protein Content

The amount of proteins was determined by measuring the nitrogen content of the material and multiplying that value by the factor 6.25 [15].

2.2.2. Zeta-Potential and Particle Size of Grass Pea Flour Suspension

The suspension was prepared dissolving the flour in distilled water at concentration of 1 mg mL^{-1}. In order to sediment the starch, the sample was kept overnight at 4 °C. After that, the sample was centrifuged at 10,000 rpm for 5 min at the temperature of 10 °C and the pellet was removed. Before the analysis, the supernatant was further filtrated with 0.45 micron filter and the pH was adjusted to 2 by using HCl 0.1 N. A titration as function of pH (from 2 to 12) was carried out to measure Zeta-potential and particle size of grass pea flour suspension by means of Zetasizer Nano-ZSP (Malvern®, Worcestershire, UK). As titrants we have used 0.01, 0.1 and 1 N NaOH solutions, respectively. All results were analyzed by using the Zetasizer software (version 7.12).

2.3. Film Forming Solutions Preparation and Characterization

2.3.1. mTGase Preparation

The enzyme solution was prepared by dissolving the commercial preparation "Activa" (containing 1% of enzyme and 99% of maltodextrins, specific activity 92 U/g) in distilled water at a concentration of 20 U mL^{-1}. The mixture was stirred for 10 min to allow the solubilization of mTGase preparation.

2.3.2. Film Forming Solution (FFS) Preparation

Flour (41.5 g) was dissolved in 500 mL of distilled water (concentration of 83 mg mL^{-1}) and the stock solution was stirred for 1 h. Afterwards the pH was adjusted from 6.5 to 9 with NaOH 1 N. Then the solution was centrifuged at 10,000 rpm for 10 min at 4 °C and the pellet was removed. The pH of supernatant was adjusted to 7 by adding HCl 1 N and the solution was centrifuged under the same conditions (described above) in order to remove additional aggregates. FFSs without mTGase were prepared by mixing 30 mL withdrawn from solution and mixed with 200 µL (corresponding to 8% of glycerol in respect to protein content) of glycerol (100 mg mL^{-1} *w/v*) and 19.8 mL of distilled water. FFSs with mTGase were prepared as previously described and by adding 1 mL of mTGase (this amount corresponds to 33 U of enzyme/g of protein). Both FFSs, treated or not with mTGase, were incubated for 2 h at 37 °C. After incubation, the pH of FFSs was adjusted to 9. The final volume of each solution was 50 mL.

2.3.3. Zeta-Potential and Particle Average Size

Zeta-potential, average particle size, and polydispersity index of the FFSs, containing or not mTGase, were analyzed using the Zetasizer Nano-ZSP. Three independent Zeta-potential measurements at pH 9 were carried out on each sample of FFSs (1 mL) introduced in the measurement vessel. Temperature was set up at 25 °C, applied voltage was 200 mV and duration of each analysis was approximately of 10 min. The software calculated mean diameter of particles, determined at pH 9 by using dynamic light scattering, and the polydispersity index, representing the relative variance in the particle size distribution. The device uses a helium-neon laser of 4 mW output power operating at the fixed wavelength of 633 nm (wavelength of laser red emission). All the results were reported as mean ± standard deviation.

2.3.4. Viscosity

Standard Ostwald capillary viscometer was used for the experiments. The viscometer was thermostated to 30.0 ± 0.1 °C in a water bath. The flow time for water was approximately 83.3 ± 0.1 s. Flow times for the FFSs (untreated and treated with mTGase) were measured in duplicate using a stopwatch. Each FFS was diluted 1:2 starting from concentration of flour of 29.3 mg mL^{-1} to 1.83 mg mL^{-1}. Specific viscosity was obtained by using the following equation:

$$\text{Specific Viscosity} = (\text{FFS flow time} - \text{water flow time})/(\text{water flow time}) \tag{1}$$

2.4. Film Preparation and Characterization

2.4.1. Film Casting

FFSs, prepared as described above, were poured in Petri's dishes and placed in a climatic chamber at 25 °C and 45% of R.H. for 48–72 h.

2.4.2. Thickness

Thickness was obtained using a micrometer (Metrocontrol Srl, Casoria, Naples, Italy, mod. H062 with the precision of ± 2 µm). The results were obtained measuring thickness in four random points, then the average and the standard deviation were calculated.

2.4.3. Opacity

The opacity of each samples was investigated reproducing the method used by Shevkani et al. [16]. This method is based on the measurement of absorbance at 600 nm (spectrophotomer UV/Vis SmartSpec 3000 Bio-Rad, Segrate, Milan, Italy) divided by the thickness (mm). All the samples (our bioplastics and commercial material used for references) were cut into pieces of 1 cm × 3 cm and they were let adhere perfectly to the wall of the cuvette.

2.4.4. Scanning Electron Microscopy (SEM)

SEM analysis of both surface and cross-section of grass pea flour-based films was carried out by using field emission scanning electron microscope (Nova NanoSem 450-FEI-Thermo Fisher, Scientific, Waltham, MA, USA). Briefly, the samples were placed on an aluminum stub by using a graphite adhesive tape. A thin coat of gold and palladium was sputtered at a current of 20 mA for 90 s. The sputter-coated samples were then introduced into the specimen chamber and the images were acquired at an accelerating voltage of 3 kV, (4.4–5.2) mm working distance, through the Everhart Thornley Detector (ETD, 450-FEI-Thermo Fisher, Scientific, Waltham, MA, USA). Two different samples of each type of films were subjected to SEM and four micrographs of each sample were taken. Micrographs of surfaces and cross-sections were obtained taking parts at 2600× magnification of the samples.

2.4.5. Mechanical Properties

Film tensile strength, elongation at break and Young's modulus were determined by using an Instron Universal Testing Instrument (model no. 5543A, Instron Engineering Corp., Norwood, MA, USA). Film sample strips (1 cm wide and 5 cm long), obtained by using a sharp razor blade, were equilibrated for 2 h at 50% RH and 25 °C in an environmental chamber, and four samples of each film type were tested. Tensile properties were measured according to the ASTM D882-97 [17]. The initial grip separation was 40 mm, and the crosshead speed was 5 mm min^{-1} in tension mode. The acquisition and elaboration of the data were made by the using the software BlueHill 2.21.

2.4.6. In Vitro Film Digestion

The films prepared in the absence and in the presence of mTGase were subjected to a three-stage in vitro digestion by using adult model [18–20], under simulated oral, gastric and duodenal physiological conditions. For our analyses, 5 mg of each type of films were incubated in 600 µL of Simulated Salivary Fluid (SSF, 150 mM of NaCl, 3 mM of urea, pH 6.9) for 5 min at 170 rpm. Afterwards the samples were subjected to gastric and duodenal digestion as described by Giosafatto et al. [18] with some modifications. Briefly, aliquots (100 µL) of Simulated Gastric Fluid (SGF, 0.15 M of NaCl, pH 2.5) were placed in 1.5 mL microcentrifuge tubes and incubated at 37 °C. 75 µL of films dissolved in SSF, the pH of which was adjusted to 2.5 with HCl 6 M, were added together with pepsin (1:20 *w/w* respect to grass pea protein content) to each of the SGF vials to start the digestion reaction. At intervals of 1, 2, 5, 10, 20, 40, 60 min, 40 µL of 0.5 M of ammonium bicarbonate (NH$_4$HCO$_3$) were added to each vial to stop the pepsin reaction. The control was set up by incubating the sample for 60 min without the protease. Duodenal digestions were performed using, as the starting material, the gastric digests after 60 min, adjusted to pH 6.5 with 0.5 M Bis-Tris HCl pH 6.5. Bile salts (sodium taurocholate and sodium glycodeoxycholate) dissolved in Simulated Duodenal Fluid (SDF, 0.15 M of NaCl at pH 6.5) were added to a final concentration of 4 mM. After equilibrating at 37 °C for 10 min, trypsin, chymotrypsin (the ratio of trypsin and chymotrypsin with test proteins was 1:400 (*w/w*) and 1:100 (*w/w*), respectively) were added to the duodenal mix. Aliquots were removed over the 120 min digestion time course and proteolysis was stopped by addition of a two-fold excess of soybean Bowmann-Birk trypsin-chymotrypsin inhibitor above that calculated to inhibit trypsin and chymotrypsinin of the digestion mix. The control was carried out by incubating the sample without the proteases for 120 min. The samples were then analyzed using the SDS-PAGE (12%) procedure described below.

2.4.7. Sodium Dodecyl Sulphate Polyacrylamide Gel Electrophoresis (SDS-PAGE)

For SDS-PAGE of FFSs, an aliquot of 5 μL of sample buffer (15 mM of Tris–HCl, pH 6.8, containing 0.5% (*w/v*) of SDS, 2.5% (*v/v*) of glycerol, 200 mM of β-mercaptoethanol, and 0.003% (*w/v*) of bromophenol blue) were added to aliquots of 20 μL of FFS (either untreated or mTGase treated) and analyzed by 12% SDS-PAGE. The SDS-PAGE of cast films was carried out by dissolving 20 mg of each film in 250 μL of sample buffer. The samples were treated at 100 °C for 5 min, and then centrifuged for 10 min at 13000 ×g. Three μL of each supernatant were analyzed by SDS-PAGE (12%). For the analysis of film digestion carried out under physiological conditions, 5 μL of sample buffer were added to 20 μL of each protolyzed film sample and analyzed by 12% SDS-PAGE.

In all cases SDS-PAGE was performed as described by Laemmli [21], at constant voltage (80 V for 2–3 h), and the proteins were stained with Coomassie Brilliant Blue R250 (Bio-Rad, Segrate, Milan, Italy). Bio-Rad Precision Protein Standards were used as molecular weight markers.

2.4.8. Densitometry Analysis

Densitometry analysis was carried out by means of Image Lab software (version 5.2.1) from Bio-Rad Laboratories. Each SDS-PAGE image was analyzed by detecting all the lanes and protein bands. Protein bands, possessing a relative molecular mass (*Mr*) of 50 kDa were used to determine the band intensity of film digested in the absence of mTGase respect to the control carried out without proteases. Protein bands >250 kDa were used to determine the band intensity of film digested in the presence of the microbial enzyme with respect to control incubated without proteolytic enzymes.

2.5. Statistical Analysis

All data were analyzed by means of JMP software 5.0 (SAS Institute, Cary, NC, USA), used for all statistical analyses. The data were subjected to analysis of variance, and the means were compared using the Tukey-Kramer HSD test. Differences were considered to be significant at $p < 0.05$.

3. Results and Discussion

3.1. Stability of Grass Pea Flour Suspension and FFSs

In order to evaluate the pH stability of grass pea flour dissolved in water at a concentration of 1 mg mL^{-1}, a titration as function of pH was carried out to measure Zeta-potential. The charge of particles depends on the solvent used [22]. Zeta-potential is a function of the surface charge of the particle, of adsorbed layer at the interface, and of the nature and composition of the surrounding suspension medium. Generally, Zeta-potential values higher than ±25 mV indicate that the solution is quite stable [22]. The data reported in Figure 1 show a moderate stability of grass pea flour suspension, in fact, the potential changes from +27 to –25 mV by varying the pH from 2 to 12. At pH 4, the suspension became unstable (0.01 ± 0.53 mV) since this pH is close to isoelectric point of grass pea proteins (globulins and albumins), which are in the range of 4–6, as also demonstrated by Romano et al. by performing two-dimensional gel electrophoresis [23]. Also, the dimension of particles was quite stable (data not shown) during the titration, being the main particle size diameter equal to roughly 200 nm of diameter for all the pHs analyzed (data not shown).

FFSs were prepared, both in the presence and the absence of mTGase, at pH 9, since, as reported in Figure 1, we have an acceptable stability at this pH (Zeta-potential = −25 mV). After the preparation, 1 mL of each solution was analyzed at Zetasizer Nano-ZSP (Malvern®, Worcestershire, UK) to confirm the stability.

In Table 1 results about average size, polydispersity index and Zeta-potential of FFSs are reported. The solutions possess a similar Zeta-potential, regardless the presence of mTGase. The average size seems to be slightly reduced in the FFS prepared in the presence of the enzyme as already reported by Porta et al. [8]. It is important to note that polydispersity index is around 0.5 indicating that the size of particles is quite uniform in both the systems.

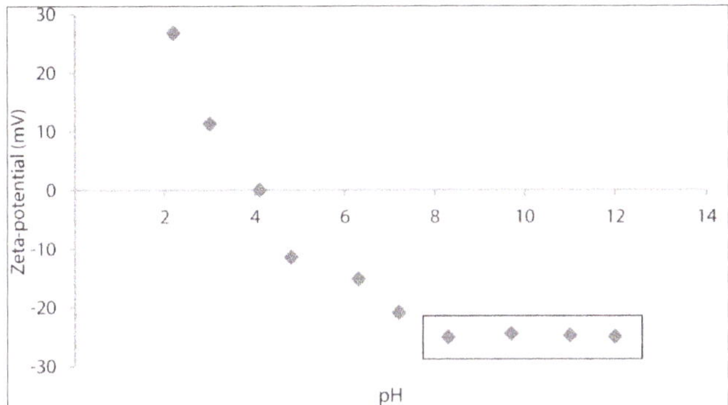

Figure 1. Zeta-potential of grass pea flour suspension as function of pH. Values in the frame represent the Zeta-potential range of stability.

Table 1. Average size, polydispersity index and Zeta-potential of FFSs treated or not by mTGase.

Sample pH 9	Average Size (d/nm)	Polydispersity Index	Zeta-Potential (mV)
FFS	139.40 ± 1.06 [a]	0.53 ± 0.01 [a]	−27.10 ± 1.90 [a]
FFS + mTGase	127.30 ± 2.50 [a]	0.57 ± 0.02 [a]	−28.00 ± 1.63 [a]

Values are mean ± standard deviation; Means followed by the same letters are not significant different (Tukey-Kramer test, $p < 0.05$).

3.2. Modification of Grass Pea Flour Proteins by Means of mTGase

Both FFSs and cast films were analyzed by means of SDS-PAGE (12%). The Figure 2 demonstrated that mTGase was able, under these experimental conditions, to modify grass pea proteins. In fact, from the gel (Figure 2) it is possible to note the formation of *Mr* polymers and the concomitant disappearance of lower *Mr* protein bands in the sample treated with mTGase both in FFSs (Figure 2A) and the solubilized films (Figure 2B), indicating that the mTGase-catalyzed reaction occurs also in the casting system. This result was also supported by viscosity analysis that demonstrated that FFS treated with mTGase has a higher viscosity than the one untreated (Supplementary Materials). An increase of viscosity is due to mTGase activity that, by forming intra and intermolecular ε-N-(γ-glutamyl)-lysine crosslinks between proteins, reinforces the network. These results are in good agreement with those obtained by Nio et al. [24], and Temiz et al. [25] that studied the gelation of casein and soybean globulins by mTGase, demonstrating that the enzyme treatment increases the viscosity of solution.

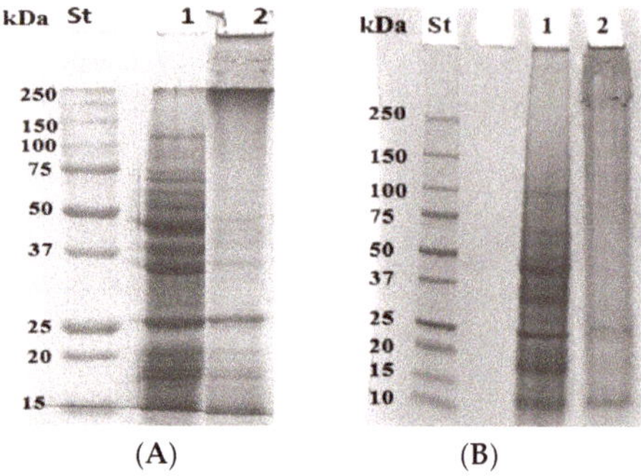

Figure 2. Panel A-SDS-PAGE of untreated (lane 1) and mTGase-treated (lane 2) FFSs. Panel B-SDS-PAGE of solubilized films cast in the absence (lane 1) and presence (lane 2) of mTGase. St, Molecular weight standards, Bio-Rad. (**A**) FFSs; (**B**) FILMS.

3.3. Opacity

As shown in Table 2, grass pea-based films, cast in the absence of mTGase, possess an opacity value of 7.74 ± 0.26 A_{600nm}/mm that is similar to the ones obtained by Shevkani et al. [16] which studied hydrocolloid edible films made up of proteins from bean (*Phaseulus vulgaris*) and pea (*Pisum sativum*). mTGase-treated films have a opacity value (4.04 ± 0.06 A_{600nm}/mm) that is statistically lower ($p < 0.05$) than the ones exhibited by grass pea-based films. The opacity was also determined in traditional commercial plastics such as cellulose triacetate (CTA) and polypropylen (PP5). As expected CTA, glossy plastic sheets used for projecting, appeared very transparent (0.53 ± 0.08 A_{600nm}/mm), whereas PP5, normally used for bakery products, macroscopically opaque, showed an opacity value equal to 32.02 ± 3.35 A_{600nm}/mm.

Table 2. Opacity of grass pea flour film cast with and without mTGase, compared to commercial plastics.

Film Features	Thickness (mm)	Opacity (mm^{-1})
Grass Pea-Based Films	0.084 ± 0.005 [b]	7.74 ± 0.26 [b]
Grass Pea-Based Films + mTGase	0.12 ± 0.02 [a]	4.04 ± 0.06 [c]
Kidney Bean-Based Films *	0.064 ± 0.002	8.9 ± 0.3
Field Pea-Based Film *	0.064 ± 0.002	7.3 ± 0.3
CTA	0.131 ± 0.001 [a]	0.54 ± 0.09 [d]
PP5	0.054 ± 0.003 [c]	32.02 ± 3.35 [a]

Values are mean $\pm$ standard deviation; Means followed by the same letters are not statistically different (Tukey-Kramer test, $p < 0.05$); * Data from Shevkani et al. [16]; CTA, cellulose triacetate; PP5, polypropylene.

3.4. Scanning Electron Microscopy (SEM)

The film both cast in the presence and absence of mTGase macroscopically appear quite handleable and flexible with a homogeneous structure. Figure 3 shows the SEM images of untreated and mTGase-treated bioplastics. As it is possible to see from Figure 3A, the surface of film cast in the absence of mTGase has a very heterogeneous structure with a high grade of roughness and deep cracks. On the other hand, film surface of films treated with mTGase appears smoother and homogeneous. This observation can be better appreciated in the cross sections of the films, shown in Figure 3B, where the untreated film is highly wrinkled, appearing not compact; instead in the presence of mTGase the film sections appear more homogeneous and uniform, with less cracks. These results reflect those obtained by Giosafatto et al. [3] and Mariniello et al. [26] that state that mTGase treatment confers a smoother and compact structure in pectin and phaseolin-based films.

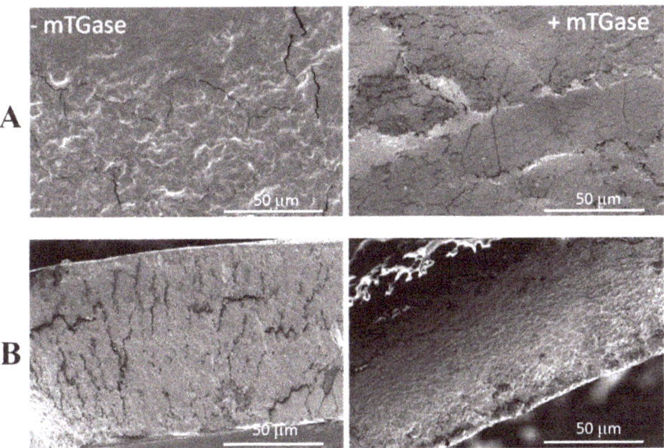

Figure 3. SEM micrographs of surface (**A**) and cross-sections (**B**) at 2600× magnification of grass pea flour-based films prepared in the absence and the presence of mTGase.

3.5. Oral, Gastric and Duodenal in Vitro Digestion of Grass Pea Flour-Based Edible Films

Gastric and duodenal digestion experiments were performed under physiological conditions in order to study the possible digestion of the films by the human gut [3,18]. As it is possible to note from SDS-PAGE (12%) shown in Figure 4A unmodified proteins are more susceptible to be digested in the gastric environment than the mTGase-crosslinked ones (Figure 4B). In fact, low *Mr* proteins occurred only following the pepsin hydrolysis of untreated grass pea proteins; on the other hand, the mTGase-catalyzed polymers seemed quite resistant and stable even after 60 min of incubation with pepsin (Figure 4B). In fact, densitometry analysis showed (lower part of Figure 4B) that mTGase-modified forms start being digested only after 20 min incubation with pepsin, and about 76% of these polymers were still present following 60 min incubation in comparison to the band intensity of control (lower part of Figure 4B), whereas the undigested proteins represented only the 36% in the samples that were not subjected to mTGase-mediated modification (lower part of Figure 4A).

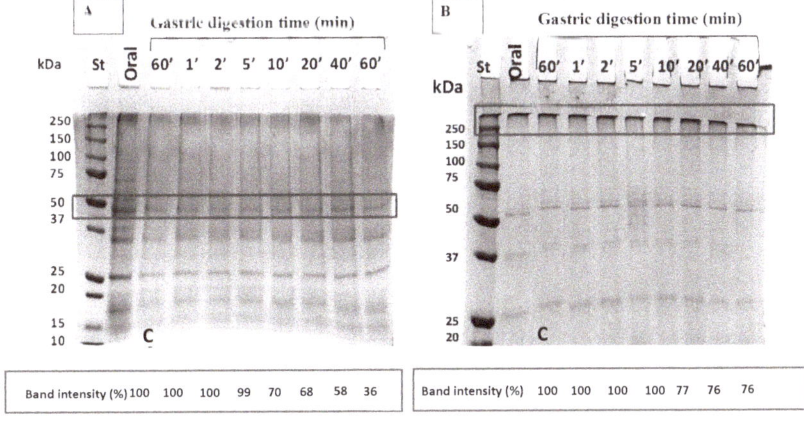

Figure 4. (**A**) Oral and gastric in vitro digestion and densitometry analysis of 50 kDa protein bands of grass pea film cast without mTGase; (**B**) Oral and gastric in vitro digestion and densitometry analysis of protein bands of >250 kDa of grass pea film cast in the presence of mTGase (33 U/g). C is control sample incubated without pepsin. St, Molecular weight standards, Bio-Rad.

The samples obtained after 60 min of pepsin digestion were further processed by recurring to trypsin and chymotrypsin, with the aim of mimicking duodenal digestion (Figure 5). We found that both unmodified (Figure 5A) and mTGase-modified (Figure 5B) were more difficult to be digested, even though, once again, the samples incubated in the absence of the crosslinking enzyme appeared more prone to be hydrolyzed by the intestinal enzymes. mTGase-derived polymers are gradually digested and after 120 min incubation (lower part of Figure 5B) with trypsin and chymotrypsin, 61% of unbroken polymers are still detectable. On the contrary, densitometry analysis of residual intact 50 kDa protein present in the unmodified grass pea flour indicated that 41% of protein was observed still intact following 120 min digestion with trypsin and chymotrypsin (lower part of Figure 5A). These results clearly indicate that the TGase-mediated intra- and inter-molecular crosslinks confer resistance to gastric and duodenal digestion as demonstrated by other proteins when modified by mTGase [18,27]. These characteristics make such materials usable as scaffolds for the incorporation of active molecules to be delivered in the intestinal tract.

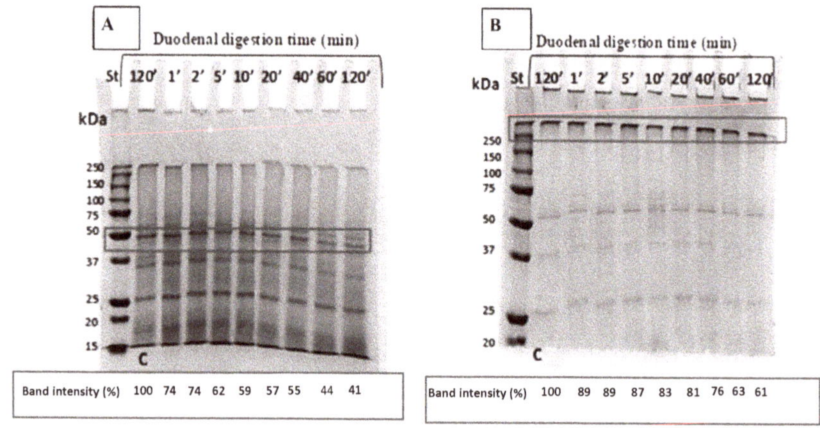

Figure 5. (**A**) Duodenal in vitro digestion and densitometry analysis of 50 kDa protein bands of grass pea film cast without mTGase; (**B**) Duodenal in vitro digestion and densitometry analysis of protein bands of >250 kDa) of grass pea film cast in the presence of mTGase (33 U/g). SDS-PAGE 12%. Molecular weight standard, Bio-Rad. C is control sample incubated without chymotrypsin and trypsin. St, Molecular weight standards, Bio-Rad.

3.6. Mechanical Properties

Tensile strength (TS), Elongation to break (EB) and Young's Modulus (YM) are shown in Table 3. As it possible to see, TS of grass pea flour-based film mTGase-untreated is lower than the one treated with mTGase. These results are in agreement with data reported by our research group [6]. The mTGase induces an increasing of TS because of the occurrence of the mTGase-catalyzed isopeptide bonds within film matrix [28–31]. Also, EB is higher for grass pea flour-based film treated with mTGase than the one performed by untreated sample. It has been reported that deamidated gluten films crosslinked by mTGase showed a gaining of EB likely due to the formation of covalent linkages by mTGase which confers more flexibility [31]. These results are also in agreement with the ones obtained by Mariniello et al. [32], and Tang et al. [33], who suggest that there is a development of a more compact and more elastic film structure after the mTGase treatment. YM data show that the films cast in the absence of mTGase are more rigid than the ones cast with mTGase, the latter possessing lower values of YM. The results reflect those reported from Porta et al. [6], that studied bitter vetch protein concentrate (BVPC) films treated or not with mTGase and affirmed that a treatment with the microbial enzyme induces an increase of resistance and a reduction of stiffness (Table 3). Moreover, from Table 3 it is possible to compare mechanical properties of grass pea flour based-films with those performed by Viscofan® and Mater Bi® [34] plastics, already available on the market and based on natural molecules. In particular, Viscofan® is obtained from collagen, cellulose and fiber-reinforced cellulose [35], whereas Mater Bi® is made up of corn starch mixed with some vegetal oils [36] in order to improve the technological features. Viscofan® has a higher value of TS and YM (Table 3) than our bioplastics prepared both in the presence and the absence of mTGase, demonstrating that this bioplastic is more mechanically resistant but more rigid than our bioplastics.

On the other hand, EB (Table 3) performed by Viscofan® is lower than that one performed by grass pea flour based-film, indicating that the latter is more extensible than the commercial bioplastic. As far as Mater Bi® is concerned, it is possible to note again that the grass pea flour-based bioplastics are less resistant, less stiff and less extensible then the starch-based one (Table 3).

Table 3. Mechanical properties of films cast in the presence and the absence of mTGase compared to commercial plastics.

Film Type	TS (MPa) Resistance	EB (%) Extensibility	YM (MPa) Stiffness
Films	0.70 ± 0.03 [b]	32.2 ± 4.4 [b]	26.2 ± 0.7 [a]
Films + mTGase	1.04 ± 0.10 [a]	59.1 ± 6.1 [a]	17.1 ± 2.8 [b]
* BVPC	1.59 ± 0.18	32.08 ± 2.52	78.14 ± 3.04
* BVPC + mTGase	2.14 ± 0.47	21.04 ± 1.29	65.13 ± 2.10
** Viscofan NDX®	36.6 ± 8.1	13.1 ± 2.9	356 ± 29
** Mater Bi (S-301)®	18.4 ± 2.7	317.9 ± 35.9	75.2 ± 2.7

Values are mean ± standard deviation; Means followed by the same letters are not significant different (Tukey-Kramer test, $p < 0.05$); * Data from Porta et al. [6]; ** Data from Porta et al. [34].

4. Conclusions

It has been demonstrated that grass pea flour suspension treated or not with mTGase in the presence of a very low amount (8%) of glycerol, used as plasticizer, is able to produce edible films. Zeta-potential and polydispersity index of the resulting FFSs do not seem to be affected by treatment with mTGase, while average protein agglomerate size appears to be slightly affected by enzyme treatment, resulting on a reduction of particle size. Optical analyses show that grass pea flour-based films are quite transparent in the presence of mTGase, the film opacity being 7 times greater than that performed by the transparent CTA and 8 times lower than the opaque PP5. Morphology studies demonstrated that mTGase confers a smoother and uniform structure as evident from the SEM micrographs of both film surface and cross-section. Digestibility analysis carried out under physiological conditions demonstrated that the grass pea flour proteins were more easily broken down by both gastric and duodenal proteolytic enzymes when the bioplastics were prepared in the absence of mTGase, whereas, the enzyme was able to produce high molecular weight polymers that resulted very resistant to the hydrolysis. Finally, mechanical analyses showed that the bioplastics prepared in the presence of mTGase were more resistant, more extensible and less rigid that the ones prepared in the absence of the enzyme. Further studies will be devoted to assess barrier properties toward O_2, CO_2 and water vapor permeability

Supplementary Materials: The following are available online at http://www.mdpi.com/2079-6412/8/12/435/s1, Figure S1: Specific viscosity of grass pea flour FFSs prepared in the absence and the presence of mTGase.

Author Contributions: Conceptualization, C.V.L.G. and L.M.; Methodology, A.D, A.A.-A and V.R.; Software, M.E.; Validation, C.V.L.G. and L.M.; Formal Analysis, L.M.; Resources, C.V.L.G. and L.M.; Data Curation, A.A.-A and M.E..; Writing-Original Draft Preparation, C.V.L.G.; Writing-Review & Editing, C.V.L.G. and L.M.; Visualization, A.D.; Supervision, L.M.; Project Administration, C.V.L.G.; Funding Acquisition, C.V.L.G. and L.M.

Funding: This work was supported by "MINISTERO DELLE POLITICHE AGRICOLE, ALIMENTARI, FORESTALI E DEL TURISMO (MIPAAFT) (contributi per il finanziamento dei progetti innovativi relativi alla ricerca ed allo sviluppo tecnologico nel campo della "shelf life" dei prodotti alimentari e al confezionamento dei medesimi, finalizzati alla limitazione degli sprechi alimentari nonché per il finanziamento dei progetti di servizio civile, CUP J57G17000190001).

Acknowledgments: We are grateful to Maria Fenderico for her helpful technical assistance.

Conflicts of Interest: The authors declare that they do not have any conflicts of interests.

References

1. Geyer, R.; Jambeck, J.R.; Law, K.L. Production, use, and fate of all plastics ever made. *Sci. Adv.* **2017**, *3*, 7–11. [CrossRef] [PubMed]

2. Sabbah, M.; Di Pierro, P.; Cammarota, M.; Dell'Olmo, E.; Arciello, A.; Porta, R. Development and properties of new chitosan-based films plasticized with spermidine and/or glycerol. *Food Hydrocoll.* **2019**, *87*, 245–252. [CrossRef]

3. Giosafatto, C.V.L.; Di Pierro, P.; Gunning, P.; Mackie, A.; Porta, R.; Mariniello, L. Characterization of citrus pectin edible films containing transglutaminase-modified phaseolin. *Carbohydr. Polym.* **2014**, *106*, 200–208. [CrossRef] [PubMed]

4. Giosafatto, C.V.L.; Di Pierro, P.; Gunning, A.P.; Mackie, A.; Porta, R.; Mariniello, L. Trehalose containing hydrocolloid edible films prepared in the presence of transglutaminase. *Biopolymers* **2014**, *101*, 931–937. [CrossRef] [PubMed]

5. Materials European Bioplastics V. Available online: https://www.european-bioplastics.org/bioplastics/materials/ (accessed on 1 October 2018).

6. Porta, R.; Di Pierro, P.; Rossi-Marquez, G.; Mariniello, L.; Kadivar, M.; Arabestani, A. Microstructure and properties of bitter vetch (*Vicia ervilia*) protein films reinforced by microbial transglutaminase. *Food Hydrocoll.* **2015**, *50*, 102–107. [CrossRef]

7. Falguera, V.; Quintero, J.P.; Jiménez, A.; Muñoz, J.A.; Ibarz, A. Edible films and coatings: Structures, active functions and trends in their use. *Trends Food Sci. Technol.* **2011**, *22*, 292–303. [CrossRef]

8. Porta, R.; Di Pierro, P.; Sabbah, M.; Regalado-Gonzales, C.; Mariniello, L.; Kadivar, M.; Arabestani, A. Blend films of pectin and bitter vetch (*Vicia ervilia*) proteins: Properties and effect of transglutaminase. *Innov. Food Sci. Emerg. Technol.* **2016**, *36*, 245–251. [CrossRef]

9. Rossi Marquez, G.; Di Pierro, P.; Mariniello, L.; Esposito, M.; Giosafatto, C.V.L.; Porta, R. Fresh-cut fruit and vegetable coatings by transglutaminase-crosslinked whey protein/pectin edible films. *LWT Food Sci. Technol.* **2017**, *75*, 124–130. [CrossRef]

10. Campbell, C.G. *Grass Pea, Lathyrus Sativus L.*; Promoting the conservation and use of underutilized and neglected crops; IPGRI: Rome, Italy, 1997; Volume 18, pp. 1–91.

11. Al-Asmar, A.; Naviglio, D.; Giosafatto, C.V.L.; Mariniello, L. Hydrocolloid-based coatings are effective at reducing acrylamide and oil content of french fries. *Coatings* **2018**, *8*, 147. [CrossRef]

12. Kieliszek, M.; Misiewicz, A. Microbial transglutaminase and its application in the food industry. A review. *Folia Microbiol.* **2014**, *59*, 241–250. [CrossRef] [PubMed]

13. Sorrentino, A.; Giosafatto, C.V.L.; Sirangelo, I.; De Simone, C.; Di Pierro, P.; Porta, R.; Mariniello, L. Higher susceptibility to amyloid fibril formation of the recombinant ovine prion protein modified by transglutaminase. *Biochim. Biophys. Acta.* **2012**, *1822*, 1509–1515. [CrossRef] [PubMed]

14. Porta, R.; Giosafatto, C.V.L.; Di Pierro, P.; Sorrentino, A.; Mariniello, L. Transglutaminase-mediated modification of ovomucoid. Effects on its trypsin inhibitory activity and antigenic properties. *Amino Acids.* **2013**, *44*, 285–292. [CrossRef] [PubMed]

15. Kjeldahl, J. Neuemethodezurbestimmung des stickstoffs in organischenkörpern. *Zeitschrift für Analytische Chemie* **1883**, *22*, 366–382. [CrossRef]

16. Shevkani, K.; Singh, N. Relationship between protein characteristics and film-forming properties of kidney bean, field pea and amaranth protein isolates. *Int. J. Food Sci. Technol.* **2015**, *50*, 1033–1043. [CrossRef]

17. *ASTM D882-97 Standard Test Method for Tensile Properties of Thin Plastic Sheeting*; ASTM: Philadelphia, PA, USA, 1997.

18. Giosafatto, C.V.L.; Rigby, N.M.; Wellner, N.; Ridout, M.; Husband, F.; Mackie, A.R. Microbial transglutaminase-mediated modification of ovalbumin. *Food Hydrocoll.* **2012**, *26*, 261–267. [CrossRef]

19. Minekus, M.; Alminger, M.; Alvito, P.; Balance, S.; Bohn, T.; Bourlieu, C.; Carrière, F.; Boutrou, R.; Corredig, M.; Dupont, D.; et al. A standardised static in vitro digestion method suitable for food—An international consensus. *Food Funct.* **2014**, *5*, 1113–1124. [CrossRef] [PubMed]

20. Bourlieu, C.; Ménard, O.; Bouzerzour, K.; Mandalari, G.; Macierzanka, A.; Mackie, A.R.; Dupont, D. Specificity of Infant Digestive Conditions: Some Clues for Developing Relevant In Vitro Models. *Crit. Rev. Food Sci. Nutr.* **2014**, *54*, 1427–1457. [CrossRef] [PubMed]

21. Laemmli, U.K. Cleavage of structural proteins during the assembly of the head of Bacteriophage T4. *Nature* **1970**, *227*, 680–985. [CrossRef] [PubMed]

22. Bhattacharjee, S. DLS and Zeta potential—What they are and what they are not? *J. Controlled Release* **2016**, *235*, 337–351. [CrossRef] [PubMed]

23. Romano, A.; Giosafatto, C.V.L.; Al-Asmar, A.; Masi, P.; Aponte, M.; Mariniello, L. Grass pea (*Lathyrus sativus*) flour: Microstructure, physico-chemical properties and in vitro digestion. *Eur. Food Res. Technol.* **2018**, 1–8. [CrossRef]

24. Nio, N.; Motoki, M.; Takinami, K. Gelation of Casein and Soybean Globulins by Transglutaminase. *Agric. Biol. Chem.* **1985**, *49*, 2283–2286.

25. Temiz, H.; Dağyıldız, K. Effects of Microbial Transglutaminase on Physicochemical, Microbial and Sensorial Properties of Kefir Produced by Using Mixture Cow's and Soymilk. *Korean J. Food Sci. Anim. Resour.* **2017**, *37*, 606–616. [CrossRef] [PubMed]

26. Mariniello, L.; Giosafatto, C.V.L.; Di Pierro, P.; Sorrentino, A.; Porta, R. Synthesis and resistance to in vitro proteolysis of transglutaminase cross-linked phaseolin, the major storage protein from *Phaseolus vulgaris*. *J. Agric. Food Chem.* **2007**, *55*, 4717–4721. [CrossRef] [PubMed]

27. Monogioudi, E.; Faccio, G.; Lille, M.; Poutanen, K.; Mattinen, M. Effect of enzymatic crosslinking of β-casein on proteolysis by pepsin. *Food Hydrocoll.* **2011**, *25*, 71–81. [CrossRef]

28. Motoki, M.; Aso, H.; Seguro, K.; Nio, N. αs1-Caseinfilm prepared using transglutaminase. *Agric. Biol. Chem.* **1987**, *51*, 993–996.

29. Mahmoud, R.; Savello, P.A. Solubility and hydrolyzability of films produced by transglutaminase catalytic cross-linking of whey protein. *J. Dairy Sci.* **1993**, *76*, 29–35. [CrossRef]

30. Yildirim, M.; Hettiarachchy, N.S. Biopolymers Produced by Cross-linking Soybean 11S Globulin with Whey Proteins using Transglutaminase. *J. Food Sci.* **1997**, *62*, 270–275. [CrossRef]

31. Larré, C.; Desserme, C.; Barbot, J.; Gueguen, J. Properties of deamidated gluten films enzymatically cross-linked. *J. Agric. Food Chem.* **2000**, *48*, 5444–5449. [CrossRef] [PubMed]

32. Mariniello, L.; Di Pierro, P.; Esposito, C.; Sorrentino, A.; Masi, P.; Porta, R. Preparation and mechanical properties of edible pectin-soy flour films obtained in the absence or presence of transglutaminase. *J. Biotechnol.* **2003**, *102*, 191–198. [CrossRef]

33. Tang, C.H.; Jiang, Y.; Wen, Q.B.; Yang, X.Q. Effect of transglutaminase treatment on the properties of cast films of soy protein isolates. *J. Biotechnol.* **2005**, *120*, 296–307. [CrossRef] [PubMed]

34. Porta, R.; Di Pierro, P.; Roviello, V.; Sabbah, M. Tuning the functional properties of bitter vetch (*Vicia ervilia*) protein films grafted with spermidine. *Int. J. Mol. Sci.* **2017**, *18*, 2658. [CrossRef] [PubMed]

35. Products and markets—Viscofan. Available online: http://www.viscofan.com/products-and-markets (accessed on 1 October 2018).

36. Uso delle risorse—Materbi. Available online: http://materbi.com/uso-delle-risorse/ (accessed on 1 October 2018).

Review

A Guide to and Review of the Use of Multiwavelength Raman Spectroscopy for Characterizing Defective Aromatic Carbon Solids: from Graphene to Amorphous Carbons

Alexandre Merlen [1], Josephus Gerardus Buijnsters [2] and Cedric Pardanaud [3,*]

[1] Institut Matériaux Microélectronique Nanoscience de Provence, IM2NP, UMR CNRS 7334, Universités d'Aix Marseille et de Toulon, site de l'Université de Toulon, Toulon CS 60584, France; merlen@univ-tln.fr

[2] Department of Precision and Microsystems Engineering, Research Group of Micro and Nano Engineering, Delft University of Technology, Mekelweg 2, 2628 CD Delft, The Netherlands; J.G.Buijnsters@tudelft.nl

[3] Laboratoire PIIM, Aix-Marseille Université, CNRS, UMR 7345, Marseille 13397, France

* Correspondence: cedric.pardanaud@univ-amu.fr; Tel.: +33-4-91-28-27-07

Academic Editor: Alessandro Lavacchi
Received: 27 July 2017; Accepted: 11 September 2017; Published: 25 September 2017

Abstract: sp^2 hybridized carbons constitute a broad class of solid phases composed primarily of elemental carbon and can be either synthetic or naturally occurring. Some examples are graphite, chars, soot, graphene, carbon nanotubes, pyrolytic carbon, and diamond-like carbon. They vary from highly ordered to completely disordered solids and detailed knowledge of their internal structure and composition is of utmost importance for the scientific and engineering communities working with these materials. Multiwavelength Raman spectroscopy has proven to be a very powerful and non-destructive tool for the characterization of carbons containing both aromatic domains and defects and has been widely used since the 1980s. Depending on the material studied, some specific spectroscopic parameters (e.g., band position, full width at half maximum, relative intensity ratio between two bands) are used to characterize defects. This paper is addressed first to (but not limited to) the newcomer in the field, who needs to be guided due to the vast literature on the subject, in order to understand the physics at play when dealing with Raman spectroscopy of graphene-based solids. We also give historical aspects on the development of the Raman spectroscopy technique and on its application to sp^2 hybridized carbons, which are generally not presented in the literature. We review the way Raman spectroscopy is used for sp^2 based carbon samples containing defects. As graphene is the building block for all these materials, we try to bridge these two worlds by also reviewing the use of Raman spectroscopy in the characterization of graphene and nanographenes (e.g., nanotubes, nanoribbons, nanocones, bombarded graphene). Counterintuitively, because of the Dirac cones in the electronic structure of graphene, Raman spectra are driven by electronic properties: Phonons and electrons being coupled by the double resonance mechanism. This justifies the use of multiwavelength Raman spectroscopy to better characterize these materials. We conclude with the possible influence of both phonon confinement and curvature of aromatic planes on the shape of Raman spectra, and discuss samples to be studied in the future with some complementary technique (e.g., high resolution transmission electron microscopy) in order to disentangle the influence of structure and defects.

Keywords: multiwavelength Raman spectroscopy; carbon solids; graphene; disordered carbon; amorphous carbon; nanocarbons

---

1. Introduction

Raman spectroscopy is an inelastic light scattering process that allows to identify and characterize the structure of molecules from gas to solid phase, from amorphous to crystals. It is created by a fluctuating electric dipole caused by both the incident light beam and by the elementary excitations of the scattered media: E.g., ro-vibrations of free molecules, phonons in crystals, impurities, and local vibrational modes. In material sciences, it is used routinely, since the 1970s, to characterize carbon-based materials, ranging from very well organized carbons such as four coordinated diamond; to three coordinated aromatic carbons such as graphene [1,2], nanotubes [3], and nanoribbons, down to amorphous carbons [4]. The latter materials are disordered carbon solids containing a mixture of tri- and tetravalent bonds [5], with or without hetero atoms [6]. In between the two extremities of highly ordered and the very disordered three coordinated aromatic carbons, nanographites which display a local order at the nanometric scale (nanographites can refer here to soots, coals, pyrolitic graphite, implanted graphene/graphite, etc.), are also covered by this spectroscopic technique.

Researchers new to the field are in general astonished by the huge amount of papers found in a first raw bibliographic search. The main aim of this review paper is to help them in identifying how Raman spectroscopy aids in studying different forms of aromatic carbons containing defects. To emphasize the role of Raman spectroscopy played in this aromatic carbon community, a quick bibliographic search on Web of Science returns more than 1700 publications having the keyword "graphite" in the title and the keyword "Raman" in the topic. This number increases up to 9000 items when replacing "graphite" with "graphene" in the title, which makes a huge number of publications. The number of publications remains high (close to 900) if both "graphene" and "Raman" are considered as keywords in the title, implying that they are intrinsically correlated. This strong correlation is still significant for amorphous carbons as the amount of publications reaches 220 with the keywords "Raman + amorphous carbon" or "diamond like carbon" both in the title. We will detail the origin of this correlation below.

At this stage the most relevant questions are: How is Raman spectroscopy generally used (or how can it be used) for characterizing carbon-based materials, and what can we learn from the Raman spectra? The answer to these questions is not easy to give as it depends on the goal of the study: Basic characterization, or deeper fundamental study, or both. What is sure is that in all cases Raman spectroscopy gives information on defects. Most often in the scientific literature it is used to confirm that the good allotrope has been obtained after a given preparatory process (e.g., ion implantation, deposition varying relevant parameters like pressure or substrate temperature, mechanical modification by milling) or to quantify the amount of defects or structure deformation introduced after a given transformation of the pristine sample. It can also be used to give a rough estimation of the stored hydrogen content [6,7], to monitor chemical changes under some physicochemical process, to determine mechanical stress or stress release, to characterize the electronic properties, diameter of carbon nanotubes, coupling between a carbon phase and another environment, and so forth. Occasionally, Raman spectroscopy is employed to obtain more fundamental information on the material properties, such as the Grüneisen parameter [8]. In this paper, we review the most important uses of multiwavelength Raman spectroscopy of sp^2 based carbon samples containing defects to answer the questions detailed above for these peculiar materials. We emphasize the role played by the laser wavelength used because of resonance effects that can be positive (i.e., allow to obtain additional information) or negative (i.e., introduction of unwanted experimental biases due to the wavelength dependency of the Raman cross sections for sp^2 or sp^3 carbons [9], merging of bands due to dispersion behavior caused by resonance effects, etc.).

In Section 2, we give an introductive and historical background of Raman spectroscopy in general and then applied to carbons. As the history of Raman spectroscopy starts at the same time as the beginnings of quantum mechanics, we have decided to give some details on the latter as well since generally it is skipped from specialized text books and review papers. Section 2 is split into three subparts, one giving the historical context, another giving some basics on the Raman effect, and the last one is more focused on Raman spectroscopy applied in the study of carbon solids. In Section 3, we give results (basically correlation between Raman spectroscopic parameters such as band intensity ratios, band position and band width, for different laser wavelength) related to different kinds of aromatic containing carbons that range from disordered graphene to amorphous carbons. The aim of Section 3 is to give a concrete and practical view on how Raman spectroscopy can be used to classify the nanostructure under investigation and its defectiveness. The guiding principle of this review section is the increase of complexity of the samples through the pages. In Section 4, we highlight the role of phonon confinement for a variety of nanocarbons and conclude our review.

2. Raman Spectroscopy of Carbon Solids: Basics

2.1. A Brief History of Raman Spectroscopy

The historical milestones (experimental, theoretical, and instrumental) on the Raman effect applied to carbon have been summarized in Figure 1. Below, we give more details. First light scattering experiments were performed in 1922 by Brillouin [10] and in 1923 by Compton, who used X-rays [11]. During the six following years, Raman was involved in 53 communications focused on scattering processes in liquids [12] that led him and his students to observe a new effect in the optical region. In 1928, Raman [13] finally realized he observed the analogue of the Compton effect: An inelastic light scattering process, but in the visible range of radiation. Two groups of unresolved bands were observed: One at a higher wavelength compared to the wavelength of the incident light (Stokes lines, more intense) and one at a lower wavelength (anti-Stokes lines, less intense). The denomination as Stokes or anti-Stokes lines is due to Woods [12] who noticed that, phenomenologically, this Raman effect gave a shift in wavelength as does fluorescence, a phenomenon discovered by Stokes in 1852 on fluorite, CaF_2 [14]. Some German researchers doubted the discovery of the Raman effect as they failed to reproduce it, but Sommerfeld played a major role in the acceptance of the Raman effect by the international community [15,16]. The history of this discovery is discussed in detail in Singh et al. [17]. The inelastic scattering of light in the visible range was then found very promising for the study of molecular structures because the process allows to obtain infrared and far-infrared information (ro-vibrational states) using and detecting light in the visible range. Before the most efficient detector at that time (photography) was used, the Raman effect was first observed by coupling a spectroscope with the naked eye [15].

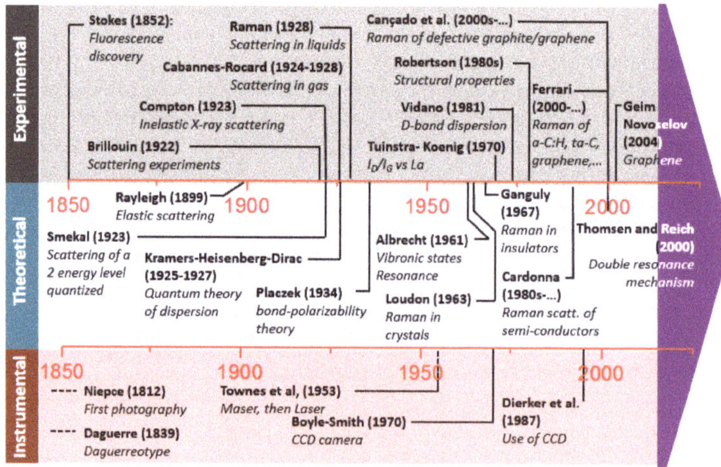

Figure 1. Quick chronology of the Raman effect.

The theoretical background at the basis of the understanding of the Raman effect started to be established before the experimental discovery of the effect itself: In 1923, Smekal quantized a two-energy level system [18]. Two years later, Kramers and Heisenberg [19] obtained the expression of the scattering cross section of an electromagnetic wave by an atom described by quantum mechanics. In 1927, Dirac derived the same expression by quantizing both the matter and light, creating quantum electrodynamics [20]. Nowadays, the theory is generally labelled the Kramers–Heisenberg–Dirac (KHD) theory. In 1932, Breit reviewed this quantum theory of dispersion (Among other points, we learn in his paper the well-known fact that both the Schrödinger and matrix mechanics from Born, Heisenberg, and Jordan were tested to obtain the Kramers–Heisenberg formula of dispersion, and that they both lead to the same result.) [21]. In 1934, Placzek [22] introduced the bond-polarizability theory of Raman scattering. This approach is still useful nowadays as it allows "easy" manipulation of Raman intensities. It is based on a time-dependent perturbation theory and on some assumptions, among others that the nuclei of the molecules are fixed and that the system is in its ground electronic state, which prevent the theory from being used for resonance effects. The main advantages of this semi-classical theory is that point group theory can be used for deriving selection rules based on symmetry considerations. In 1961, Albrecht reported multiple hypothesis, introducing vibronic states and allowing his theory to be used for normal and resonance Raman scattering [23]. In 1956, Born, who is well known to first give the interpretation of the wave function squared absolute value, co-wrote a very detailed book about the dynamics of crystals in which some sections are dedicated to the Raman effect [24]. In 1964, Loudon reviewed the knowledge acquired on the theory adapted to crystals; the theory (he contributed to create in 1963) focusing in his paper on cubic, axial, and biaxial crystals, using the first order perturbation theory applied to a system with three systems interacting: The radiation, the electron, and the lattice (phonon) [25]. In his paper, he also reviewed crystal excitations involved in the Raman effect which are not only phonons. In 1967, Ganguly and Birman developed the theory of lattice Raman scattering in insulators [26]. Reviewing the period from the seventies to the present days is a complex task due to the increasing number of papers published in this period. This explosion of works related to Raman spectroscopy can be explained by the invention of the laser, which provides monochromatic intense photons. A bibliographic search with the key words "Raman + spectroscopy" in the title returns more than 21,000 papers. This number falls to 3200 if the research is restricted to the material science field only. In the 1970s, the number of papers published per year was just close to 5, whereas currently about 140 papers are published each year. The two series *Light Scatterings in Solids* [27] and *Recent Advances in Linear and Non-Linear Raman Spectroscopy* [28] will be helpful to the readers who want to follow the evolution of this field in more detail. *The Journal of Raman Spectroscopy* (published by John Wiley & Sons, Inc.) is a dedicated journal that publishes in the field. Raman spectroscopy is now routinely used in many labs to characterize many kinds of solids, transparent or absorbent, thick or monoatomically thin, and so forth. It is coupled with an optical microscope that focalizes the laser beam to a restricted sampling area/volume and helps in collecting light more efficiently after it got scattered with matter, at the micrometer lateral resolution. The reader will take advantage in reading the reference from Gouadec and Colomban [29] which is very well adapted to efficiently learn both the theoretical and experimental basics and which illustrates these basics on well-chosen examples.

The evolution of this field of research has been correlated to the evolution of the experimental techniques. It reached its apogee in the 1940s, studying first molecules in liquids and then in gas (first measurements where done unsuccessfully in gas phase from 1924 to 1928 by Rocard [30,31]). Due to both low Raman scattering cross sections (see below) and absorption of light, studying crystals was not easy until the advent of the laser in the 1960s (see Figure 1). Only transparent samples like diamond and CdS, with a large volume probed, are reported in the period 1930–1960 [32,33]. In 1928, Mandelstam and Landsberg intended to measure Brillouin spectra of quartz, but instead they observed faint new lines with an unexpected shift, which were in fact the corresponding Raman lines of quartz [34]. Lasers, contrary to the mercury lamp previously used, offered many advantages such as: High power, monochromaticity, and coherence, thus opening the era of studying solids. In parallel, progress had been made in electronics so that photomultipliers were used first before the CCD (charged coupled device) cameras [35] which were invented in 1970, based on semiconductors technology arranged in arrays. The CCD camera was applied first in the field of Raman spectroscopy for solids in 1987 for characterizing ultrathin organized layers of organic films [36] without using specific molecules in which resonance effects enhance the Raman signature, as was done before by Rabolt et al. [37]. More information on the instrumental aspect can be found in the work by Adar et al. [38]. The early history of the Raman effect can also be found in a review by Long [39]. Laser coupling to nanometric metal nanoparticles or atomic force microscopy tips [40] allows reaching nanometric resolution, but this is another topic and will not be covered in this paper. Before discussing more details on the use of Raman spectroscopy for characterizing carbon allotropes, we first provide some basic theoretical knowledge on Raman spectroscopy.

2.2. Basic Knowledge on Raman and Resonance Raman Spectroscopy

The aim of the following section is to give the main physical ideas behind normal and resonant Raman scattering, and not to give a complete lesson on Raman spectroscopy theory, including cross section calculations/band intensities in the case of normal and resonant Raman scattering. For a deeper learning, we refer to the works of Rocard and Long [30,41] for the basics and applications in material sciences and [41] for the full quantum theory applied to free molecules. The review by Born and Huang [24] displays, among other useful developments, a full description of calculations leading to the Placzek's formalism. For solids, the studies by Cardona et al. [42,43] are highly recommended.

2.2.1. Experimental Set-Up

The typical experimental set-up applied for Raman spectroscopy measurements, with a backscattering geometry, is presented in Figure 2a. Briefly, a laser beam is aligned by a set of mirrors and driven to an objective that focalizes it on the sample placed on a motorized XY stage. Depending on the set-up, different laser sources (laser wavelength from λ_0 = 244–1064 nm) with their corresponding optics can be found. Depending on the community, either the wavelength or the laser energy is used to display spectroscopic data. They are related by the expression (see Equation (1)):

$$E_0(\text{eV}) = 2.41 \frac{514.5}{\lambda_0(\text{nm})} \qquad (1)$$

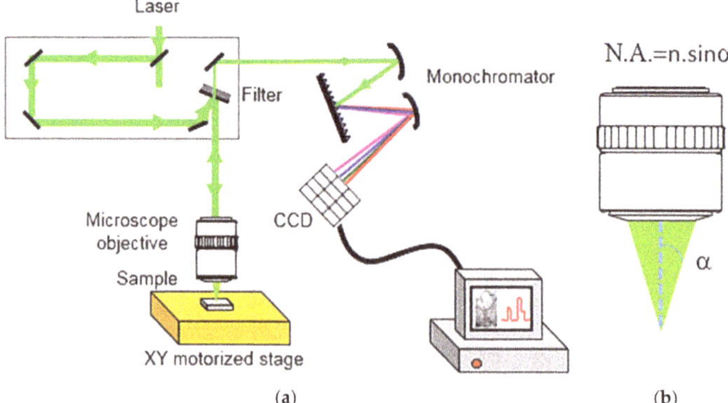

Figure 2. Experimental set-up with a backscattering geometry: (**a**) Details of the system; (**b**) Definition of the numerical aperture (N.A.), with *n* the index of refraction of the medium in which the lens is working and α the maximal half-angle of the cone of light that can enter or exit the lens.

After light and matter have interacted in the sample, the photons (either reflected, elastically scattered, inelastically scattered, or other photons coming from competing processes such as fluorescence) are collected by the same objective and driven to a filter that diminishes the intensity of the elastic photons. Generally, the cut-off frequency of this filter is close to 50–100 cm^{-1}, meaning that modes with a lower wavenumber will not be detected. Photons other than the elastic ones are then driven to a monochromator in which light dispersion occurs. Light is finally spread on a CCD camera, converted in an electronic signal that is recorded on a computer. The spot radius on the sample, R, is $\approx \frac{0.6\lambda_0}{N.A.}$, where λ_0 is the laser wavelength and N.A. the numerical aperture as defined in Figure 2b, n being the refractive index separating the sample and the optics. According to the Rayleigh criterion, R is also the lateral resolution, as it is the smallest distance between two points that can be probed. The XY stage is generally motorized in order to work in a mapping mode: The stage moves to different x,y-positions at which spectra are recorded to check spatial inhomogeneity at the micron scale. The vertical resolution, labelled Δz, is generally of the order of the micron, if the material is transparent. In addition, it can be adjusted using confocal mode. Note that it can be lower if the material under investigation is absorbing the laser light. More details on the experimental operation of Raman spectroscopy can be found in the paper by Gouadec and Colomban [29].

2.2.2. Conservation Rules

Raman photons are created by a fluctuating electric-dipole in the scattering medium, by the simultaneous action of the incident light beam and the elementary excitations of the solid, leading to an induced polarization moment $\vec{P}$. To describe this interaction, three ways can be used. The easiest one describes classically both electromagnetic field and matter. The second one describes, classically, the electromagnetic field but quantifies matter. The third one quantifies both.

Let us start by introducing the classical monochromatic electric field of the incident laser $\vec{E} = \vec{E_0} \cos(\omega_0 t - \vec{k_0} \times \vec{r})$, E_0 being its amplitude, ω_0 its frequency, $\vec{k_0}$ its plane wave propagation vector (which is $k_0 = \omega_0/c$ in vacuum, c being the speed of light) and $\vec{r}$ the 3D position in space. The elementary excitation of the crystal, called the phonon, has a crystal momentum called $\vec{q}$, and a corresponding frequency ω_q, for each value of $\vec{q}$. As the system is isolated, two conservation rules (on total energy and momentum) occur, leading respectively to Equation (2) and (3):

$$\omega_0 = \omega_S + \omega_q \tag{2}$$

$$\vec{k_0} = \vec{k_S} + \vec{q} \tag{3}$$

where ω_S and $\vec{k_S}$ are the scattered light frequency and wavevector, respectively. Due to these conservation rules, the geometry of the experiment normally determines orientation and magnitude of the scattering wave vector. In current experiments, the backscattering geometry (i.e., $\theta = 180°$, see Figure 3 for the definition of this angle) is one of the mostly routinely used in labs. The displacement of a peculiar ion in the unit cell around its rest position is given by $u(\vec{r}, t) \propto Q(\omega_q, t)e^{i(\omega_q t - \vec{q} \times \vec{r})}$, where $Q(\omega_q, t)$ is the phonon coordinate. The phonon coordinate is a linear combination of bond lengths and bond angles and is associated to the normal modes of vibration. Whatever the crystal symmetry, $|\vec{q}|$ varies from 0 to a value $|\vec{q}_{max}|$ which is of the order of $1/a$, a being the typical lattice parameter, close to 1 Å. In the visible range of radiation, considering the conservation rule on momentum, and whatever the value of θ, as the ratio between $|\vec{q}_{max}|$ and $|\vec{k_0} - \vec{k_S}|$ is close to 100, it means that necessarily the phonon that will satisfy the conservation rule will be close to $|\vec{q}| \approx 0$, i.e., close to the center of the Brillouin zone. For higher order processes (i.e., involving more than just one phonon), the wave vector conservation rule becomes Equation (4):

$$\vec{k_0} = \vec{k_S} + \sum_i \vec{q_i} \tag{4}$$

As a consequence, not only phonons at the center of the Brillouin zone can contribute now.

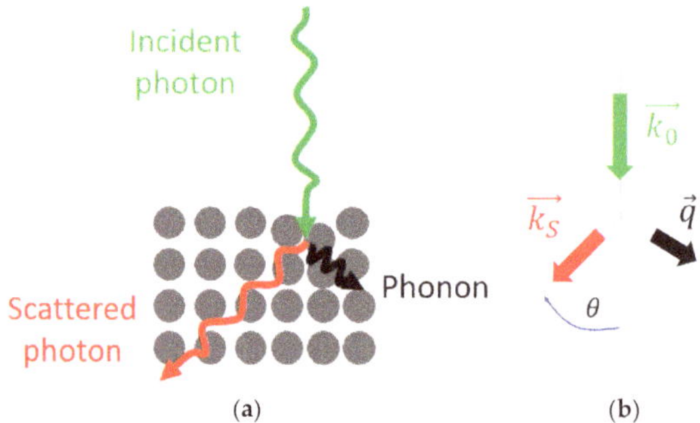

Figure 3. Scattering geometry: (**a**) Photon/phonon interaction; (**b**) Corresponding momentum.

2.2.3. Classical Expressions for Molecules

Before considering the more complex case of solids, we introduce some basics of Raman theory for molecules. The incident electromagnetic field induces a dipolar momentum in an electronic system, the ρ-component ($\rho = x$, y, z) being given by $P_\rho \approx \sum_\sigma \alpha_{\rho\sigma} E_\sigma$, with the polarizability tensor $\alpha_{\rho\sigma}$, (higher order terms, related to hyper-Raman spectroscopy, are not mentioned here). The components of $\alpha_{\rho\sigma}$ can be approximated by a Taylor expansion. For simplicity, let us forget about the ρ and σ indexes coding for directions in space, and let us assume we consider a linear molecule with a normal vibrational coordinate $Q(t)$ characterized by an intrinsic vibrational frequency ω_{vib}. Then, $\alpha \approx \alpha^{(0)} + \left(\frac{\partial\alpha}{\partial Q}\right)_{Q=Q_0} \times Q_0 \cos(\omega_{\text{vib}}t)$. The term $\left(\frac{\partial\alpha}{\partial Q}\right)_{Q=Q_0}$ is sometimes noted R and called the first order Raman tensor. It gives the coupling strength of the nuclear and electronic coordinates. By including both the expression of the incident electric field and the polarizability elements expansion, one obtains by using the product of two cosines the following expression (see Equation (5)):

$$P \approx \alpha^{(0)} \cos(\omega_0 t) + \frac{E_0 Q_0}{2} R \times \left(\cos([\omega_0 - \omega_{\text{vib}}]t) + \cos([\omega_0 + \omega_{\text{vib}}]t)\right) \tag{5}$$

in which we have omitted the $\vec{k_0} \times \vec{r}$ term, for simplicity, and the random phase of the nuclear mode of vibration acquired during the scattering process. This expression contains three terms: The first one has the same frequency as the incident laser and is interpreted as due to elastic Rayleigh scattering, the second and third ones having their frequencies lowered or increased by the frequency ω_{vib}. These latter terms are respectively interpreted as due to the inelastic Stokes and anti-Stokes Raman scattering. These processes are represented in an energy diagram in Figure 4a–c, respectively. For the Stokes photon, and far from the oscillating dipole, the amplitude of the electric field is given by classical electrodynamics as (see Equation (6)):

$$E_{\text{Stokes}} = \frac{(\omega_0 - \omega_{\text{vib}})^2}{4\pi\varepsilon_0 c^2} |P_{\text{Stokes}}| \frac{e^{i k_S r}}{r} \sin\varphi \tag{6}$$

φ being the orientation given from the dipole axis and ε_0 the permittivity of vacuum. The outgoing flux of energy is given by the Poynting vector: $S_{\text{Stokes}} = \frac{\varepsilon_0 c}{2}|E_{\text{Stokes}}|^2$. Integrating over the unit sphere, this flux gives the total energy radiated by the Stokes induced dipole (see Equation (7)):

$$\sigma_{\text{Stokes}} = \frac{(\omega_0 - \omega_{\text{vib}})^4}{12\pi\varepsilon_0 c^3}(Q_0 E_0)^2 |R|^2 \tag{7}$$

The classical theory then predicts the fourth power (sign − is replaced by + if we consider anti-Stokes), the incident beam intensity E_0^2 and the $|R|^2$ dependencies. However, it fails to reproduce resonance effects, Stokes, and anti-Stokes intensity ratio behavior with temperature, etc.

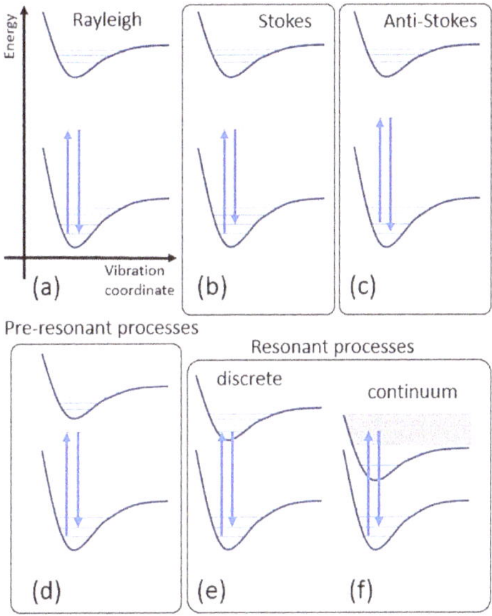

Figure 4. Jablonski diagram of Rayleigh scattering (**a**), non-resonant Stokes (**b**) and anti-Stokes (**c**) Raman scattering, and pre-resonant (**d**) and resonant Raman scatterings (**e**,**f**).

2.2.4. Semi-Classical Expression for Molecules

When matter is treated quantum mechanically, some of these behaviors are better reproduced. We only consider the first order induced dipole here. During the scattering process, one incident photon (of energy $\hbar\omega_0$, $\hbar$ being the Planck constant divided by 2π) is annihilated, and one photon and one phonon are created. Below, all the results are obtained in the framework of time-dependent perturbation theory applied for a molecule (details in Chapter 4 of [11]), limited here to the first order of expansion [22]. Let us define the unperturbed initial and final states of the molecule considered: $|i\rangle$ and $|f\rangle$ respectively. The ρ-component ($\rho = x, y, z$) of the first order induced dipole is $(P_\rho)_{fi} = \sum_\sigma (\alpha_{\rho\sigma})_{fi} E_\sigma$, where $(\alpha_{\rho\sigma})_{fi} = \langle f | \alpha_{\rho\sigma} | i \rangle$ is the transition polarizability, as given in Equation (8):

$$(\alpha_{\rho\sigma})_{fi} = \frac{1}{\hbar} \sum_{\substack{r \\ r \neq i,f}} \left(\frac{\langle f | \hat{p}_\rho | r \rangle \langle r | \hat{p}_\sigma | i \rangle}{\omega_{ri} - \omega_0 - i\Gamma_r} + \frac{\langle f | \hat{p}_\sigma | r \rangle \langle r | \hat{p}_\rho | i \rangle}{\omega_{rf} + \omega_0 + i\Gamma_r} \right) \tag{8}$$

The summation r runs over all the states $|r\rangle$ of the molecule. $\omega_{ri} = \omega_r - \omega_i$, $\hbar\omega_i$ and $\hbar\omega_f$ being the energies of the initial and final states, respectively. Γ_r is related to the lifetime of the level r (damping factor). The numerator $\langle f | \hat{p}_\rho | r \rangle \langle r | \hat{p}_\sigma | i \rangle$ is the product of two transition electric dipole terms: One for a transition from $|i\rangle$ to $|r\rangle$ (absorption) and one from $|r\rangle$ to $|f\rangle$ (emission). Term one and two under the sum will not play the same role depending on the comparison between ω_0 and $\omega_{ri,f}$: The first term can increase a lot if $\omega_0 \approx \omega_{ri}$, being predominant in the case of resonance condition, whereas this is not the case for the second one that is still present even far from resonance conditions. In general, all the pathways connecting $|i\rangle$ and $|f\rangle$ with non-null transition electric dipole terms must be considered in the summation. Representing this scattering in an energy diagram is simplified by introducing a virtual state and its corresponding virtual level of energy (i.e., not a stationary state coming from the resolution of the Schrodinger's equation), as it is done in Figure 4b,c (the virtual level is marked by a dashed line). The case where $\omega_0 \ll \omega_{ri}$ is represented in Figure 4b (Stokes) and c (anti-Stokes), whereas the case $\omega_0 \approx \omega_{ri}$ is represented in Figure 4d–f corresponding respectively to pre-resonance, discrete resonance and continuum resonance Raman scatterings. By considering the Born–Oppenheimer condition (electron and nucleus motions are not coupled), one has the vibronic wavefunction that can be written $|r\rangle = |e_r\rangle|v_r\rangle$ with the corresponding quantum number v_r for vibration and the energy $\hbar\omega_r = \hbar\omega_{er} + \hbar\omega_{vr}$. In the ground electronic state, the expression of $(\alpha_{\rho\sigma})_{fi}$ (see Equation (8)) does not change a lot. It is labelled "the A-term" in the Albrecht denomination [23]. However, if one introduces a perturbation calculated by considering electron–nucleus interaction (i.e., the electronic Hamiltonian is expanded in the nuclear displacements around the equilibrium position Q_0), then other additional terms (called B, C, and D terms) appear in the previous expression of $(\alpha_{\rho\sigma})_{fi}$. More information about the meaning of these terms can be found in the paper by Long [41].

As the electric dipole moments can be expanded in a Taylor series over the normal vibrational coordinates, one can obtain the following expression (Equation (9)):

$$\langle v_f|\alpha_{\rho\sigma}|v_i\rangle \approx \alpha_{\rho\sigma}{}^{(0)}\langle v_f|v_i\rangle + \sum_k \left(\frac{\partial\alpha_{\rho\sigma}}{\partial Q_k}\right)_{Q=Q_0} \times \langle v_f|Q_k|v_i\rangle + \ldots \tag{9}$$

with $\langle v_f|Q_k|v_i\rangle = \sqrt{v_i+1}$ if $v_f = v_i + 1$, $\sqrt{v_i}$ if $v_f = v_i - 1$, or 0 in the other cases, noting the elements of the Raman tensor as $R_{\rho\sigma}^{(k)} \propto \left(\frac{\partial\alpha_{\rho\sigma}}{\partial Q_k}\right)_{Q=Q_0}$. We can notice that the vibrational mode will be Raman active only if this last term is different from zero. This condition defines the so-called "selection rules". They are directly related to the symmetry of the system. See [44] for further details. Considering that the molecule is not oriented preferentially, the total energy radiated by the Raman effect is given by the expression (Equation (10)):

$$\sigma_{\text{Stokes}-fi} \propto (\omega_0 - \omega_{\text{vib}})^4 E_0^2 \left|\sum_{\rho\sigma}(\alpha_{\rho\sigma})_{fi}\right|^2 \tag{10}$$

where $|(\alpha_{\rho\sigma})_{fi}|^2$ is composed of k products in the form of $\left|\left(\frac{\partial\alpha_{\rho\sigma}}{\partial Q_k}\right)_{Q=Q_0}\right|^2 \times (v_i + 1)$ for Stokes and $\left|\left(\frac{\partial\alpha_{\rho\sigma}}{\partial Q_k}\right)_{Q=Q_0}\right|^2 \times (v_i)$ for anti-Stokes processes.

2.2.5. Raman Effect in Crystals

For well oriented crystals, the same kind of expressions holds but with some changes. First, the vibrational quantum number is replaced by the Bose factor $n = [\exp(\hbar\omega/kT) - 1]^{-1}$, k being the Boltzmann constant and T the temperature expressed in K. Second, for monocrystals, one must take care of the incident and scattered polarization directions. In general, one has the differential cross section expressed as (Equation (11)):

$$\frac{d\sigma}{d\Omega} \propto \left|\sum_{\rho\sigma} e_{0\rho}\alpha_{\rho\sigma}e_{S\sigma}\right|^2 \tag{11}$$

where $d\Omega$ is a solid angle, $e_{0\rho}$ and $e_{s\sigma}$ are the ρth and σth components of respectively the incident $\vec{e_0}$ and scattered $\vec{e_S}$ elementary polarization vectors. It sometimes can be found as: $d\sigma/d\Omega \propto |\vec{e_0} \times [\alpha] \times \vec{e_S}|^2$, $d\sigma/d\Omega \propto |\vec{e_0} \times R \times \vec{e_S}|^2$, or $d\sigma/d\Omega \propto |\sum_{\rho\sigma} e_{0\rho}R_{\rho\sigma}e_{S\sigma}|^2$. For single crystals, by measuring the dependence of the scattered intensity on $\vec{e_0}$ and $\vec{e_S}$, one can deduce the symmetry of the Raman tensor and hence the symmetry of the corresponding Raman-active phonon [45].

To determine the cross sections experimentally, one needs first a reference with a known Raman cross section. In order to deduce the cross section relatively to the reference material, one has to correct from electromagnetic biases such as reflection losses which can be different from the reference material and the sample to analyze (due to specific refractive index) or interference effects [46–48]. The reader interested in the method can read the detailed work of Klar et al. that was performed on graphene [49].

2.3. Basic Properties of Graphene and Related Materials

Due to its valency, carbon exists under several allotropic forms—such as diamond, graphite, graphene, nanotubes, and fullerene—and many other forms will likely be discovered in the future [50]. Plasma and nanoscale plasma/surface interactions are processes responsible for a large number of them (we do not pretend to list all these interactions; they can be found in the review by Ostrikov et al. [51]). Roughly speaking, two families exist: The "sp^3 family" (with the tetra-coordinated diamond), and the "sp^2 family" with the graphene as the model of aromatics, with many possibilities in between the two families [52]. Amorphous carbons can mix aromatic sp^2 and sp^3 carbons, as well as heteroatoms.

Graphene is a monolayer thick crystal organized as displayed in Figure 5a. The hybridization between one s-orbital and two p-orbitals leads to a trigonal planar structure with three in-plane sp^2 σ bonds and one out-of-plane π bond. The σ bonds are responsible of the robustness of the lattice, whereas the π bond, by binding covalently with a neighbor π bond, is responsible of the electronic conduction. Two atoms per unit cell are necessary to reproduce the crystal, and the two vectors, $a_1 = (3a/2, a\sqrt{3}/2)$ and $a_2 = (3a/2, -a\sqrt{3}/2)$, with a = 1.42 Å, are displayed in Figure 5a [53]. The reciprocal-lattice vectors are b_1 and b_2. In this momentum space, three specific points of the Brillouin zone are the Γ, M and K points. The lower part of Figure 5b represents the electronic structure in vicinity of the K point, where the π and π* bands meet with a zero-gap energy. The dispersion of the π and π* bands is $E_\pm \approx \pm v_F q$, where v_F is the Fermi velocity, and q the momentum measured relatively to the K point [54]. The shape of the electronic bands forms a cone which is called a Dirac cone, because of the linear dispersion curve. It is because of this peculiar point that graphene is called a semi-metal. A review by Neto et al. presents the basic theoretical aspects of graphene's peculiar electronic properties [53]. Note that the sp^2 behavior can be partially modified by some mechanisms, such as chemical adsorption of hydrogen on top of C atoms, that modify curvature and then induce a mixed sp^2/sp^3 state [55].

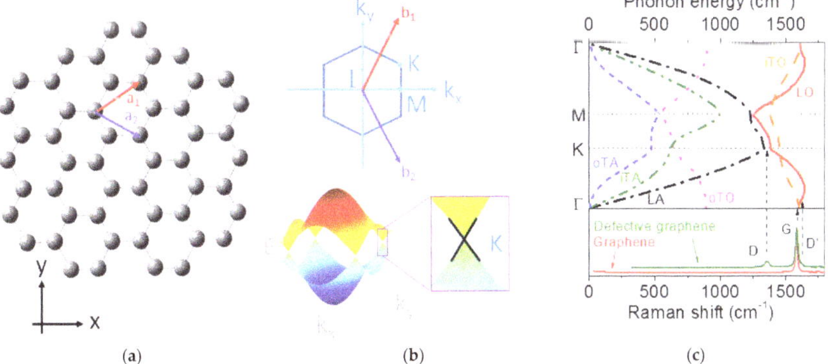

Figure 5. Honeycomb lattice of graphene. (**a**) Unit-cell vectors (a1 and a2). (**b**) Electronic structure in the reciprocal lattice. Γ, K, and M are high symmetry points in this space. (**c**) Phonon dispersion curve and Raman spectra (*LO*, *LA*, *iTO*, *iTA*, *oTO*, and *oTA*. O and A refer to optic and acoustic phonon branches, L and T refer to longitudinal and transversal, and i and o refer to in plane or out of plane, respectively).

The upper part of Figure 5c displays the phonon dispersion [56] for a Γ-*M*-*K*-Γ trajectory in the Brillouin zone (the figure is generally presented with the phonon energy horizontally). As there are two atoms in the unit cell, there are 3×2 phonon branches. "*O*" and "*A*" stand for optical (three branches) and acoustic (three branches) phonon branches, acoustic branches being close to zero at the center of the Brillouin zone. "*L*" and "*T*" stand for lateral or transverse vibrations and "*i-*" or "*o-*" stand for in plane or out of plane. The lower part of Figure 5c displays a typical Raman spectrum for defect free and defective graphene. In this spectral range, one can observe the presence of one band (called the G band) at 1582 cm^{-1} for the defect free sample, whereas one can see two extra bands (the D and D' bands) for the defective sample. As it has been discussed in Section 2.2.2, not all the phonons give rise to a band in the Raman spectrum. We give more information about the Raman spectra of graphene related materials in the next section.

2.4. Raman Spectra of Graphene, Graphite, and Disordered Carbons

This section is separated in two subsections. The first one gives usable information about the origin and behavior of the bands often encountered for aromatic based materials. The second section deals with a historical overview of the findings about Raman spectroscopy in the field, in a linear way.

2.4.1. Basics of Raman Spectroscopy for Graphene and Graphite

Raman spectra of a wide variety of disordered carbons are displayed in Figure 6 in order to overview what varies and how much in terms of band intensities, width, position, etc. The Raman spectra displayed in Figure 6 are obtained from a highly oriented pyrolytic graphite, nanographites, and amorphous carbon (we do not specify the kind of synthesis here as we consider only the main trends in the Raman spectra). Note that a complete discussion about what is called "intensity" in the literature when speaking about band fitting procedures, is given in Section 2.6.

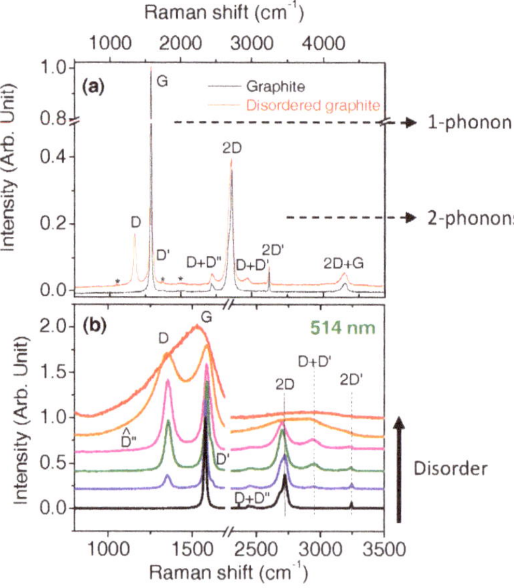

Figure 6. Raman spectra of graphite and disordered carbons recorded at 514 nm. (**a**) Attribution of the main bands in the first, second, and third order regions. (**b**) Same for a variety of more disordered carbons.

Graphite belongs to the $P6_3/mmc$ (D_{6h}^4) space group. If considering only a graphene plane, at the Γ point of the Brillouin zone, there are six normal modes that possess only one mode (doubly degenerate in plane) with a E_{2g} representation, which is Raman active (see [57]). Its wavenumber is at 1582 cm^{-1} and it gives rise to the so-called "G band". Its width is close to 15 cm^{-1} and it is mainly due to an electron–phonon coupling interaction [58]. By combining two graphene planes to build graphite, one can obtain another Raman active mode that gives rise to a band at 42 cm^{-1}, which cannot be measured by standard set-ups as the one presented here. Then for standard set-up operation, only one band should be detected. However, this is not the case: For pure graphite, for example (see Figure 6a), one extra band with a high intensity is detected (one-third of the G band intensity, at 2720 cm^{-1}). This band is composed of several bands for graphite and few layers graphene, but has a Lorentzian shape for monolayer graphene and disordered graphite in which stacking in the c direction is not like in graphite [1,59]. Its intensity, compared to the G band can vary from 3 to 1/3 from graphene to graphite, respectively. When disorder increases, this band broadens, overlapping with other bands and nearly disappears. For amorphous carbon, the intensity compared to that of the G band is lower than 6% (see Figure 6b). There are also several weaker bands (2–5% of the G band intensity, at 2450, 3240, 4300 cm^{-1}) and even weaker bands, marked by stars ($\approx$0.4% of the G band intensity at 1750 cm^{-1}). Most of the weak bands are listed in the paper by Kawashima and Katagiri [60]. As proposed by the international consensus, these bands are due to the double resonance mechanism which is described in detail in Section 2.5. A defective graphite presents other bands that can be as intense as the G band at 1350 and 1615 cm^{-1} (see Figure 6). These bands are activated by defects due to the breaking of the crystal symmetry that relax the Raman selection rules. They are called the "D and D' bands", respectively. The "D" stands for "defect" and has nothing to do with the diamond band that lies at 1332 cm^{-1}. The intense bands lying below 1640 cm^{-1} are due to first order phonons (see the phonon relation dispersion in Figure 5c). In the range $\approx$2000–3000 cm^{-1}, they are due to two-phonon processes and named 2D, D + D', 2D', and so forth. For higher wavenumbers, they are due to third order processes, etc. When increasing the order of the process, the intensity is generally diminished because of the cross section which becomes less and less probable. Considering the less intense bands, the D" band is present in the shoulder of the D band of very defective samples. It is always needed when fitting (see Section 2.6). Other bands are the 2450 cm^{-1} band, which has been attributed recently to a D + D" band by Couzi et al. [61], the D + D' (in literature, the wrong D + G label is often found [62]), the 2D' bands and the 2D + G band. Even weaker bands are present and marked by stars in Figure 6a. The 1750 cm^{-1} band is called the M band and has been understood in the framework of the double resonance mechanism [63].

In Figure 6b, one can see that when disorder increases, the bands broaden, and the relative intensity of the bands changes: The D band increases with disorder and then decreases when being close to an amorphous carbon. These two behaviors have been related to the coherence length L_a (obtained from structural analysis), with the two historical formulae (Equations (12) [64] and (13) [4,5]):

$$\text{For } L_a > 2 \text{ nm}: \quad \frac{I_D}{I_G} = \frac{c}{L_a} \text{ (Tuinstra and Koenig relation)} \tag{12}$$

$$\text{For } L_a < 2 \text{ nm}: \quad \frac{I_D}{I_G} = c'L^2 \text{ (Ferrari's relation)} \tag{13}$$

c and c' are parameters that depend on the fourth power of the laser wavelength [4,65,66] and that can vary from one sample to the other [65,67]. Why I_D/I_G behaves like $1/L_a$ for $L_a > 2$ nm is also justified in Section 3.1. Since 2010, an upgrade was published that allows a better understanding of the information retrieved from the intensity ratio. It is presented in Section 3.3. Note that instead of I_D/I_G, another spectral parameter that is often used is proportional to $I_D/I_G \times E_0(\text{eV})^4$, E_0 being the laser energy. This is because the c and c' parameters evolve as the fourth power of the laser wavelength [66,68]. It allows to compare results from different wavelength on a same plot.

One important fact about band parameters behavior is that some bands disperse, which normally does not happen according to constant eigenenergy's values obtained by quantum mechanics. This, again, is due to the double resonance mechanism, detailed in Section 2.5. Here, we briefly describe what is observed. The D band disperses linearly with the energy of the laser used, the slope being close to 50 cm^{-1}·eV^{-1}. This is a consequence of the double resonance mechanism. This has been observed for many aromatic based carbons, with different amounts of disorder (see Figure 7), an offset being present from one sample to the other (in the range $\pm$ 15 cm^{-1}). The D-band dispersion is useful for differencing an aromatic based sample from a diamond based sample because the 1332 cm^{-1} band in the latter case does not disperse. The D' band also displays a dispersion (not shown here) but it is less significant than that of the D band.

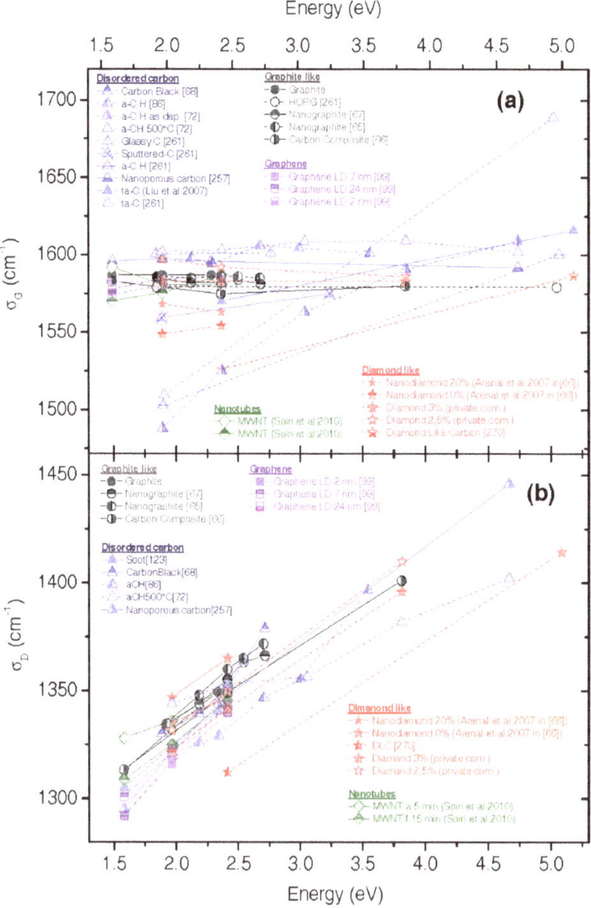

Figure 7. Dispersion relations of the (**a**) G band and (**b**) D band for many disordered samples (data derived from literature as indicated).

The 2D band has twice the slope of the D band (not shown here). Several other weaker bands also display a dispersive behavior, which can be explained by the double resonance mechanism [61,69]. The G band is not due to the so called double resonance mechanism and for perfect graphite and graphene, it does not display dispersion, lying at 1582 cm^{-1}. However, its position can change with the state of disorder in the material, in the range 1590–1600 cm^{-1} (for nanocrystalline graphite) down to 1520 cm^{-1} for amorphous carbons. This band position is sensitive to clustering of the sp^2 phase, bond disorder, presence of sp^2 rings and/or chains, presence of sp^3 carbons and the way they are coupled to aromatic carbons [70]. Due to the electronic resonance Raman process (see Figure 4e,f), the G band can display a dispersion behavior, as displayed in Figure 7b, driven by the size of the aromatic domains (from roughly 0.5 eV up to few eVs for few aromatic carbon clusters [71]). This is particularly true for amorphous carbons. To our knowledge, this dispersion parameter is generally not used to better characterize amorphous carbons with small aromatic size skeleton, but should be used more often such as was done in the work by Lajaunie et al. [72].

The width of the G band (Γ_G) can be used to have an idea about the amount of defects (even if depending on the sample there are some differences, as can be seen in Part 3 with doped graphene, for example). It varies generally from 10 up to 200 cm^{-1} from graphene to amorphous carbons, but can be influenced by doping. It was found to vary roughly as a power law of L_a [5] for a large variety of samples, but it was found to vary as $1/L_a$ for samples with L_a in the range 20–65 nm [65]. Using only Γ_G is not enough to better characterize the sample analyzed, since the I_D/I_G parameter can be different for two kinds of samples whereas Γ_G is the same.

2.4.2. Historical Aspects

The following section is meant to provide an intermezzo, but detailed description of historical highlights in Raman spectroscopy of aromatic carbons. This section is less essential to the general reading of this paper, and therefore we refer to Section 2.5 on the double resonance mechanism for a continued subject reading.

Figure 8 displays the most important historical breakthroughs in Raman spectroscopy of aromatic carbons. First Raman spectra from a carbon were obtained on diamond in 1930. A band was observed at 1332 cm^{-1} [73], with a better understanding of the first and second orders reported in 1970 [74]. The phonon spectrum of graphite was modelled and obtained for the first time in 1965 [75,76]. Only two modes are Raman active according to the point group symmetry analysis. In 1970, Tuinstra and Koenig [64] detected with Raman spectroscopy only one band at 1575 cm^{-1} for graphite and attributed it to a doubly degenerate deformation vibration of hexagonal rings, corresponding to the E_{2g} mode (The true frequency admitted today is 1582 cm^{-1} [77], but at that time the authors did not realize they used a high power (300 mW instead of the few mW or even less for absorbent materials) that can increase the equilibrium temperature of the sample, resulting in a downshift of the bands.) Moreover, they compared graphite with more disordered carbons (i.e., activated charcoal, lampblack, and vitreous carbon) and found a new band close to 1355 cm^{-1}. They noted that this band is close to the 1332 cm^{-1} band of diamond, but ruled out the fact that the 1355 cm^{-1} band is due to sp^3 carbons. They also showed that the relative height ratio between the 1355 cm^{-1} band (labelled D band, for "defect-induced band") and the 1575 cm^{-1} band (labelled G band for "graphite allowed band") is correlated to the crystallite size, L_a, retrieved from X-ray diffraction measurements. This is the so called Tuinstra–Koenig relation. In 1978, Tsu et al. observed three behaviors that are nowadays used to characterize defects and electronic structure of aromatic carbons [77]. First, they observed a new mode at 1627 cm^{-1} that they attributed wrongly to a splitting of the E_{2g} mode. Then, they saw a high shift of the 1575 cm^{-1} band (up to 1590 cm^{-1}) and finally they observed the presence of a band at 2742 cm^{-1}, twice the frequency of the 1370 cm^{-1} band (also high shifted) even if the 1370 cm^{-1} band is not present in the spectrum. Nowadays, we label the 1627 and 2742 cm^{-1} bands as the D' and 2D band, respectively. Using a different polarization configuration, Vidano et al. [78] related the D, G, D', and 2D bands to in-plane vibrations and noticed the composite behavior of the 2D band, prophesying that once well understood, this behavior could help in better characterizing disorder and structural imperfection. Indeed, the shape and intensity of this 2D band was found to be very dependent on the electronic structure and the number of stacked layers in a multilayer graphene sample [1], and to the quality of the layer stacking in nano graphite by Cançado et al. [59], for example. The spectroscopic parameters of all these bands were noticed to be very dependent on the structure of the samples and then, to give a quick and good idea of the degree of graphitization, a thermal treatment was performed on different carbon fibers, coals, and pyro carbons between 450 and 3000 °C [78–80] and later, on amorphous carbons by Ferrari and Robertson [5].

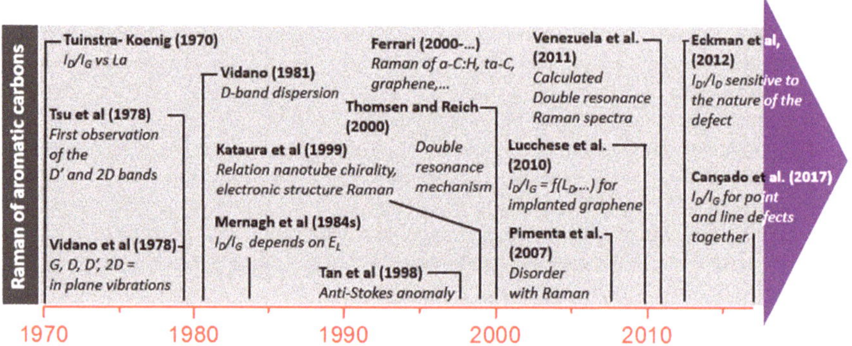

Figure 8. Chronology of the Raman effect applied to aromatic carbon.

D Band and Combination Band Dispersions

In 1979, Nemanich and Sollins attributed the 2D band plus weaker bands appearing at 2450 and 3248 cm^{-1} to vibrational density of states features, using calculated phonon dispersion curves of graphite [81]. However, this was not completely satisfactory as the 2D band, even when the graphite sample is defect free, is of the same order of magnitude as the G band, whereas it should depend on the defect density according to this hypothesis (Before the physical origin of the 2D band was well understood, it was named the G' band, exactly because it was always observed in the spectra together with the Raman allowed G band. The D* band can also be found in the literature. Be careful, the denomination D* is used since 2016 for another band. See in the paragraph text for details.). Part of the problem was solved when Vidano, in 1981, noticed that the D and 2D-band frequencies were dependent on the laser wavelength (the D band varying from 1360 down to 1330 cm^{-1} when using laser wavelengths from 488 nm up to 647 nm, and a slope which is twice for the 2D band) [82], hiding a vibronic resonance behavior. The same kind of dispersion for the D and 2D bands was observed on other graphites [83,84], carbon blacks [68], hydrogenated amorphous carbons [85], and later on carbon nanotubes [86,87] and most recently on graphene [62] (and references therein). The slope is close to 50 cm^{-1}/eV for the D band and close to the double for the 2D band. Dispersion of weaker bands, mainly due to combination modes such as 2D + G, 2D + 2G, etc., were also reported in [4,5,60].

In 1984, Mernagh et al. also noticed that the I_D/I_G ratio was depending on the laser wavelength [68], which is another proof that the D band arises from a resonant effect (if not, the ratio should be constant because of the $\approx \omega_0{}^4$ dependency in the non-resonant Raman cross section affecting all non-resonant bands). In 1989, Knight and White determined the c value appearing in the Tuinstra and Koenig relation at 4.4 nm and that was done with 514 nm lasers [88] before the work of Cançado et al. [65]. In 1998, Tan et al. studied thermal effects on graphite and ion implanted graphite [89]. By comparing Stokes and anti-Stokes spectra, the D and 2D bands were found shifted differently for a given temperature, without clear explanation [89]. The origin of the dispersive effect of the D and 2D band were tentatively given by Pócsik et al. [90] and Matthews et al. [91] in 1998 and 1999, respectively. The slopes of the dispersions were reproduced in their work by considering a coupling between electrons and phonons with the same wave vector near the K point of the Brillouin zone (and called the $k \approx q$ quasi selection rule). However, this coupling was introduced ad hoc and moreover the authors failed to reproduce the puzzling anti-Stokes behavior. Thomsen and Reich went a step further by calculating the Raman cross section of graphite in a double resonance process [92] (completed by Saito et al. [93]), and were able to reproduce both the D and 2D dispersion relations together with the Stokes/anti-Stokes shifts. This double resonance mechanism is due to a heritage of the way Raman spectroscopy is treated in semiconductors and in which Cardonna played a major role [27,42,45]. Because of conservation rules of energy and momentum, the double resonance process, involving photons, electron-hole pairs and phonons, and described a little bit more in Section 3, selectively enhances a peculiar phonon wavevector (close to the K-point in the Brillouin zone) and then a phonon frequency [57]. For the D-band, a phonon and a defect are involved. For the 2D-band, only two phonons (without defect) are necessary. This explains why the 2D-band is always visible, even for defect free samples. The Stokes/anti-Stokes differences were understood, as the double resonance mechanism does not involve the same phonons during creation (Stokes) or annihilation (anti-Stokes) processes. In 2007, Pimenta et al. published a paper prophesying that being able to reproduce the resonant Raman behaviors will allow to better characterize disorder in nano-graphite based systems [94].

Graphene was experimentally first obtained in 2004 [95] and most of its fantastic electronic properties, related to the Dirac cone, were reviewed soon after [53,96]. The ability of Raman spectroscopy to study these properties is presented in the work by Ferrari and Basko [62]. Jorio, Cançado, and Lucchese et al., in a series of papers [97–102], were able to characterize and distinguish the influence of 0D and 1D defects on graphene. Venezuela et al. published a theoretical paper in 2011 [69] in which they calculated the double resonant Raman spectra of defective graphene, and reproduced the dispersion of D, D′, 2D bands and weaker ones (which are combination bands but not necessarily, some bands being attributed to acoustic branches). (The names of the bands do not necessarily respect the names given by other authors in the literature. For example, on the one hand the D″ band in their paper can be labelled D4 in other papers such as [61]. On the other hand, Venezuela et al. labelled bands D3, D4, D5, etc. Note that we did not use their labelling here. The supplementary information of [62] gives (among other things) valuable information on the history of this nomenclature.) Weaker bands, due to two phonon scatterings, were also observed and understood in the framework of the double resonance mechanism [103,104]. In 2012, Eckmann et al. showed that the intensity of the D′ band compared to the D band, which is not sensitive to the amount of defects, is however sensitive to the nature of the defect (sp^3, vacancy, edge) [105]. A bibliometric search with the key words "double resonance" + "graphite" + "Raman" in the abstract returns at present 122 papers, concerning not only graphene but also other allotropes like carbon nanotubes, meaning this double resonance mechanism is well established. In fact, it is so well established that it is now used on other isoelectronic and structural analogs of graphene such as 2D dichalcogenides (ME_2, M = metal, E = S, Se or Te) to interpret multiwavelength Raman spectra [106–108].

However, two things have to be noticed at this point about understanding Raman cross sections: First, we have to mention the existence of an alternative approach and second we need to introduce a possible recent breakthrough. Thus, first: Another approach was performed to model the Raman spectrum of graphite based materials. It started in 2002 by Castiglioni et al. [109–116]. This approach is based on calculating resonant Raman cross sections of disconnected aromatic molecules of different sizes, governed by interactions between π electrons and the nuclei. This approach is also able to reproduce the D-band dispersion. Even if the double resonant mechanism is now commonly used and admitted, it does not mean the molecular approach has ceased today. For example, the group of Castiglioni recently published a paper on a polycyclic aromatic hydrocarbon resembling a longitudinally confined graphene ribbon with armchair edge [117]. Their technique could be applied in the future because it can help in studying confined and nanoshaped graphene. As a proof, the authors were able to reproduce the intensity of the G + D combination band which could be used as an experimental measure of confinement in graphitic materials, as they claimed. The second thing to be noticed, the possible breakthrough mentioned above, is that in 2016 the group of Heller et al. re-interpreted the theory of graphene Raman scattering using the Kramers–Heisenberg–Dirac theory without needing to introduce the double resonance mechanism [118]. Therefore, the authors were able to describe with a second order perturbation theory (double resonance mechanism is a fourth order perturbation theory [45]) the following characteristics: The bands' dispersive behavior, defect sensitivity, Stokes/anti-Stokes anomalies, intensities, etc. Many other effects have now to be reproduced in the framework of this theory, especially concerning recent advances on UV Raman to probe the dispersion of graphene far away from the K point of the Brillouin zone (see at the end of Section 3.1), before being able to admit that this theory is better than the double resonance mechanism theory. We have to mention that Placzeck's theory also works fine to reproduce coupling effects between layers in a multi-layer graphene, being able to reproduce the out-of-plane mode variation with the number of stacked layers, in the range 10–50 cm^{-1} [119].

For more defective samples (such as carbon blacks, soots, nanographite, and amorphous carbons) other bands were reported many times in the literature, already since 1985. The D_3 and D_4 bands are examples of such bands. (In that nomenclature, the D_1 and D_2 bands are the D and D' bands lying at 1350 and 1620 cm^{-1}. In the literature, the D_3 band is sometimes called the A-band and the D_4 band is occasionally referred to as the TPA band (which stands for transpolyacetylene). They are found at $\approx$1500 and $\approx$1200 cm^{-1}, respectively [66,120–123]. They are generally difficult to observe as they fall in the D-G region, and for very defective materials, the D and G bands are broad ($\Gamma \approx$100 cm^{-1} or so), so that only the D_4 band can be seen as a shoulder on the overall spectrum. The D_4 band has been recently understood in the framework of the double resonance mechanism too [61]. More precisely, using a large variety of disordered graphitic carbon materials, Couzi et al. [61] have shown with the help of multiwavelength spectroscopy that this band is in fact composed of three sub-bands (the D*, D**, and D'') that disperse differently. However, the authors failed in relating the intensity of these sub-bands to the L_a parameter, essentially because the main factors governing the resonant Raman process (and thus the corresponding band intensities) are related to the nature of defects (point defect, edge defect, staking fault, curved or twisted planes, etc.). Note that the D'' at 1100 cm^{-1} in the visible range has been introduced by Venezuela et al. [69] in 2011. We give more information on the D'' band in the last paragraph of Section 3.4. The Raman spectra of amorphous carbons can be seen as simpler, because only a broad asymmetric band is seen close to 1500 cm^{-1}. However, this is incorrect, and for several reasons. First, many different kinds of amorphous carbon exist: sp^2 dominated ones (a-C), sp^3 dominated ones (ta-C, t referring to tetrahedral), one containing hetero-atoms such as H (a-C:H, ta-C:H, ...) or N. Their structure and properties are related but widely varying [124–126]. Second, as there is some aromatic carbon embedded in their structure, some resonance occurs [127]. Ferrari and Robertson studied and reviewed the Raman behavior of such materials in the 2000s [4,5,128]. Contrary to the Tuinstra and Koenig relation, the Ferrari's relation says that when the size of the aromatic domains is large, then the D band is intense. This relation was supposed to be connected to the Tuinstra and Koenig relation for an L value of 2 nm, imposing a relation between c and c', but this connection was found empirically. More than a decade later, the evolution of I_D/I_G on a large scale of L was understood, and the change of slope was found close to 1 nm [101]. It was done by bombarding a graphene layer varying the flux of incident ions creating the surface coverage of "0D" defects in the honey comb structure. The average distance between these 0D defects, L_D, was then introduced and a relation between I_D/I_G and L_D found. Very recently, the group of Cançado et al. disentangled the contribution of graphene samples containing 0D (point) and 1D (line) defects, by giving I_D/I_G in function of L_a and L_D [97]. More details on the way ion bombardment can help to better understand Raman spectra are given in Section 3.3. Just note that very recent improvements have been done by the same group in order to distinguish point defects and line defects using multiwavelength Raman spectroscopy [97].

The historical approach in this review is mainly focused on Raman spectroscopy of graphene based samples. However, one should note that other allotropes played an important role in the historical development of Raman spectroscopy of carbons as well, which we rapidly cite here. C$_{60}$ has been discovered by Kroto in 1985 [129]. Vibronic resonance effects in C$_{60}$ were evidenced 5–6 years later (taking into account only the A term from Albrecht's theory and reproducing the two orders of magnitude enhancement) [130–132]. Carbon nanotubes (CNT) were studied intensively since 1991, after their discovery (or rediscovery [133]) by Iijima [134]. Raman spectra of CNT not only contain G, D, and 2D bands, but also the well-known radial breathing modes (RBM). As Raman spectroscopy is a resonant process, these RBM frequencies and intensities depend on the nanotube electronic structure which is itself driven by the way the graphene plates are rolled. The Kataura plot is a tool that can give the intensity of a given mode in function of the laser used to perform the Raman spectroscopic measurements [135]. Raman spectroscopy can thus be seen as a power tool to distinguish between different nanotubes, the symmetry aspects of which are reviewed in the review paper by Barros et al. [136].

2.5. Brief Introduction to the Double Resonance Mechanism

Below, we give the basics of the double resonance mechanism which is a fourth order perturbation theory. The reader who also wants to learn about the details of this mechanism is referred to the following papers [2,137–139].

The Raman cross section contains sums of terms like the following one in Equation (14) (Γ terms, accounting for relaxations processes, have been omitted in the denominator, for simplicity):

$$\frac{\langle i|H_{\text{e-Radiation}}|a\rangle\langle a|H_{\text{e-phonon or defect}}|b\rangle\langle b|H_{\text{e-phonon or defect}}|c\rangle\langle c|H_{\text{e-Radiation}}|f\rangle}{(\hbar\omega_0 - \hbar\omega_{\text{a}\leftarrow\text{i}}))(\hbar\omega_0 - [\hbar\omega_{\text{b}\leftarrow\text{i}} - \hbar\omega_{\text{vib}}])(\hbar\omega_0 - [\hbar\omega_{\text{c}\leftarrow\text{i}} - \hbar\omega_{\text{vib}}])} \tag{14}$$

Here, $H_{\text{e-Radiation}}$ and $H_{\text{e-phonon or defect}}$ are the Hamiltonians describing the interaction between electrons and light, and electron and phonons (or defects), respectively. $|i\rangle$ and $|f\rangle$ are the initial and final electronic states, $|a\rangle$, $|b\rangle$, and $|c\rangle$ are intermediate states corresponding to the intermediate steps of the double resonance mechanism. In the reciprocal space, when the valence and conduction bands cross at the K-point (see Figure 5b), an incoming photon with the energy $\hbar\omega_0$ and wave vector $\vec{k_0}$ can always excite an electron/hole pair in the vicinity of this crossing point. This is how the double resonance mechanism starts: An electron-hole pair is excited by the incoming photon that is absorbed. Then, the electron is scattered by a phonon or a defect with a vibrational energy $\hbar\omega_{\text{vib}}$ and wavevector $\vec{q}$.

Many different scatterings can occur (see Figure 9), however only the resonant one will modify the Raman cross section by minimizing its denominator. Other scatterings are possible (not displayed on Figure 9, see Figure 2 of [62] for a complete description). After the scattering process between the excited electron and the phonon, the electron has a $\vec{k_0} + \vec{q}$ wavevector. The electron is then scattered back by a phonon or a defect, and finally the electron-hole pair recombines, emitting a new photon with the energy $\hbar\omega_0 - \hbar\omega_{\text{vib}}$, if one considers the Stokes process. The first resonances occur from $|i\rangle$ to $|a\rangle$, and the second resonance occurs from $|a\rangle$ to $|b\rangle$. The third and fourth steps are not resonant. Changing the energy of the incident photon will select another phonon that will maximize the Raman cross section, leading to the dispersion of the D and 2D bands. The intravalley process, using a scattering with a defect and with a phonon, will give rise to the D' band. The intervalley process between K and K' points in the Brillouin zone will give rise to the D band if the excited electron is scattered by a phonon and a defect, or to the 2D band if the scattering by the defect is replaced by a scattering back with another phonon.

Then, for the bands appearing because of the intervalley process, we have a set of two coupled quantities (electron and phonon momentums + dispersion relation) that obey the quasi selection rule $q \approx 2k$, meaning the phonon with wavevector q will couple preferentially with the electronic state that has the 2k wavevector measured from the K point. Using multiwavelength Raman spectroscopy data allows to follow the iLO and iTO phonon energies along the dispersion relations, exploring the Brillouin zone, which is strictly forbidden for conventional Raman spectroscopy (see for example Section 3.1 and Figure 12 in the article by Malard et al. [139]).

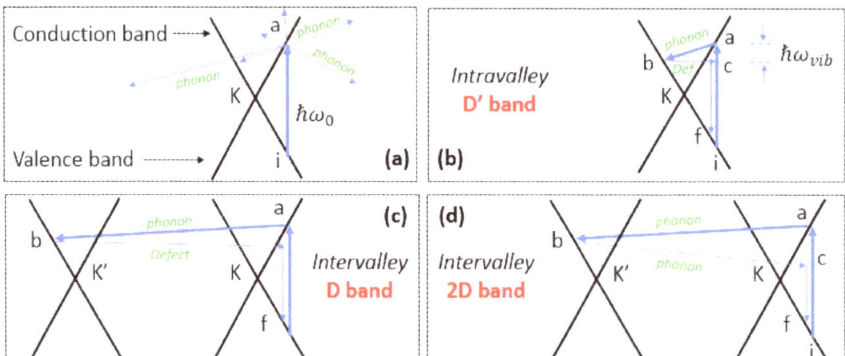

Figure 9. Double resonant scattering for the valence and conduction bands, adapted from [92]. (a) Electron-hole excitation followed by non-resonant phonon scattering. (b) Intravalley process giving rise to the D′ band: Phonon (resonant) + defect scattering. (c,d) Intervalley processes giving rise to D and 2D processes with phonon (resonant) + defect and double phonon processes, respectively.

The quasi selection rule cited above can be very useful for quantum calculations: A procedure for calculations can be to identify first the different scattering configurations between electrons and phonons by using analytical laws describing the electronic structure and phonon dispersion relation in order to perform Density Functional calculations in a second step, as they are more time consuming. 2D integration in the Brillouin zone is then necessary to take into account contributions from all the directions. Quantum interferences can occur, as the cross section is proportional to the square of the absolute value of a sum of terms integrated on the Brillouin zone (see [140]).

2.6. Intensity, Band Profiles, and Models for Fitting Spectra of Aromatic Carbons

If we do not take into account the instrumental transfer function that can be negligible in many cases (If the natural width is comparable to the width of the instrumental transfer function, which is generally a Gaussian function, then the intensity of the band is a convolution between the natural line shape and the instrumental function. Depending on the grating used, the instrumental width varies but is in general close to 1 cm^{-1}.), the total intensity of one phonon mode with a wavevector q_0 and a frequency $\omega(q_0)$, in a perfect crystal, is spread on a symmetric profile which is Lorentzian, see Equation (15): (If one considers only the dominant term of $(\alpha_{\rho\sigma})_{fi}$ (Equation (8) in Section 2.2.4) and applies the square of the modulus to calculate the Raman intensity, one will find: $|1/(\omega_{ri} - \omega_0 - i\Gamma_r)|^2 = 1/[(\omega ri - \omega_0)^2 + (\Gamma_r)^2])$

$$I(\omega) \propto \frac{1}{(\omega - \omega(q_0))^2 + (\Gamma_0)^2} \tag{15}$$

The full width at half maximum of this band, Γ_0, is the inverse of the phonon lifetime, with electron–phonon and electron–electron interactions that can contribute. For the G band of graphene it was shown that the dominant term that determines the width is the electron–phonon coupling that leads to 11 cm^{-1} at room temperature [58,141]. At the other extreme, the material is amorphous. The analysis of the profile of Raman modes of amorphous carbon has led to many studies [142,143] (and references from Ferrari et al., see Section 3.6). The profile of a Raman mode is related to the neighboring of the vibrating molecule or atoms. Any variation in this neighboring will affect the width of the Raman mode. For single crystals, all atoms have equivalent environment and the associated Raman bands are sharp. This is evidently not the case for an amorphous material where each atom has a specific environment and the ranges of values of bond angles and bond distances are wider than in a crystalline material. This leads to a broadening of the bands, which is effectively observed for amorphous carbons. Usually, a Gaussian function is used to fit the Raman bands, even if, as mentioned by Wang et al. [144], there is no clear justification for using this function.

In between amorphous materials and perfect crystals, defects play a role in the disturbed Raman scattering process. The double resonance mechanism involving defects is an example, but it is not the only one. Another mechanism that can influence the inelastic scattered light is confinement at the nanoscale. Defects can localize the wavefunction of the phonon, leading to a delocalization of its momentum in the Brillouin zone, due to the Heisenberg principle, relaxing the $|\vec{q}| \approx 0$ rule. Then, because of the phonon dispersion relation $\omega(\vec{q})$, it allows to explore away from the Γ point, new frequencies appearing in the Raman spectrum. Richter et al. [145] succeeded in understanding asymmetric profiles observed on the band at 520 cm^{-1} in silicon by following this approach, the phonon confinement model. They proposed to multiply the phonon wavefunction by a Gaussian function that localizes the phonon [145]. It can account for linewidth increase and wavenumber decrease. The intensity of the band is written as (see Equation (16)):

$$I(\omega) \propto \int_{BZ} \frac{W(\vec{q})e^{-(\vec{q}-\vec{q_0})^2 L^2/4}}{\left(\omega - \omega(\vec{q})\right)^2 + (\Gamma_0/2)^2} d^2\vec{q} \tag{16}$$

By choosing a good weighting function W, and by describing correctly the phonon branches in the Brillouin zone *(BZ)* and integrating the above expression, one can obtain the influence of phonon confinement on the shape of the spectrum. This was done for graphene by Puech et al. in 2016 [146]. *LO* and *TO* phonons introduced a second G band component at a lower wavelength (close to 1550 cm^{-1}), but also broadened all the bands

Another kind of profile, which is asymmetric, has to be mentioned: The Breit–Wigner–Fano (BWF) profile (see Equation (17)):

$$I(\omega) \propto \frac{1 + \left[(\omega - \omega(q_0))/q\Gamma_0\right]^2}{1 + \left[(\omega - \omega(q_0))/\Gamma_0\right]^2} \tag{17}$$

where the q parameter is a real number (here, q has nothing to do with the $\vec{q}$ momentum, not enough letters being possible to choose in the many alphabets usable). $1/q$ accounts for an interaction between a phonon and a continuum of states. It is generally used to fit the G bands of metallic nanotubes. Sometimes it can be used just because it is convenient to use it, e.g., to account for confinement effects or an unresolved band lying in the wing of the G band. Then, the term $1/q$ has no physical meaning.

2.7. Procedures for Fitting of the First Order Region

To fit what is often called the first order region (1000–1700 cm^{-1}), a large variety of procedures have been reported in the literature. Here, we just mention some of them (more details will be given in the corresponding sub sections of Section 3):

- One band: The G band is fitted by a Lorentzian if symmetric, and by a BWF if not symmetric.
- Three bands: The G, D, and D′ bands are fitted by Lorentzians. D and D′ bands are sometimes labelled D_1 and D_2, respectively. The D′ band is less intense than the D band by an order of magnitude and can be forgotten when the D and is much less intense than the G band.
- Four bands: The G, D, and D′ bands are fitted by Lorentzians and a Gaussian band is added in the redshift wing of the G band (close to 1500 cm^{-1}). This band is sometimes called the D_3 band, in other cases it is called the A band.
- Five bands: Same as the four bands model, but adding another band around 1200 cm^{-1}, which is sometimes found Lorentzian, otherwise found Gaussian. This band is called the D_4 band, or D″ since the theoretical work of Venezuela et al. [69].

- Six bands: Same as the five bands model, but with another distinct band close to 1150 cm^{-1}. This band is generally more easily seen using red laser (633/785 nm) instead of green/blue lasers.
- Recent developments [61] have revealed that the bands around 1100–1200 cm^{-1} are in fact three bands (D″, D*, and D**) with different dispersion behavior (see Figure 6).
- Occasionally, no D′ band is observed (possibly merged with the G band so that authors do not try to decompose each component).
- Sometimes, the D and G bands which are Lorentzian are accompanied by two other broader bands (Gaussian or of different line shape) that are red-shifted compared to the D and G bands. The term amorphous component can often be found as well.

We advise the reader to carefully check for the relative intensity ratio notation used in the literature. In some papers, for example, the ratios of two band intensities are the area ratios (noted A), sometimes it is the maximum intensity (noted I) and elsewhere it is reported without fitting (noted H). Jorio et al. [147] performed a comparison between A_D/A_G and I_D/I_G for bombarded graphene and found that I_D/I_G should be used instead of A_D/A_G.

2.8. Examples of How Comparison with First-Principle Calculations Can Help

Systems are sometimes so complex that the use of first-principle calculations are more than needed, helping in retrieving quantitative data of the structure or kind of defects. Lattice-dynamic behavior of a crystal affects physical properties such as phonon dispersion, which (as we know) can be compared to Raman spectra for some points of the Brillouin zone. Our aim in this part is not to detail the theory from the first-principle, but to illustrate the use of the theory and which information can be retrieved. The reader who is interested in more details about the basics of the lattice vibration theory is advised to read the seminal paper by Born and Huang [24], and the reader who would like to know more about the density functional (perturbation) theory, the methodology, and its approximations could read the review paper by Baroni et al. [148]. Such theories are implemented in open source codes like QUANTUM-ESPRESSO [149] or licensed codes like the Vienna Ab Initio Simulation Package (VASP).

As a first example, Mohiuddin et al. applied an external strain on a graphene plane both experimentally (observing the splitting of the E_{2G} Raman active mode) and theoretically [8]. The good match between both experiment and theory strengthens the physical description of the graphene on the basis of the calculation. Going a step further is possible as well: Bonini et al. [141] were able to determine from first principle the thermal properties of graphene and graphite, enhancing the role of electron–phonon and phonon–phonon interactions and compared them with the band shift and band width (σ_G and Γ_G) evolution with temperature. Γ_G being essentially due to electron-phonon coupling at room temperature, whereas the band width of the infrared active mode is narrower, only caused by phonon–phonon interactions. Next example is about identifying defects signature, which is a holy grail in the field [69], and will be discussed in much more detail in Sections 3.3–3.6 (some issues can be found in the review article by Ferrari and Basko [62]). Here, we mention the fact that for carbon nanotubes, the works by Saidi et al. [150] have revealed that di-vacancies and other defects influence the non-resonant G bands more than the radial breathing modes. These defects can also affect the intensity of the resonant Raman bands, allowing them to be identified by this way [151,152]. As a final example, we would like to mention graphene oxide, which is a heavily oxidized carbon. The G band of graphene oxide has been found blue shifted experimentally, compared to graphene. Its explanation was obtained with the help of first principles [153]: Graphene oxide can be composed of sp^2 areas surrounded by an alternating pattern of single-double carbon bonds.

3. Raman Spectroscopy of Different Aromatic Carbons

In this section, we review the way multiwavelength Raman spectroscopy is used in literature to better characterize aromatic based carbons, by focusing on more and more disordered aromatic carbons. The aim is to give a concrete and practical view on how Raman spectroscopy can be used to classify the nanostructure. Figure 10 displays the Raman cross sections of typical graphene based materials compared to some other relevant molecules or reference materials.

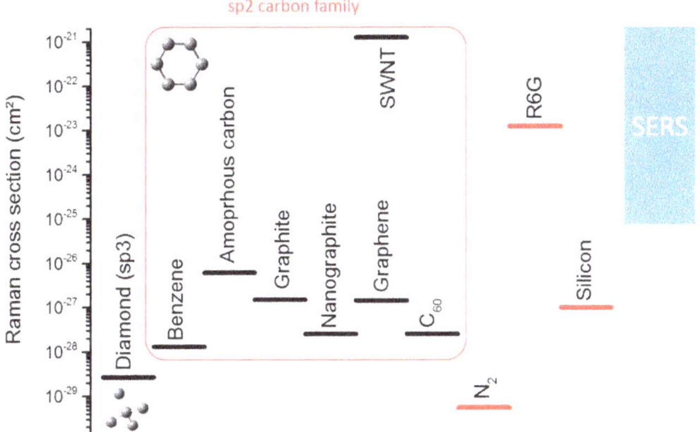

Figure 10. Raman cross section for carbonaceous materials compared to reference materials. Sources: Diamond [154], graphene [49], nanographite [65], benzene [155], silicon [156], SERS [157]. Note that Raman cross section of solids can sometimes be difficult to obtain because of substrate effects that can give rise to interference effects that depend on the thicknesses and optical indexes, as it was shown for graphene [48,158] and earlier for amorphous carbon [46].

We start with graphene in Section 3.1, and will continue with what we call nanographenes in Section 3.2 (nanotubes, nanohorns, nanocones, nanofibers), followed by bombarded graphene in Section 3.3, very defective carbons (e.g., coal, soot, black carbons) in Section 3.4 and graphite intercalated compounds in Section 3.5. We will finish with amorphous carbons in Section 3.6.

3.1. Graphene

Graphene has many astonishing properties (that are reviewed by Peres [159]). One of them is the existence of Dirac cones crossing at the Fermi level at the *K* point of the Brillouin zone (see Figure 5b), and that determine mostly the transport properties in graphene. As there is some electronic resonance with incoming photons, these cones also determine the properties of Raman spectra, which can appear counterintuitively. The main thing to keep in mind about Raman spectroscopy of graphene is that it is then driven by the electron properties—how they move, interfere, and interact—affects the Raman spectra. Disorder can modify the electronic properties, and then can find finger prints in the Raman spectra because of the resonance mentioned above. Among other defects, one can find: Adsorbed species, folded regions, rippling, vacancies, topological defects (such as Stone–Wales defects), charged impurities on wafer [159], and so forth.

Graphene is the building block of nanocarbons: Staking individual (graphene) layers will give rise to multi-layer graphene and eventually graphite. Rolling it, results in the formation of carbon nanotubes. Creating point defects or linear defects (edge or grain boundaries) can account for different processes such as ion implantation or crystal growth under thermal treatment of amorphous carbons, for example. If one wants to understand the Raman spectrum of aromatic carbon containing samples, one has to understand first the Raman spectrum of graphene. Many reviews can be found on Raman spectroscopy applied to graphene, but we advise the reader to first use the work of Ferrari [62] and its useful 13 pages of supplementary information, or the work of Beams et al. [160]. Below, we give some trends before focusing, in Section 3.2, on some specific nanoforms and on few kinds of defective graphene samples.

Graphene without defect gives rise to two main bands: The symmetry allowed G band and another band, the 2D band. For multilayer graphene, useful information can be found in the work of Malard et al. [139]. The G band is due to the E_{2g} phonon at the center of the Brillouin zone (and called the Γ-point). The 2D band is due to TO phonons around the K point in the Brillouin zone, and is active due to the double resonance mechanism [57], as presented briefly in Section 2.5. As an illustration, multi-layer graphene was found to display characteristic Raman spectra, especially in the 2D spectral region [1]. As it has different electronic structures close to the K point, and because the double resonance mechanism connects phonons to the electronic structure, the position, shape (composed of several overlapped bands), and intensity of the 2D band(s) can be used to distinguish from monolayer up to 5–10 stacked layers. The relative intensity ratio between the 2D and G bands was also found to be dependent on the number of layers: I_{2D}/I_G is close to 3 for monolayer graphene, and falls down to 0.3 for highly oriented pyrolytic graphite (HOPG). Then, the Raman plot σ_{2D} versus I_{2D}/I_G can be used to rapidly have an idea of the quality of the graphene samples handled. Figure 11 gives such an illustration for two wavelengths: 514 and 633 nm. The HOPG samples have been cleaved by the tape method to obtain multilayer graphene flakes that were deposited on a silicon substrate. Raman spectra have been obtained from all the flakes. For 514 nm, HOPG is situated at (2725; 0.3) and the monolayer is situated at (2686; 3). One can see that, with this method, many intermediates are obtained. The broadening around the guide for the eyes lines can be due to stacking faults that modify the electronic structure of the multilayers, and thus the 2D shape and intensity [161–163]. One has to note that σ_G slightly depends also on the number of layers but the shift compared to HOPG is no more than 5–6 cm^{-1} [164]. The width of the G band, due to electron–phonon coupling has been evaluated to be 11.5 cm^{-1} [58], phonon–phonon scattering being responsible to 4–5 cm^{-1} extra broadening found experimentally. For graphite and multilayer graphene, we mention the existence of a C band, which is sensitive to coupling between layers: It lies at 44 cm^{-1} for graphite and shifts regularly down to 31 cm^{-1} for bilayer graphene [165].

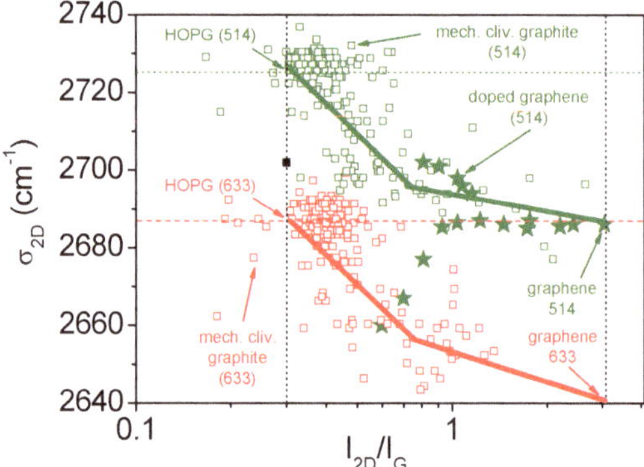

Figure 11. 2D band position as function of I_{2D}/I_G for laser wavelengths of 514 and 633 nm. Graphene frequencies were obtained from [1]. Empty squares are from mechanically cleaved graphene flakes (original data). Stars are from [160] (doped with carrier density from -3×10^{13} cm^{-2} to $+4 \times 10^{13}$ cm^{-2}), the downshift corresponds to positive doping, whereas upshift corresponds to negative doping.

As another example, dopant impurities can modify the Raman spectrum of graphene [166–168]. Playing with the electron doping (i.e., changing the position of the Fermi level) modifies the Raman spectrum: The band positions and intensities are modified [169]. In Figure 11, the doping effect on a single layer graphene has been added: Doping with electrons or holes can increase or decrease σ_{2D}, and both diminish I_{2D}/I_G. The effects of doping have been reviewed in 2007 [137] and in 2015 [160]. We report that changing the carrier density also affects the D band spectroscopic parameters [170].

Impurities adsorbed on graphene can also modify carrier mobility and thus the Raman spectrum of graphene [171], changing the D band intensity.

Substrate effects on the position of the 2D and G bands (downshift, splitting) have been reported in the works of Das and Calizo et al. [164,172,173]. Strain can be such effect [174]. If that is the case, Frank et al. demonstrated that using different wavelength gives very different 2D band shape. Among others, doping can be an important and unintentional effect as it can change the position of the 2D band, as shown by Yang et al., [175], for graphene grown on SiC, or by Das et al., [176], for other substrates. However, as doping also affects the G band position and width, checking G band anomalies can be used to detect such an unwanted effect. Roughly, for electron or hole concentrations higher than 10^{13} cm^{-2}, the G band width is diminished by a factor of 2 (i.e., close to 7–8 cm^{-1}) [176]. G band splitting and intensity enhancement (i.e., Surface Enhanced Raman Scattering) were observed for multilayer graphene in contact with Ag and Au [177]. The splitting and intensity enhancement were found to be higher for single layers. As the 2D band intensity was found diminished by the Ag deposition, an n-type doping effect was concluded. A p-doping effect was deduced for Au deposition because of the 2D band shift. If the reader is interested in the detection of graphene combined with the SERS effect, a further reading of the 2013 review can be found in the work of Xu et al. [178]. A good way to differentiate between doping effect and subtract effect is to plot σ_G versus σ_{2D} of the analyzed samples (this differentiation is possible because strain affects the band position by changing bond length and angles between bonds, whereas doping affects the electron–phonon coupling [179]).

Depending on the optical constants of the substrate on which the graphene layer is deposited, the substrate (layer) thickness, the numerical aperture of the objective used or the wavelength of light used for performing the micro Raman spectroscopy analysis, the absolute intensity of the G and 2D bands can be altered due to electromagnetic interference effects so that their relative intensity ratio is changed in appearance [48,158]. For example, for a SiO$_2$ substrate layer thickness in the range of 225–375 nm, I_{2D}/I_G appears to vary from 2 to 8 instead of being constant at $\approx$3. A refined analysis was performed by Klar et al. [49] who were interested in the absolute intensities of the G and 2D bands for multilayer graphene, checking numerical aperture effects, wavevector polarization, substrate effect, etc.

We have to mention a breakthrough due to deep UV Raman spectroscopy that has been used only since very recently (2015) on graphene [180]. The advantage is that, due to the double resonance mechanism, one gets far away from the K point in the Brillouin zone, exploring a larger part of the dispersion relations, for example near the M point of the Brillouin zone. The $(\hbar\omega_0)^4$ wavelength dependency of the G band intensity was confirmed for 266 nm (4.7 eV), but a surprising $(\hbar\omega_0)^{-1}$ relationship was found for the 2D band. It would lead to the conclusion that I_{2D}/I_G is proportional to $(\hbar\omega_0)^5$. On nanographites [65], Cançado et al., who observed a L_a dependency, did not observe any $(\hbar\omega_0)$ dependency, but their data were recorded in the visible range, in a small range of wavelength. Tyborski et al. observed that when the 2D band disappears, the two phonon density of states rises in the 2600–3300 cm^{-1} spectral region [181]. Among the many things they deduced from their analysis, taking into account symmetry arguments, they were able to deduce from this spectral region, the maximum frequency of the LO phonon at 1626 cm^{-1}, and its frequency at the M point at 1408 cm^{-1}. Saito et al. performed predictive calculations up to 177 nm (7 eV) [182]. They also determined the asymmetry parameter evolution with $\hbar\omega_0$, appearing in the Breit–Wigner–Fano profile of metallic nanotubes (see Section 3.2.3 for more details) what has never been down before, to our knowledge. Finally, Raman spectroscopic measurements from UV up to 325 nm were performed on multilayer graphene in order to test the model predictions on the number of bands shaping the 2D band of multilayer graphene (due to the complex electronic structure close to the K point). The 2D band profile was found to change: For monolayer graphene, three sub bands appear at 325 nm [183]. The one with the lowest frequency (2800 cm^{-1}), supposed to have the same origin as the G* band, has an intensity enhanced when the number of layer increases, whereas the one with the highest frequency (2900 cm^{-1}) display another behavior. This experimental result is found to be in agreement with previous calculations announcing three strong bands and a weak band caused by four electron-hole scattering processes involved in the double resonance mechanism [184]. To conclude about UV Raman spectroscopy, it allows deeper exploration of the Brillouin zone.

Finally, Raman spectroscopy can also be used to determine mechanical properties [185], study strain effects [186], and characterize grain boundaries [187,188] of graphene grown on a substrate. In addition, it is used to deduce the thermal conductivity of suspended graphene [189], and to monitor electrostatic deflection of suspended graphene [190].

3.2. Graphene based Nanoforms

3.2.1. Nomenclature

A huge number of different carbon nanoforms are now produced around the world such as single and multiwall nanotubes, bamboo nanotubes, nanotube forests, fullerenes, nano-onions, nanocones, stacked nanocones, scrolled graphene, nanofibers, nanowalls, nanosheets, and nanoplates. Graphene is definitely their building block, and by applying a transformation, such as stacking, cutting, circularly wrapping, scrolling, coiling, screwing, etc., all the other forms can be obtained. In 2012, the editors of the journal *Carbon* decided to propose a nomenclature to classify all these sp^2 carbon nanoforms [191]. To help researchers in their bibliographic studies, they proposed to classify all the known forms in three families: Molecular forms (0D), cylindrical nanoforms (1D), layered nanoforms (2D). Another family should be included, to our opinion, and called "graphenic carbon materials", as was reported in [192]. This fourth family contains graphite, carbon fibers, and could also include all amorphous carbons that are based on aromatic rings, but at a local order. If one considers the addition of oxygen to the carbon nanoforms, an alternative approach to categorize them was proposed in the work of Wick et al. [193], based on their toxicologic identification.

For some of these families, the state of the art is not yet enough advanced to propose a bijection between all the members of the above families and their Raman spectra. Below, we present the understanding and main trends about Raman spectroscopy of some of the nanoforms and, where possible, the links between different kinds of nanoforms.

3.2.2. Nanodomains, Nanoribbons

Graphene edges, labelled as 1D defects, exist in two configurations which are named zigzag and armchairs, related to the shape of the terminal carbons of the hexagons (see Figure 12a). Nanoribbons are delimited by these edges, which give them peculiar electronic properties (See for example Section 7 in the work of Mohr et al. [140].), forming gaps in the electronic dispersion relations due to spatial confinement of the electron wavefunction. This can affect their Raman spectra, again due to resonance.

The first Raman study on edges was performed by Cançado et al. in 2004 on step-like edges on graphite [194]. Roughly 2 microns away from a step, no D band was observed whereas a D band and a D′ band were observed on both zigzag and armchair configurations. The intensity of the D′ band was found non-sensitive to these two configurations, whereas the D band was found to be more intense for armchair edges than zigzag edges. The explanation of these two behaviors is that as the D′ band is an intravalley process (see Section 2.4), momentum conservation upon the double resonance mechanism can be satisfied whatever the direction in the Brillouin zone, whereas for the D band, which is an intervalley process, it cannot be satisfied for zigzag edges (see [139,194] for more details). You et al. [195] have also investigated this point in 2008. They observed that a piece of graphene with two edges separated by 90° have necessarily different chirality (this is the case if the angles are 30°, 90°, or 150°, whereas when it is 60° or 120° the chirality is the same, due to geometry). They also observed that the D band intensity is weaker for zigzag edges. Note that because graphene edges are sometimes not perfect at the micrometric scale (the size of the laser beam once focalized on the surface of a sample), especially when graphene is obtained by mechanical cleavage, it can be difficult to always conclude on the zigzag or armchair edge origin with Raman spectroscopy [196]. In 2014, Islam et al. took into account non-ideal edges orientation at a scale smaller than the micrometric size of the spot (see Figure 1 of [197]). Raman spectroscopy is now used to control the properties of edges of individual grains grown by chemical vapor deposition on large scales (few tens of microns) [187].

Because of a lack of translation symmetry, the edges do not have the same environment as in between edges of the ribbons, thus their vibrational frequencies should be different. Bands lying at 1450 and 1530 cm^{-1} have been observed in 2010 by Ren et al. and attributed to localized edge phonon modes of zigzag and armchair configurations terminated by H atoms [198]. These bands were observed only with red laser but not with green laser, meaning a resonance behavior may happen. Vibrational density of states calculations taking into account different sizes and shapes of nanoribbons were performed by Mazzamuto et al. in 2011 [199]. Among other features, out of plane edge localized modes were predicted at 630 cm^{-1} for armchair nanoribbons, whereas in plane edge localized modes were found at 480 cm^{-1} for zigzag nanoribbons [199]. Non-resonant Raman spectra were also studied in the same period by Saito et al. [200]. They found that, depending on the edge configuration and polarization directions of the incident and scattered photons relatively to the edge orientation, a symmetry selection rule for phonons appeared. Very recently, the group of Castiglioni published a paper on a polycyclic aromatic hydrocarbon resembling a longitudinally confined graphene ribbon with armchair edge and found the presence of G and D bands, low wavenumber bands, and combination modes in the 2500–3250 cm^{-1} spectral region like in graphene, with intensity very sensitive to laser wavelength due to resonance effect [117]. Radial breathing like modes, looking like the one found in nanotubes (see Section 3.2.3), have also been predicted and detected [201,202]. Verzhbitskiy et al. have shown that below 1000 cm^{-1}, the spectral region is very sensitive to the edge morphology and functionalization [203], as can be seen in Figure 12b, as is found for the radial breathing modes of carbon nanotubes (see below) or poly aromatic hydrogenated molecules [117]. Moreover, the D band dispersion is found to vary from 10 to 30 cm^{-1}/eV, lower than the classical 50 cm^{-1}/eV, allowing to use multiwavelength Raman spectroscopy as a detector of such nanostructures. Very recently, the Raman spectroscopy of nanoribbons has been reviewed by Casiraghi and Prezzi [204].

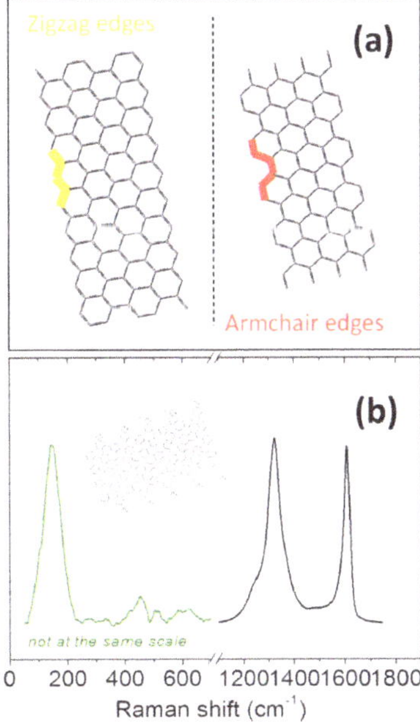

Figure 12. Nanoribbons. (**a**) Zigzag and armchair nanoribbons; (**b**) Typical spectra of a real nanoribbon, as studied by Casiraghi et al. [204].

3.2.3. Nanotubes

Basically, a single wall carbon nanotube (SWNT) can be seen as a rolled piece of graphene (see inset of Figure 13). The way this piece is rolled is named the chirality of the tube [205,206]. This chirality defines several properties of the tube: Its geometry (diameter) but also its electronic properties (band gap). Due to their quasi unidimensional structure (quantum confinement) the electronic density of states of SWNTs exhibits Van Hove singularities, which are singularities associated to a very high density of electronic states. Therefore, the chirality defines the allowed electronic transition between Van Hove singularities with a photonic excitation. In addition, the chirality also governs the band gap and a tube can be either metallic or semi-conductor. As a result, it is possible to plot all the allowed electronic transitions with the diameter of the SWNT. This plot is called the Kataura plot [135]. As Raman scattering for SWNTs obeys a resonant mechanism, the Kataura plot directly indicates which kind of tube is resonant for a given excitation wavelength. The final recorded Raman signal will come only from the resonating tubes. As a consequence, the Raman cross section of SWNTs is extremely high (contrary to their IR signal) [207], and since the pioneering work by Rao et al. [87], Raman spectroscopy has been extensively used for the study of SWNTs [208]. The typical first order Raman spectrum of SWNTs is divided in three parts:

- The low frequency region (typically between 100 and 300 cm^{-1}), which is associated to radial breathing modes (RBM), see Figure 13. The frequency of these modes is directly related to the diameter of the tubes. One can find a review on RBM not only limited to nanotubes published recently by Ghavanloo et al. [209];
- The D band (around 1300 cm^{-1}), related to defects (as for graphite and graphene);
- The G band (around 1550 cm^{-1}), also similar to the G band of graphite and graphene.

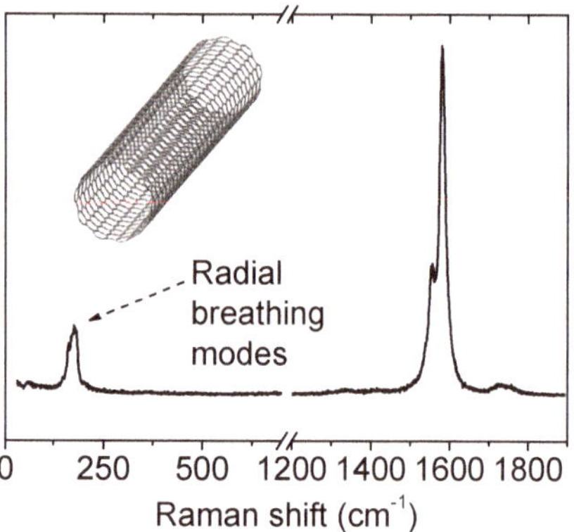

Figure 13. Example of carbon nanotube Raman spectrum (λ_0 = 514 nm).

Nevertheless, in the case of SWNTs the profile is related to their electronic properties. As mentioned above, it is possible to determine which kind of tube is resonant for a given excitation wavelength using the Kataura plot, and then resonance allows to identify the kind of nanotube probed [210]. If the resonant tubes are metallic, the G band exhibits an asymmetric profile named Breit–Wigner–Fano (BWF). This feature is not present for semi-conducting tubes and is coming from a specific interaction between the phonons and the electronic continuum [211]. Saito et al. performed calculations on the shape of the Breit–Wigner–Fano profile for different excitation energies, taking into account the double resonance mechanism and found that the shape is modified [182].

To conclude this part, rings of single nanotubes with different diameters have been synthetized since 2006 [212]. Ring size-dependent Raman G-band splitting in carbon nanotubes has been reported and attributed to the resonance condition changes caused by additional curvatures in rings instead of attributed to bond length change induced by curvature/strain (ring diameter ranging from 129 to 372 nm generates strain in the range 0.3–1.3%) [213]. When deconvoluted, the sub bands were found to lie at σ_{G1} = 1535 cm^{-1}, σ_{G2} = 1553 cm^{-1}, σ_{G3} = 1563 cm^{-1}, σ_{G4} = 1575 cm^{-1}, σ_{G5} = 1593 cm^{-1} and σ_{G6} = 1603 cm^{-1}. This splitting rises for strains up to roughly 1%. Such splitting was found on strained graphene as well [8].

3.2.4. Fullerenes

Here we briefly review the main papers dealing about the Raman spectroscopy of fullerenes. First, fullerene are mixture of pentagonal and hexagonal rings which leads to a curved shape, like a soccer ball, the name of footballene being encountered sometimes, referring to C_{60}. Since their discovery in 1985, once vaporizing graphite using laser light [129], fullerenes have been extensively studied using Raman spectroscopy. Among others, C_{60}, C_{70}, C_{80}, C_{84} have been detected, up to C_{400} [214]. The paper of Dresselhaus in 1996 [215], and then the one of Kuzmany et al. in 2004 [216] both review the Raman spectroscopy of C_{60} and related materials, from pristine C_{60} up to peapods, which are single wall nanotubes filled by C_{60} molecules [216]. Contrary to the other sp^2 carbon forms presented in our review, fullerenes exhibit specific Raman modes that can be easily identified. For example, as shown in Figure 14, in the case of C_{60}, the modes labelled Hg(1) to Hg(8) rise at 272, 433, 709, 772, 1099, 1252, 1425, and 1575 cm^{-1}, respectively, and the modes labelled Ag(1) and Ag(2) rise at 496 and 1470 cm^{-1}. The intensity of the Ag(2) mode at 1470 cm^{-1} is much more intense than the other modes due to a vibronic coupling that enhances its Raman intensity for E_0 close to 2.6 eV [130]. This band intensity can be used as a probe of the coupling between C_{60} and its environment, like was shown by Bardelang et al. [217], where C_{60} has been introduced in an open framework. Due to different selection rules, the corresponding modes for C_{70} are different [218]. Both C_{60} and C_{70} can be polymerized under UV radiation [219], leading to the rise of low vibrational modes at 118 cm^{-1}, which correspond to bonds between C_{60} in the solid phase. Note that heating C_{60} thin films can lead to the formation of highly disordered nanographites [220].

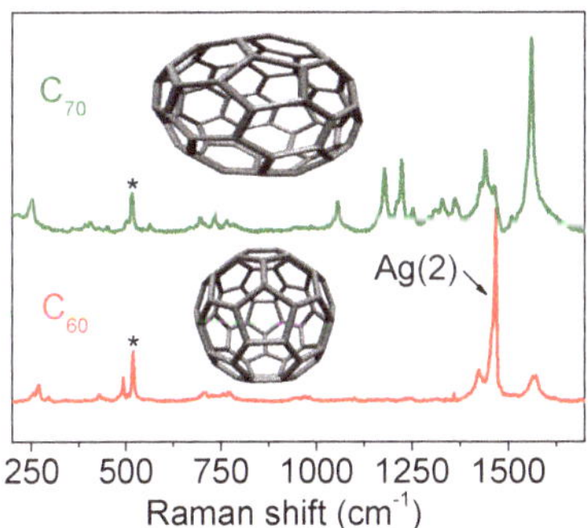

Figure 14. Raman spectra obtained from fullerenes C_{60} and C_{70}. The band at 520 cm^{-1} and marked by a star is due to the underlying silicon wafer. λ_0 = 514 nm.

3.2.5. Nanocones

Nanocones, also called nanohorns, are typically 2–5 nm in diameter and 40–50 nm in length, looking like needles, the number of pentagons at their tips piloting their shape, as for fullerenes [221]. They were first obtained by Iijima et al. in 1999 [222], and Raman spectra performed at that time were looking like Raman spectra in between nanographite and amorphous carbons. Depending on the synthesis conditions, four different types have been identified, and labelled because of their shape observed by transmission electron microscopy: Dahlia-like, bud-like, seed-like, and petal-like [223].

A typical spectrum of a nanocone is displayed in Figure 15. By increasing the pressure up to 8 GPa, some types debundle whereas others change from one type to the other, and new promising configurations, of interest for future functionalization, have been found [223]. Calculations by Sasaki et al. [224] predict the existence of a topological D band, shifted by 50 cm^{-1} and being non dispersive with the incoming photon energy. However, to our knowledge, such a shift has not yet been observed.

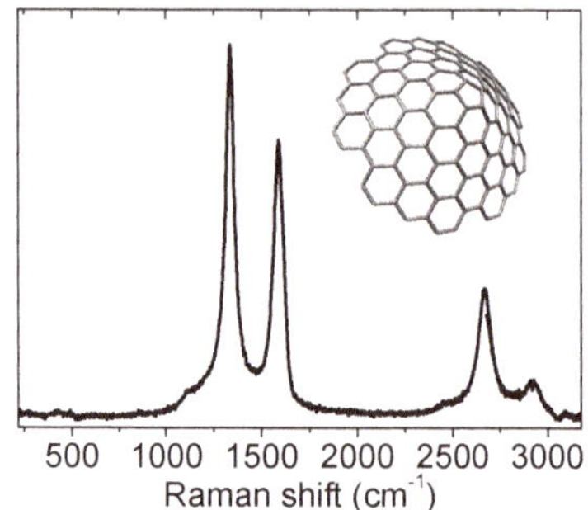

Figure 15. Typical Raman spectrum of a nanocone. The insert is an illustration.

In situ Raman spectroscopy reported done by Pena-Alvarez et al. [223], has revealed several interesting things: The G band high shifts with stress, a band rises close to 1510 cm^{-1} (looking like the G band of an amorphous carbon), all the bands increase their width, and the most important thing: The width of the D band increases much more than the G band. Exactly all these effects have been found in a large variety of disordered samples that are not nanocones (ranging from nanographite to amorphous carbon) exhibiting very typical, similar, and often encountered Raman spectra, looking like the orange one in Figure 6b [66].

In a previous study [66], we have used a simple parametric model to describe the relation between the spectroscopic parameters I_D/I_G, Γ_G, Γ_D, and the presence of a band close to 1500 cm^{-1}, for a huge variety of disordered carbons and using three wavelengths from 325 to 633 nm (based on empirical laws relating these parameters to L_a plus the assumption that the $(\hbar\omega_0)^4$ dependency affecting I_D/I_G also prevails for the Ferrari relation). To reproduce the spectra well, an unknown source of broadening of the D band was absolutely needed for 514 and 633 nm, but not for 325 nm, highlighting a resonance effect. Except the resonance effect which was not studied by Pena-Alvarez et al. [223], exactly all these effects have been observed in stressed nanocones as can be seen in Figure 16. In this figure, we compare the width of the G band (Γ_G) versus I_D/I_G scaled by the laser wavelength (λ_0) for the large variety of samples analyzed by Pardanaud et al. [66], (referred here to "disordered carbons"), heated amorphous carbons (work of Pardanaud et al. [225,226]), bombarded graphite (work of Niwase et al. [227]) and the nanocones from the work of Pena-Alvarez et al. [223]. The grey lines are the models based on the Tuinstra and Ferrari relations (see the work of Pardanaud et al. [66], for more details) and stand for disordered graphite and amorphous carbons, respectively. All these data draw a common pattern that has to do with the so-called "amorphization trajectory" of Ferrari [4], presented in more details in Section 3.6. Briefly, starting from a perfect graphite, on the left corner of the figure, both Γ_G and I_D/I_G increase, Γ_G being linear with I_D/I_G. At a certain point (in the range I_D/I_G = 0.6–3), I_D/I_G starts to decrease whereas Γ_G continues to increase, changing the trajectory in this plot. This happens when

- both the Ferrari relation and Tuinstra and Koenig's law meet, and/or
- a new set of bands close to 1200 and 1500 cm^{-1} appears, and/or
- the D bands broaden more than excepted.

This change of trajectory is observed for both disordered carbons and nanocones where the compressive stress has been increased up to 8 GPa.

Widths are generally related to lifetimes, here the decay lifetime of the *iTO* phonon involved in the D band, due to confinement effect induced by external compressive stress, can be the cause of the band width increase. The rise of the 1510 cm^{-1} band when increasing the compressive stress, can be understood qualitatively as a phonon confinement effect too [146]. Note that for L_a = 2 nm, Puech et al. were able to produce a peak which is close to 1550 cm^{-1}, 40 cm^{-1} away from the detected position. However, in their calculations they used flat graphite, whereas nanocones are not flat. Bonds, and then of course band positions, are however affected by curvature [8] so that a combined effect of phonon confinement plus curvature may explain the existence of this band at 1510 cm^{-1}. That band was not observed on disordered samples studied by Pardanaud et al. [66], using UV, due to resonance effects. We believe that in situ multiwavelength Raman spectroscopy coupled to skeleton analysis of TEM images (like the one performed by Oschatz et al. [228], or by Da Costa et al. [229]) performed on stressed nanocones could be the next insight to pave the way between 2D ordered and 3D disordered aromatic carbons, especially by taking advantage of resonance effects.

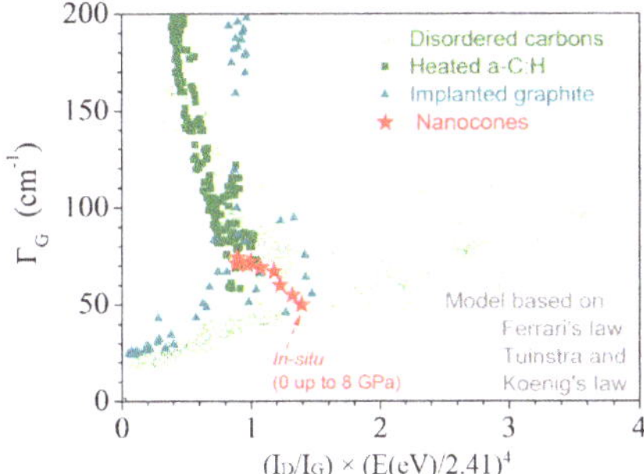

Figure 16. Γ_G vs. I_D/I_G plot for a large variety of disordered graphitic samples. Grey lines are obtained from a fit (more explanation in the work of Pardanaud et al. [66]. Data obtained from disordered carbons (514 and 325 nm data) were also obtained from this paper. Data of implanted HOPG (triangles) were taken from [230] and from [231], respectively, and those of thermally heated amorphous carbons were taken from [225]. Data belonging to nanocones under high pressure are given in green stars, and were extracted from [223].

3.3. Disordered Graphene as a Reference for More Disordered Carbons

Structural defects that may appear during a growth process or some processing can modify local properties and make new properties varying from one kind of defect to another [232]. As an example, defective graphenes composed of randomly oriented domains, called amorphous 2D materials, have been observed in 2011 under electron beams [233]. Here, we focus on the Raman spectra of point defects (0D), line defects (1D), and Stone–Wales defects. Writing "disorder", or "defect", one immediately thinks about the D and D′ bands, not only the G and 2D bands. Even though the shape of the G and 2D bands can also be changed by disorder (we will see that in Section 3.4), we will now focus on the D and D′ bands that become activated by defects, whereas they are forbidden by selection rules for perfect graphene planes. We just mention that the rise of these bands subsequently to ion irradiation was first studied in the 1980s and 1990s on graphite [227,234–238], giving an insight to the production and behavior of defects in graphite.

Coming back to graphene, in 2012 Eckmann et al. [105] showed that the intensity of the D′ band compared to the D band ($I_D/I_{D'}$) is very effective to discriminate between different kinds of defects: $I_D/I_{D'}$ = 3.5, the minimum value, is found for boundaries, whereas $I_D/I_{D'}$ is close to 7 for vacancy like defects and up to 13 for sp^3 defects. The $I_D/I_{D'}$ ratio is interesting because, according to the resonance Raman theory, it is not sensitive to the number of defects but only to the type of defects. Until today, we do not know if all defects activate these D and D′ bands. To answer this question, one has to introduce in a controlled way many kinds of defects in graphene and record their corresponding Raman spectra. Since 2006, this task has been challenged mainly by the group of Jorio and Cançado by studying 0D [98,99,101] and 1D [65,67] defects of graphene separately, and together very recently [97], and also with 0D defects in multilayer graphene [99,100]. Eckmann et al. also challenged this point, by comparing sp^3-C, vacancies, and substitutional boron atoms, thereby highlighting the role played by the D′ band [239].

In 2006 and 2007, Cançado et al. worked first on nanographites with crystallites delimited by 1D defects (and obtained by heating diamond like carbons), and clarified the Tuinstra and Koenig relation, highlighting the $(\hbar\omega_0)^4$ dependency of the I_D/I_G ratio for 1D defects [65,67,240]. Moreover, they have shown that the width of the D, G , D′, and 2D bands verify a relation which is: $\Gamma = A + B/L_a$, A and B being constants (A = 11 cm^{-1} for the G band, for example, close to the 11.5 cm^{-1} calculated for the perfect graphite crystal [58]). As a direct consequence, I_D/I_G and Γ are linear if plotted one against the other for these nanographites, with L_a in the range 20–65 nm. This is what is evidenced in the bottom of Figure 16. Another direct consequence is that Γ_G and Γ_D evolve linearly too, with a slope close to 1.

In 2010, Lucchese et al. [101] introduced their well-known "local activation model" in order to reproduce the spectra obtained on graphene samples bombarded at different fluences by 90 eV Ar$^+$ ions. They found the following expression (see Equation (18)) for 0D defects by writing the evolution equation of the S and A regions as shown in Figure 17a,b

$$\frac{I_D}{I_G} = C_A \frac{r_A{}^2 - r_S{}^2}{r_A{}^2 - 2r_S{}^2}\left[e^{-\frac{\pi r_S^2}{L_D^2}} - e^{-\frac{\pi(r_A^2 - r_S^2)}{L_D^2}}\right] + C_S\left[1 - e^{-\pi r_S^2/L_D^2}\right] \tag{18}$$

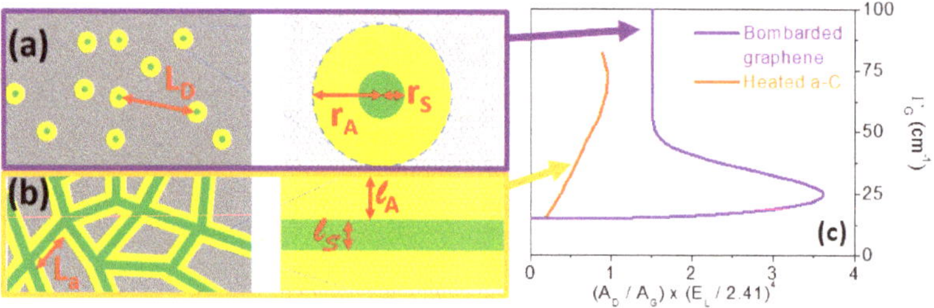

Figure 17. 0D and 1D local activation models. (**a**,**b**) Definition of the structurally damaged regions and activated regions, with L_D, the distance between defects, and L_a, the size of crystallites surrounded by edges, for respectively the 0D and 1D defects; (**c**) Γ_G as a function of A_D/A_G.

Here, r_S (see Figure 17a) is the radius of the structurally disordered area around the defect, and r_A the radius of the activated region (i.e., the region in which the selection rules are broken leading to the intervalley double resonance mechanism and giving rise to a D band). C_A and C_S are constants whose origins are discussed in [101] (C_A is related to the Raman cross sections, C_S is related to the geometry of the defect), L_D is the average distance between defects. In the regime where the quantity of defects is small (i.e., when the mean distance between defects is large compared to $r_A - r_S$), a first order Taylor expansion of the above equation reduces to Equation (19):

$$\frac{I_D}{I_G} \approx C_A\pi\frac{r_A{}^2 - r_S{}^2}{L_D^2} + C_S\pi\frac{\pi r_S^2}{L_D^2} \tag{19}$$

In their study [101], the defect was a vacancy and the parameters found were: $r_A = 3$ nm, $r_S = 1$ nm, $C_A = 4.2$, and $C_S \approx 0$. For a low number of defects, the I_D/I_G ratio behaves like $1/L_D{}^2$, which better fits the data than the Tuinstra and Koenig $1/L_a$ relation. In another study [239], no dependency with the type and number of defects (such as sp^3-C) was obtained using this ratio (but a difference was noticed from 3D materials, see Figure 3 and its discussion in that study). However, comparing D and D' intensities with the type and number of defects leads to a strong correlation. In the low concentration regime, one can find Equation (20):

$$\frac{I_D}{I_{D'}} \approx \frac{C_{A,D}(r_{A,D}{}^2 - r_{S,D}{}^2)}{C_{A,D'}(r_{A,D'}{}^2 - r_{S,D'}{}^2) + C_{S,D'}r_{S,D'}{}^2} \tag{20}$$

with the subscript D or D' to refer to the D or D' bands. One can see that the $I_D/I_{D'}$ ratio strongly depends on the $C_{S,D'}$ coefficient, which is found to be equal to 0.1 for sp^3 defects whereas it is 1.2 for vacancy defects.

Ribeiro-Soares et al. studied 1D defects in graphene [102]. They heated a diamond-like carbon from 1200 to 2800 °C. They obtained an expression for the I_D/I_G intensity ratio [102], by noting that two sets of two bands have to be considered: A first set of D and G bands occurring from the activated region, and one set of D and G bands, downshifted due to softening of the phonon modes, occurring from highly disordered areas. Let us thus consider a structurally damaged ribbon (width l_S), and activated region (width l_A) and the mean size of a crystallite L_a (see Figure 17b). The evaluation of the intensity coming from the structurally damaged ribbon is purely geometric, whereas that of the ones from the activated region is geometric plus an exponential function introduced to account for the localization of the scattering giving rise to the D band. The final expression is then (see Equation (21):

$$\frac{I_D}{I_G} = \frac{C_{A,D}l_A(L_a - 2l_S)\left[1 - e^{-2(L_a - 2l_S)/l_A}\right] + 4C_{S,D}l_S(L_a - l_S)}{C_{A,G}(L_a - 2l_S)^2 + 4C_{S,G}l_S(L_a - l_S)} \tag{21}$$

The C constants are related to the cross sections and contain the wavelength dependency. The values of l_S and l_A were found to be 1.4 nm and 4 nm, respectively, and in very good agreement with Scanning Tunneling Microscopy (STM) data. When $L_a \gg l_S$ (i.e., the size of the structurally disordered region is negligible compared to the size of the crystallites), one obtains Equation (22):

$$\frac{I_D}{I_G} \approx \left(\frac{C_{A,D}l_A + 4C_{S,D}l_S}{C_{A,G}}\right)\frac{1}{L_a} \tag{22}$$

which perfectly reproduces the Tuinstra and Koenig expression obtained experimentally in 1970. The group of Jorio and Cançado produced recently a work that aims to disentangle the contribution of 0D and 1D defects coexisting on graphene related materials [97]. Figure 17c displays for both 0D and 1D defects the plot of Γ_G as function of the A_D/A_G ratio scaled with the laser energy, A being the band area (see comment in Section 2.7).

We now focus on another kind of defects named topological defects. These are non-hexagonal arrangements of carbon atoms in the graphene lattice. They are reported for most of the graphene based materials such as SWNTs, graphene [241,242] and graphite [243]. This kind of defect introduces a curvature in the flat geometry of a graphene layer. Probably one of the most studied topological defects is displayed in Figure 18: The Stone–Thrower–Wales (STW) defect (44 papers found using a bibliometric tool focused on abstracts), which corresponds to two pentagon-heptagon pairs. It is not easy to determine the precise frequency associated to the Raman signature of these defects. Theoretical calculations with SWNTs reported that the characteristic frequency of the STW defects should lie between 1820 and 1962 cm^{-1}, depending on the local curvature of the tube [244]. On the other hand, SERS measurements performed with SWNTs have shown two modes at 1139 and 1183 cm^{-1} associated to STW defects [245]. Theoretical calculations predict, among other things, a softening of the G band by -26 cm^{-1}, and a hardening of the D band by $+13$ cm^{-1} [246]. A way to detect their Raman signature would be to use SERS, as suggested in the work of Itoh et al. [247].

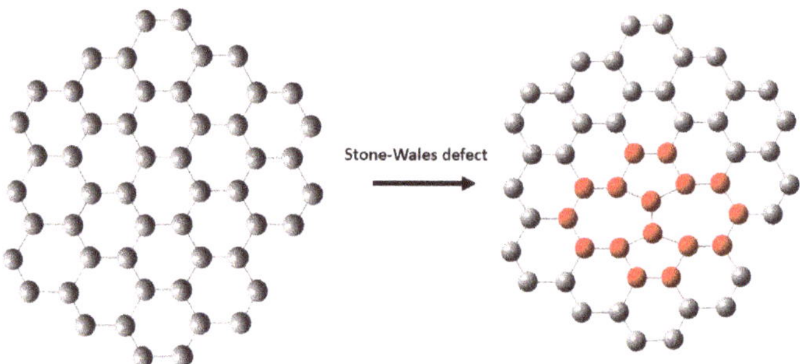

Figure 18. Stone–Wales defect.

3.4. Very Defective Carbons: Pyrocarbons, Coals, and Soots

The first reported work on very defective carbons is from Tuinstra and Koenig who found the relation relating the intensities of the D and G bands for activated charcoal, lampblack, and vitreous carbon, as reported above. In 1984, Lespade et al. [79] reported the Raman spectra of many disordered carbons such as pyrocarbons and fibers, using heat treatment to change the structure of these materials that were found to be graphitizable and not graphitizable. Four graphitization indexes were found: The frequency of the G band, its width, the intensity ratio I_D/I_G, and the width of the 2D band. Among others, they introduced a graphitization trajectory in the Γ_{2D} vs. Γ_G plot, distinguishing between 2D and 3D order [79]. The behavior of the 2D band according to a 3D order (related to the stacking order) was elucidated in more details 24 years later by Cançado et al. who heated diamond-like carbons in the range 2200–2700 °C [59]. Among other things, they evidenced a contribution of the band coming from 3D graphite and a contribution from graphene, and were able to relate their relative intensity ratio to the size of the crystallites in the c direction up to roughly 200 nm. Today, the number and origin of the 2D sub-bands have been understood for multilayer graphene [1] in the framework of the double resonance mechanism and more complex things can occur such as folding [248], misorientation [162], and stacking faults that can modify the intensities and shape. All these processes may also contribute to the intensity ratio spreading displayed in Figure 11. Table 1 in the paper by Larouche et al. [249] resumes the main spectral indicators that are used in the literature to deduce information about the structure of very defective carbons. They presented the influence of curvature by introducing the distance parameter L_{eq}, which is the average continuous graphene length including tortuosity of graphenic planes (see Figure 19). They showed that this tortuosity is very well correlated to the 2D band width. Note that they defined tortuosity as the number of phonons produced at the K point divided by the number of phonons produced at the Γ point of the Brillouin zone. We advise the reader that tortuosity is generally derived differently, from the analysis of HRTEM images [250,251]. The main message of this paper though is that the 2D band width is a good parameter to gain an insight into the tortuosity.

Table 1. Characteristics of ta-C:H and various a-C:H sub-types.

Types	sp^3 (at.%)	H (at.%)	E_g (eV)	Hardness (GPa)	Density (g/cm^3)
ta-C:H	70	25–30	2–2.5	>20	2.4
PLCH	70	40–60	2–4	soft	<1.2
DLCH	40–60	20–40	1–2	<20	2.0
GLCH	<30	<20	<1	soft	1.6
GLCHH	30	30–40	>1	soft	1.3

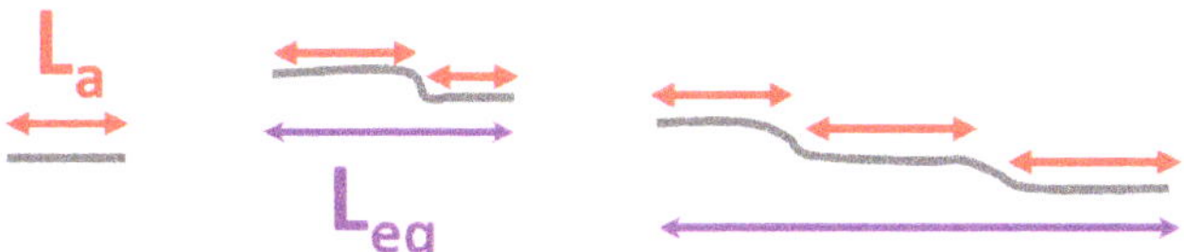

Figure 19. Tortuosity as defined by Larouche et al. [249]. The domains represented by red arrows (L_a) are aromatic domains without defects. The domains represented by purple arrows (L_{eq}) are composed of several aromatic domains, in-between there are curved graphene sites.

We now focus on soots and pyrocarbons (see a typical Raman spectrum on Figure 20). First of all, what is the difference between soots and pyrocarbons? When processing hydrocarbons (varying pressure, P, and temperature, T, in the reactor), gas phase nucleation leads to soots, whereas heterogeneous nucleation on the substrate leads to the formation of pyrocarbons, which can contain up to 5 at. % of hydrogen [252]. The Raman spectroscopy of soots (e.g., diesel soots, spark discharge soots, commercial black carbons) has been investigated in detail in 2005 [122]. To fit all the data in the 1000–1700 cm^{-1} spectral region, five bands were necessary: D (called D_1), D' (called D_2), and G bands, plus two less intense bands, the D_3 and D_4 bands, lying at 1500 and close to 1180 cm^{-1} (dependent on the wavelength of the laser), respectively. The best fit found was a four Lorentzian fit plus a Gaussian lineshape for the D_3 band. The need of such a band was first proposed by Rouzaud et al. [253]. In 2009, Brunetto et al. bombarded some aliphatic and aromatic dominated soots produced in flames to mimic the effect of irradiation encountered in the primitive solar nebulae [254]. The authors used a σ_G vs. Γ_G plot to compare their Raman spectra to the one collected on meteorites, interplanetary dusts and the grains collected from the Wild 2 comets. They found them very similar, allowing them to suppose that the irradiation played a major role in the processing of the carbon materials at the beginning of the solar system. Recently, the multiwavelength analysis is being used more than in the past to give better refined information on soots. For example, in 2011, combining temperature programmed oxidation and multiwavelength Raman spectroscopy (514–785 nm), Schmidt et al. noticed that a structure/reactivity relation from different soots can be obtained [255]. In 2015, in the framework of soot formation, Russo et al. evidenced that by changing the wavelength of the laser they were able to relate the origin of the fuel molecule and the structure of the soot [256]. The G band was found to display an asymmetric profile, and a D5 band (attributed to the presence of olefinic chains lying in grain boundaries), in the range 1100–1200 cm^{-1}, was found to be more intense using 633 nm instead of green or blue lasers. Ess et al. were able to relate the structural change of soots under heating under oxygen atmosphere up to 600 °C and relate it to the organic carbon content [257].

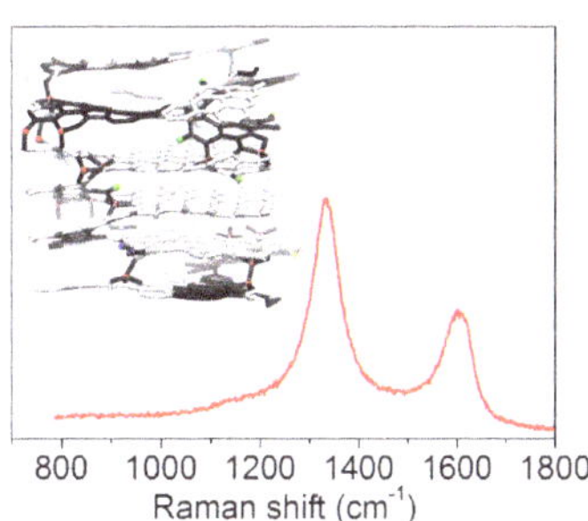

Figure 20. Raman spectrum obtained from a pyrocarbon. The inset represents a rough laminar pyrocarbon with defects (**black bonds**), sp^3 bonds (**blue**), *sp* bonds (**red**), and hydrogen bonds (**green**). ($\lambda_0 = 633$ nm).

Pyrocarbons are generally deposited on substrates by cracking hydrocarbons at temperatures higher than 900 °C, using chemical vapor deposition processes. Using infiltration process, pyrocarbons are used mainly in carbon/carbon fiber composites, whose aim is to withstand mechanical stress at high temperature [252]. A large variety of pyrocarbon textures exist. Among them, we can cite rough laminar (columnar structures), smooth laminar, and regenerative laminar [258]. High resolution TEM shows columnar and wavy structures (produced by pentagons during the deposition process [259]). Their TEM fringes length was found to be 1.6, 2.3, and 2.6 nm, respectively. In particular, the Raman D band width was found to be a good spectroscopic criterion to distinguish them. Bourrat et al. [258] found Γ_D = 80, 85, and 110 cm^{-1} for rough, smooth, and regenerative laminar pyrocarbon, respectively. The ranges were refined later: Γ_D = 80–90 cm^{-1} for rough laminar, Γ_D = 90–130 cm^{-1} for smooth laminar, and Γ_D = 170–200 cm^{-1} for regenerative laminar [123,260]. The interpretation is that Γ_D is sensitive to curvature effects, as is suggested by TEM analysis. Γ_G is spread on a narrower range, i.e., from 55 to 70 cm^{-1}.

We finish this section by focusing on the D″ band, sometimes called D$_4$ band and lying in between 1100 and 1200 cm^{-1}. It has been used for a long time to fit spectra of disordered samples (especially soot materials) and amorphous carbons with some local order (i.e., those containing sp^2 aromatic domains). It is sometimes called the TPA band (for trans-polyacetylene), as it lies close to the mode found in nanodiamonds [261]. It has been recently considered in the fitting procedure of nanoporous carbons [262]. The history of this band and its relation to the 2450 cm^{-1} band, first reported in the work of Nemanich, [81], for defectless graphite, is important to mention here because it illustrates the fact that studying disordered materials helps in understanding the spectra of well-ordered materials. The D″ band origin has been understood by Venezuela et al. [69] in 2011 on perfect graphene (it is mainly due to phonons associated to the KΓ direction in the Brillouin zone). In 2013, May et al. [263] were able to calculate the shape of the 2450 cm^{-1} band in the framework of the double resonance mechanism on perfect graphene involving *TO* and *LA* phonons close to the *K* point. They were able to reproduce its dispersion and change of shape depending on the laser wavelength. In 2016, Zhou et al. [183] observed on perfect graphene the laser sensitivity of the D″ and D + D″ bands in the UV range. In 2014, Herzinger et al. [264] created defective graphene and nanotubes by bombardment with high energy ions, and characterized the dispersion behavior of this D″ band. Couzi et al. [61] determined on defective aromatic carbons (graphite nanoplatelets, heat-treated glassy carbons, pyrograph nanofilaments, and multiwall nanotubes) the behavior of the D″ band, identifying two new bands (D* and D**) lying close to the D″ band, but with positive dispersion behavior, and reproduced the dispersion of the D + D″ in the near UV. We have to note that the D*, D**, and D″ bands do overlap, creating the well-known D$_4$ band often used to fit Raman spectra of soots. Very recently, the D$_4$ and D$_3$ bands were observed on nanotubes and partially exfoliated by acid treatment [265].

3.5. Graphite Intercalated Compounds

In this part, we briefly present some results about a kind of material that must be cited because it displays a large variety of structures and because it has great application potential in different fields (such as energy storage, superconductivity, nano-medicine, etc. [266] and references therein): Graphite intercalated compounds (GICs). GICs are multilayered materials sufficiently ordered to exhibit staging in which the number of graphitic layers in between adjacent intercalants can be varied in a controlled way. The interlayer spacing can be tuned from 0.34 up to 1 nm [267]. The denomination n-stage GICs can be found in [268], n defining the constant number of graphene layers between any nearest pair of intercalant layers. More details about the staging can also be found in [266]. The diminution of the interlayer spacing distance reduced the van der Waals interaction between planes so that it can be envisaged as a route to form graphene and nano-ribbons [269]. The list of metals and small molecules that can be embedded in between graphene planes is huge and not exhaustive here: Alkali-metals (K, Li, Cs, Rb); alkali-earth-metals (Be, Ca, Ba); halogens; C$_{60}$ [270]; FeCl$_3$ [271]; H$_2$SO$_4$ and HSO$_4^-$ [272]; AsF$_5$, HNO$_3$, and SbCl$_5$ [273]; etc. Intercalants can be electron donors or electron acceptors. Thus, charge transfer and strain are intrinsic effects in GICs, and Raman spectroscopy is sensitive to both. Raman study of highly staged GICs allowed to disentangle both effects [274]. A seminal work by Solin presenting an overview of the Raman spectroscopy of these compounds was published in 1980 [275]. A more recent work on a combination of Raman spectroscopy and ab initio calculations of GICs by Chacon-Torres et al. was reported in 2014 [266]. Inner graphene layers (also called interior layers) and outer graphene layers (also called graphene layers bounded by intercalants) can be differentiated by the G line splitting, with σ_G of the blue component that can reach values as high as 1636 cm^{-1} [266,273]. Superconductivity has been observed in many GICs (with the highest critical temperature found at a relatively high temperature, 11.5 K, for CaC_6, see references [266]) and multiwavelength Raman spectroscopy is a central characterization tool because superconductivity may be due to electron–phonon interaction and mediated by phonons [276]. Multiwavelength Raman spectroscopy also allows a direct measurement of the Fermi level by observing the Pauli blocking (which results in the 2D band intensity vanishing by tuning the wavelength of the laser) [271].

3.6. Amorphous and Diamond-Like Carbons

Most of the work cited on Raman spectroscopy of amorphous carbons [277] comes from the thousands of times cited papers by Ferrari et al. Their four landmark papers were published from 2000 to 2005 and combine all together a comprehensive view of the understanding about Raman spectroscopy of amorphous carbons [4–6,70], based on many other papers that we will not cite all here. The study of Raman spectroscopy of amorphous carbons was completed in 2015 by the work of Zhang et al. [278], as we will see soon hereafter.

To begin with, amorphous carbons are generally containing sp^3, sp^2 carbons, and heteroatoms (such as hydrogen). sp^3 carbons determine the mechanical properties (hardness, density), whereas the sp^2 aromatic clusters determine the optical properties (energy gap), mainly due to the π/π^* bonds with the energy gap in the IR–visible–UV range, depending on their size. Adding hydrogen, and organizing the structure by heating the sample, can change the optical properties together with the structure [7,124–126,225,279–282] and Raman spectroscopy can help in checking the changes. For example, hydrogen changes the electronic structure, which leads to a Raman resonance mechanism that can be observed, helping in quantifying the amount of hydrogen in the amorphous carbons [283]. Depending on the amount of aromatic/aliphatic sp^2/sp^3 carbons and hydrogen atoms, several kinds of amorphous carbons can be distinguished: a-C, ta-C, a-C:H, and ta-C:H. The t stands for tetracoordinated as these carbons contain generally close to 70% of sp^3 carbons. A basic scheme of the ternary phase diagram for amorphous carbons is displayed in Figure 21. More information can be found in [70,124]. The classification is however more complex, with even more variety of samples. a-C:H types are themselves split in subgroups with different properties: PLCH (polymer-like a-C:H), DLCH (diamond-like a-C:H), GLCH (graphite-like a-C:H) and GLCHH (graphite-like a-C:H with extra hydrogen subtype), as displayed in Table 1, adapted from [278].

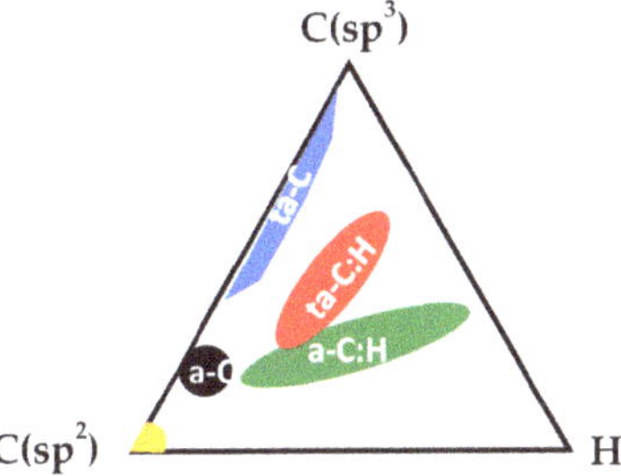

Figure 21. Basic ternary phase diagram of amorphous carbons.

Except PLCH, which only displays a photoluminescence background, Raman spectra of amorphous carbons generally display two broad overlapped G and D bands, the D band being less intense, or disappearing when the amount of sp^3 carbons is higher. The D band position displays the wavelength dispersion as in graphene but shifted depending on the kind of amorphous carbon, as shown in Figure 7b. For the G band, it depends on the local degree of order, as already displayed in Figure 7a. Figure 22 displays the Raman spectra of one a-C:H film (H being close to 30 at.%) but recorded with five different laser wavelengths, ranging from 266 to 633 nm [72]. The dependence of the shape with the laser wavelength [284] can be explained mainly by the fact that the sample is composed of a distribution of aromatic domains which display local electronic structures. The wavelength used is resonant with one kind of environment that appears stronger in the spectrum. Ferrari et al. proposed a "three stage" model that can explain the ordering of the sp^2 phase going from nanocrystalline graphite (nc-G) to highly disordered amorphous carbon, explaining the changes in the spectroscopic parameters [4]. The parameters followed are the I_D/I_G and σ_G parameters, and their changes with the laser wavelength. For example, at 514 nm, the G band position is at 1582 cm^{-1} for graphite, it upshifts to 1600 cm^{-1} for nanographite composed of only sp^2 aromatic clusters, whereas it can shift up to 1630 cm^{-1} for other forms of sp^2 carbons (chains), where sp^3 carbons are present. Then, when starting amorphization, σ_G diminishes down to 1520 cm^{-1}. In Figure 23, we illustrate that starting from an amorphous carbon, we obtain a nanographite. We plot σ_G for an a-C:H (DLCH) film that has been heated at a 15 K·min^{-1} rate, under a 1-bar Ar atmosphere, Raman spectra being recorded in situ, with ramp 1 stopped at 900 °C. A typical evolution is drawn [225] that informs us that the amorphous carbon is organizing when temperature is increased, by growth of the size of the carbon aromatic clusters. A plateau is reached at 600 °C. Ramp 2 is made on the same sample after it has first been cooled down to room temperature. One can see that the G band at room temperature is now lying at 1592 cm^{-1}. The second ramp informs us that now it behaves like graphite and no more like an amorphous carbon, with an upshift of about 10 cm^{-1}, which is typical for a nanocrystalline graphite. The diminution of σ_G with T for graphite and graphene is reversible and can be used as a contact less thermometer [172,173].

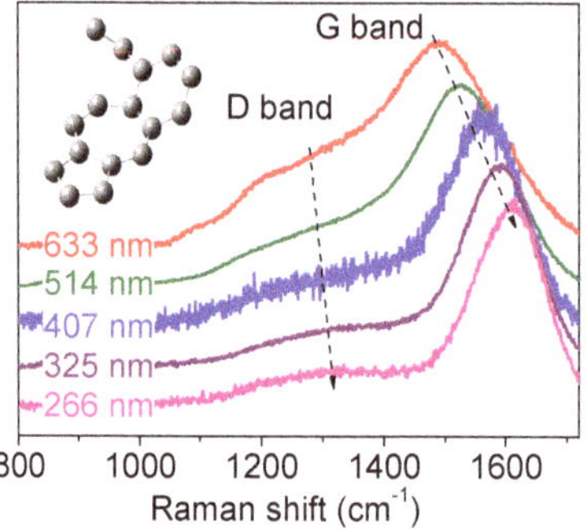

Figure 22. Raman spectra of amorphous carbon (a-C:H with about 30 at.% H) obtained with different laser wavelengths.

Next, we ask "what are the other spectroscopic parameters that can be used to characterize amorphous carbons?" The 2D band cannot be used in general as it is very low in intensity and very broad, as can be seen in Figure 6b, except when the amorphous carbon is heated. Another spectroscopic parameter has been found useful to help in determining the amount of hydrogen bonded in a-C:H: the so-called m/I_G parameter, which is the ratio between the slope of the photoluminescence background, m, in the range 800–1800 cm^{-1} divided by the G band intensity. This spectroscopic parameter gives certain qualitative information, as presented in the work of Casiraghi et al. [6], and in the work of

Buijnsters et al. [285], but is also sensitive to other defects meaning it cannot be used systematically (see the analysis on a heated a-C:H performed by comparing Raman spectroscopy with thermal desorption spectroscopy and ion beam analysis [7], and calculations at the end of the work of Rose et al. [282]). It has been found that the G band width, Γ_G, is correlated to the sp^3 content and the linear dispersion of σ_G of as deposited samples correlates to the H content [286]. Then, a good way to represent the data is to plot σ_G as a function of Γ_G, as was done for several wavelengths in [72,282]. Figure 24 displays such a plot for different kinds of heated amorphous carbons plus nanocrystalline graphite (for nc-G, ta-C, and ta-C:H data, see [66]). If one uses Γ_G as an indicator of local disorder close to sp^2 bonds in the material (which can be related to the size of the clusters [5,65] and/or to the sp^3 content close to sp^2 bonds [286]), one can use this parameter in order to have an idea of where is the sample situated in Ferrari's "three stage model". With this in mind, nc-G is more ordered than a-C:H/D which are themselves more ordered than ta-C:H and ta-C. Each kind of carbon draws its own line when heated, but all these lines tend to converge in a region close to 100 cm^{-1}. a-C, a-C:H, and a-C:D data were recorded in-situ, contrary to ta-C, ta-C:H, and nc-G, and thus appear down shifted due to dilatation and multi-phonon processes [172,173]. The presence of hydrogen systematically diminishes the position of the G band (for ta-C/ta-C:H and a-C/ta-C:H). The systematic frequency shift between a-C:H and a-C:D that tends to diminish when decreasing Γ_G (i.e., ordering the material) is partially due to an isotopical shift of C-H/C-D bonds, as the difference is in the same order as for C$_6$H$_6$ and C$_6$D$_6$ molecules [287].

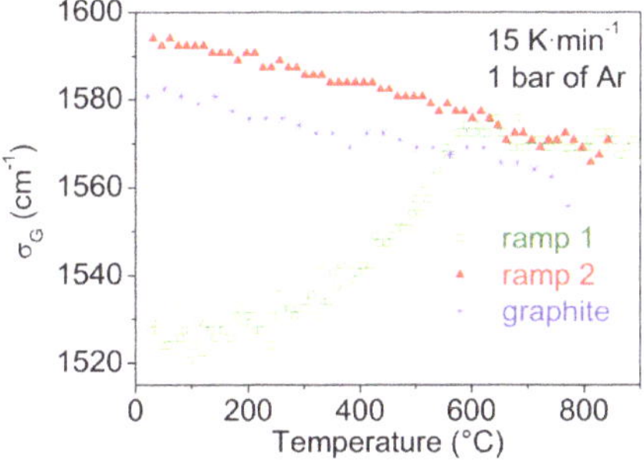

Figure 23. Temperature evolution of σ_G for a DLCH sample (H/H + C = 29%). In situ measurement is done in an environmental cell, under argon atmosphere to avoid oxidation.

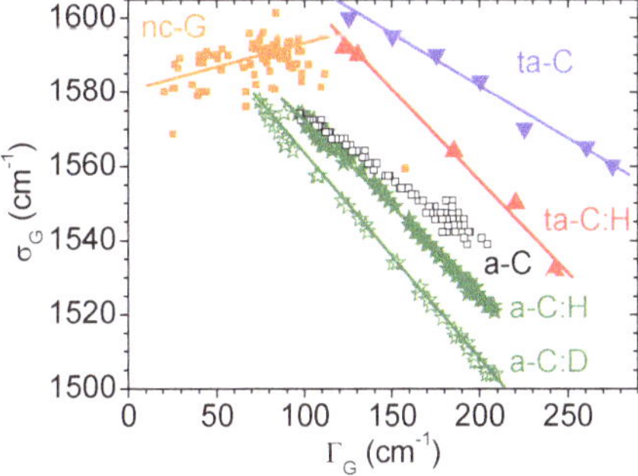

Figure 24. Plots of σ_G vs. Γ_G for nc-G and different amorphous carbons. Data were recorded with $\lambda_L = 514$ nm.

In 2015, a phenomenological model based on dispersion was proposed by Zhang et al. [278] in order to better characterize amorphous carbons by analyzing their Raman spectra recorded with several wavelengths. This model starts with the statement that the G band intensity is the sum of three kinds of sp^2 clusters (see Equation (23)):

$$I_G(\omega, \lambda_L) = I_g(\omega, \lambda_L) + I_r(\omega, \lambda_L) + I_c(\omega, \lambda_L) \tag{23}$$

where ω is the Raman frequency, λ_L the laser wavelength, and I_g, I_r and I_c are the intensities of the nc-G, fused aromatic rings and olefinic chain clusters, respectively. Without going into details here (the model is detailed in Section 2 and in the appendixes of [278]), this shows that σ_G is a weighted average of the resonant frequencies for the three types of sp^2 carbons (see Equation (24)):

$$\sigma_G = n_g \omega_g + n_r \omega_r + n_c \omega_c \tag{24}$$

where n_g, n_r, and n_c are the amount of different kind of carbons that satisfy $n_g + n_r + n_c = 1$. As ω_g, ω_r, and ω_c display a linear dispersion with λ_L [278], and as the parameters for these linear relations have been tabulated, one can in principle deduce n_g, n_r, and n_c from the G band dispersion.

4. Discussion

4.1. Role of Resonance

The resonance mechanisms at play in the Raman process of aromatic based carbons allow us to have a better insight in the study of graphenic materials, making multiwavelength Raman spectroscopy a relevant tool. So far, the well-known double resonance mechanism, based on scattering of an incoming photon by a phonon followed by a second scattering by a defect or another phonon, is the best option to explain most of the behaviors reported in the literature since the 1970s (such as the rising of bands like D*, D**, D″, D, D′, and 2D and combinations thereof), as we mentioned in this review. This mechanism can probe the phonon dispersion curve of graphene away from the gamma point of the Brillouin zone. The dispersion and intensity behaviors of these bands (i.e., the dependency of the D/D′ relative intensity ratio to the kind of defects, the relative D/G intensity ratio sensitivity to the amount of defects), the Stokes/anti-Stokes anomalies, the zigzag/armchair dependency, the doping effects are all explained in the framework of this double resonance mechanism. It must be noticed that mechanisms alternative to double resonance are also under consideration [118] but still need to be confirmed. So far, the double resonance mechanism remains considered as the most efficient model.

4.2. Role of Defects

Because of defects implied in this model, Raman spectroscopy can be used to characterize how a graphenic material is far from its crystalline state. As an illustration, Figure 25a displays the width of the D band (Γ_D) in function of the width of the G band (Γ_G), which is similar to focusing on the Γ_D/Γ_G that was done in [288]. Data shown are obtained for many graphenic materials (Another way that has already been used, is to display the Γ_D/Γ_G ratio as function of Γ_G. We then observe that the ratio evolves from 2 at $\Gamma_G = 25$ cm^{-1} down to 1 at $\Gamma_G = 75$ cm^{-1}, and following it increases up to 2.3 at $\Gamma_G = 93$ cm^{-1}, and finally stays constant till $\Gamma_G = 150$ cm^{-1}. Mallet et al. observed such a variation of Γ_D/Γ_G with L_a, reaching the lowest value at $L_a = 8$ nm (see Figure 13 in the work of Zhao et al. [271]).) (carbon fibers [289], soots [122,290,291], pyrocarbons [123,252,260], nanocones [223], a large variety of disordered carbons collected from the Tore Supra tokamak [66], nanographites from Cançado et al. [65], geothermic carbons from [292], and ion implanted graphites taken from [231]). Depending on the characteristic size of the aromatic domain, the data are grouped in three different areas: the lowest rectangle with L_a higher than roughly 10 nm, the intermediate region with L_a in the range of few to 10 nm, and the higher one with L_a close to 1 nm. Below $\Gamma_G = 50$ cm^{-1} the evolution is linear, with a slope close to 1 that was found by Cançado et al. for nanographites [65]. When Γ_G

is close to 75 cm^{-1}, the slope changes drastically: D, with Γ_D = 50 cm^{-1}, broadens extremely fast up to Γ_D = 120–150 cm^{-1}, whereas G only evolves from 50 to 75 cm^{-1}. This behavior was previously observed in [66,292] and is now also evidenced for pyrocarbons and nanocones. Note that some materials do not display the drastic change in slope, such as the data of graphite implanted by ions that are grouped around the line $\Gamma_D = 2\Gamma_G - 35$.

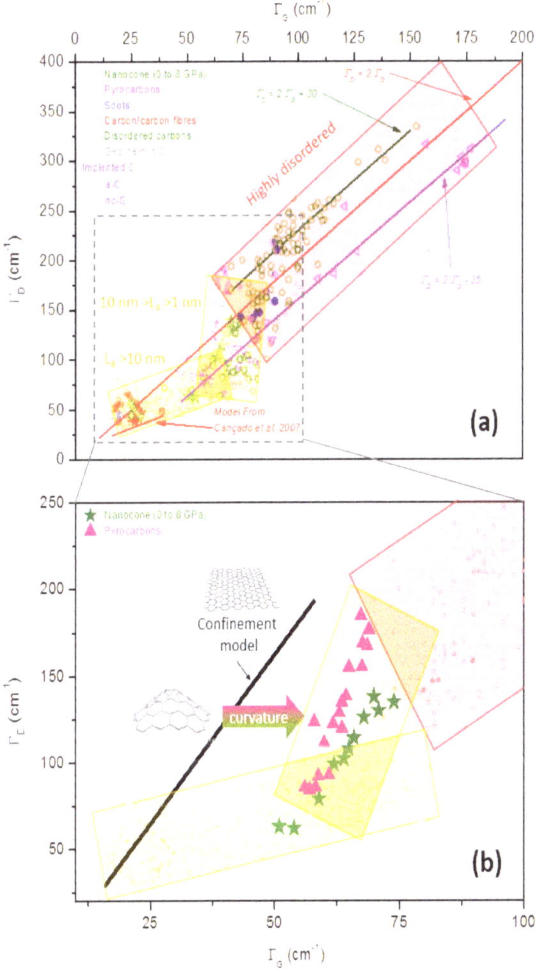

Figure 25. Γ_D versus Γ_G plot. (**a**) For a large variety of disordered aromatic carbons; (**b**) Zoom with suggestion of both phonon confinement and curvature effects.

4.3. Other Effects: Our Propositions

The strategy adopted to deals with this apparent complexity, has been to study graphene with controlled defects. Creating point defects [101] and/or linear defects [97,102] and controlling their influence on the Raman spectrum of graphene have been the milestones of a bottom-up approach aiming at understanding the Raman spectra of more complex aromatic carbons. It has been one of the best progresses made over the last few years in order to overcome the old Tuinstra and Ferrari relations that were found to give more or less the order of magnitude/trends. Nowadays, other crucial steps have to be made in order to be closer to real materials. Among other effects that must be taken into account in the understanding of the Raman spectra, there are:

- curvature effects;
- phonon confinement effects (due to poor coupling between different aromatic planes, porosities, etc.);
- and combinations thereof.

The influence of these effects on Raman spectra should be studied in detail in the future on graphene samples containing point defects and/or linear defects with controlled amounts in order to disentangle the origin of all the effects. Three good material systems for that hypothetical study could be:

- heated C_{60};
- bombarded nanoribbons of different shapes and sizes deposited on deformable surfaces;
- in situ measurements of nanocones in high pressure/temperature cells.

For the second kind of materials, we believe that in situ multiwavelength Raman spectroscopy analysis could be coupled to skeleton analysis of high resolution TEM images (like reported in Da Costa, Pre, and Farbos et al. [229,251,293] respectively) in order to relate the skeleton and the kind of defects produced under constraint and to check their influence on the Raman spectrum, via resonance effects.

Concerning what already exists in the literature about curvature and/or phonon confinement fingerprints in the Raman spectra, the analysis of pyrocarbons, which can be composed of tortuous planes of varying length, has revealed that the width of the D band is sensitive to curvature effects of the graphene sheets [123,260]. Nanocones submitted to a pressure increase up to 8 GPa display some morphological changes that are accompanied by a change in the Raman spectrum too [223]. The corresponding data are displayed in Figure 25b. Among other changes, the increase of the D band width with the pressure increase is more pronounced than for the G band leading to a rapid evolution of the D band width compared to the G band width. This rapid evolution is not seen for implanted graphite samples on which two kinds of signatures (amorphous and nanocrystalline) have been observed together [231]. An explanation may be that a synergetic effect hinders curvature of graphene planes, preferring the formation of sp^3 defects. This could be due to a characteristic of phonon confinement as evidenced by Puech et al. (data extracted from Figures 6 and 7 of the work of Puech et al. [146]). However, we know that both nanocones and pyrocarbons can be composed of curved aromatic planes and the question remains open: How can curvature affect the spectrum too? Part of the answer was obtained by observing a band shift of the G band [8], but none was obtained for defective graphene concerning the D band widths. The arrow in Figure 25b starts from the line of the confinement model and suggests that curvature produces an extra broadening of the G band. Recently, an additional source of broadening for the G band was revealed by studying pressurized graphene membranes and it was attributed to strain (if strain is higher than 1%) [294]. This should be investigated in more detail, and the question of how curvature can lead to phonon confinement answered too.

5. Conclusions

Based on the observation that many procedures are available to fit the complex Raman spectra of very defective aromatic carbons (such as soots, pyrocarbons, coals, and amorphous carbons), we decided to review the Raman spectroscopic characteristics of these materials. Basically, if someone wants to perform a structural analysis of a carbon-based sample using only Raman spectroscopy he/she has to focus on, at least:

- the ratio intensity between the D and G bands;
- the presence of additional bands (e.g., 2D, D′, etc.);
- the width of all bands.

The combination of all these parameters will give a first basic idea of the structure, namely if the sample is well-structured or strongly disordered. Nevertheless, with only a basic Raman measurement (notably if a single excitation wavelength is used) it is clear that at the present moment a more detailed analysis of the structural defects will require other complementary experimental tools, such as high-resolution transmission electron microscopy, which is a pity as Raman spectroscopy is probably the most simple, fast, and non-destructive analysis method. If possible, we therefore suggest the use of multiwavelength Raman spectroscopy, since it allows for a better characterization of defects, looking at band position, relative intensities, and width.

We can reasonably assume that each type of defect has a Raman spectral signature. The point up to now is to identify each spectral signature, which is a real experimental challenge as in most samples the different structural defects combine, thus giving rise to a complex Raman spectrum. As suggested in Figure 25 where curvature effects are evidenced in carbon nanocones and pyrocarbons, the solution is to work with well controlled samples with specific defects. This systematic approach will pave the way towards a complete guide for a multiwavelength analysis of all aromatic carbon-based species.

Acknowledgments: Cedric Pardanaud wants to thank Gerard Vignoles and Jean-Marc Leyssale for providing spectra and inset related to pyrocarbons; Miriam Peña Alvarez for the spectra obtained from nanohorns; Cinzia Casiraghi and Deborah Prezzi for spectra and inset related to nanoribbons; and Gregory Giacometti for providing Figure 5b. Finally, Cedric Pardanaud wants to acknowledge Pascale Roubin for previous scientific discussions and for being at the origin of the carbon thematic in the PIIM Lab in 2007.

Author Contributions: Cedric Pardanaud as last and corresponding author conducted the literature search. Alexandre Merlen and Cedric Pardanaud performed and analyzed own Raman measurements/data and contributed equally to the figures building. All three authors were involved in the writing and editing of the manuscript. Josephus Gerardus Buijnsters performed a deep reading at several steps of the writing process.

Conflicts of Interest: The authors declare no conflict of interest.

Abbreviations

σ_x (expressed in cm^{-1}): Band position of the band labelled x (x could be G, D, 2D, D′, …).
Γ_x (expressed in cm^{-1}): Full width at half maximum of the band labelled x.
I_x (expressed in arbitrary units related to the number of counts on the detector): height of the band labelled x.
A_x (expressed in arbitrary units related to the number of counts on the detector): integrated area of the band labelled x.

References

1. Ferrari, A.C.; Meyer, J.C.; Scardaci, V.; Casiraghi, C.; Lazzeri, M.; Mauri, F.; Piscanec, S.; Jiang, D.; Novoselov, K.S.; Roth, S.; et al. Raman spectrum of graphene and graphene layers. *Phys. Rev. Lett.* **2006**, *97*, 187401. [CrossRef] [PubMed]
2. Elias, D.C.; Nair, R.R.; Mohiuddin, T.M.G.; Morozov, S.V.; Blake, P.; Halsall, M.P.; Ferrari, A.C.; Boukhvalov, D.W.; Katsnelson, M.I.; Geim, A.K.; et al. Control of graphene's properties by reversible hydrogenation: Evidence for graphane. *Science* **2009**, *323*, 610–613. [CrossRef] [PubMed]
3. Dresselhaus, M.S.; Jorio, A.; Saito, R. Characterizing graphene, graphite, and carbon nanotubes by Raman spectroscopy. *Annu. Rev. Condens. Matter Phys.* **2010**, *1*, 89–108. [CrossRef]
4. Ferrari, A.C.; Robertson, J. Resonant Raman spectroscopy of disordered, amorphous, and diamondlike carbon. *Phys. Rev. B* **2001**, *64*, 075414. [CrossRef]
5. Ferrari, A.C.; Robertson, J. Raman spectroscopy of amorphous, nanostructured, diamond-like carbon, and nanodiamond. *Philos. Trans. R. Soc. A Math. Phys. Eng. Sci.* **2004**, *362*, 2477–2512. [CrossRef] [PubMed]
6. Casiraghi, C.; Ferrari, A.C.; Robertson, J. Raman spectroscopy of hydrogenated amorphous carbons. *Phys. Rev. B* **2005**, *72*, 085401. [CrossRef]

7. Pardanaud, C.; Martin, C.; Roubin, P.; Giacometti, G.; Hopf, C.; Schwarz-Selinger, T.; Jacob, W. Raman spectroscopy investigation of the H content of heated hard amorphous carbon layers. *Diam. Relat. Mater.* **2013**, *34*, 100–104. [CrossRef]

8. Mohiuddin, T.M.G.; Lombardo, A.; Nair, R.R.; Bonetti, A.; Savini, G.; Jalil, R.; Bonini, N.; Basko, D.M.; Galiotis, C.; Marzari, N.; et al. Uniaxial strain in graphene by Raman spectroscopy: G peak splitting, Grüneisen parameters, and sample orientation. *Phys. Rev. B* **2009**, *79*, 205433. [CrossRef]

9. Piscanec, S.; Mauri, F.; Ferrari, A.C.; Lazzeri, M.; Robertson, J. Ab initio resonant Raman spectra of diamond-like carbons. *Diam. Relat. Mater.* **2005**, *14*, 1078–1083. [CrossRef]

10. Brillouin, L. Diffusion de la lumière et des rayons X par un corps transparent homogène. Influence de l'agitation thermique. *Ann. Phys.* **1922**, *17*, 88–122. (In French) [CrossRef]

11. Compton, A. A quantum theory of the scattering of X-rays by light elements. *Phys. Rev.* **1923**, *21*, 483. [CrossRef]

12. Krishnan, R.S.; Shankar, R.K. Raman effect: History of the discovery. *J. Raman Spectrosc.* **1981**, *10*, 1–8. [CrossRef]

13. Raman, C.V. A new radiation. *Ind. J. Phys.* **1928**, *2*, 387–398.

14. Stokes, G.G. On the change of refrangibility of Light. *Philos. Trans. R. Soc.* **1852**, *142*, 463–562. [CrossRef]

15. Ramaseshan, S. The Raman effect. *Curr. Sci.* **1998**, *75*, 6.

16. Singh, R.; Riess, F. Seventy years ago—The discovery of the Raman effect as seen from German physicists. *Curr. Sci.* **1998**, *74*, 1112–1115.

17. Singh, R.C.V. Raman and the discovery of the Raman effect. *Phys. Perspect.* **2002**, *4*, 399–420. [CrossRef]

18. Smekal, A. The quantum, theory of dispersion. *Naturwissenschaften* **1923**, *11*, 873–878. [CrossRef]

19. Kramers, H.A.; Heisenberg, W. Über die streuung von strahlung durch atome. *Z. Phys.* **1925**, *31*, 681–708. (In German) [CrossRef]

20. Dirac, P.A.M. The quantum theory of dispersion. *Proc. R. Soc. Lond. A* **1927**, *114*, 710–728. [CrossRef]

21. Breit, G. Quantum theory of dispersion. *Rev. Mod. Phys.* **1932**, *4*, 504. [CrossRef]

22. Placzek, G. Rayleigh-Streuung und Raman-Effekt. In *Handbuch der Radiologie*; Marx, E.A., Ed.; Akademische Verlagsgesellschaft: Leipzig, Germany, 1934; Volume 6, p. 205. (In German)

23. Albrecht, A.C. On the theory of Raman intensities. *J. Chem. Phys.* **1961**, *34*, 1476–1484. [CrossRef]

24. Born, M.; Huang, K. *Dynamical Theory of Crystal Lattices*; The International Series of Monographs on Physics; Oxford University Press: Oxford, UK, 1956.

25. Loudon, R. The Raman effect in crystals. *Adv. Phys.* **1964**, *13*, 423–482. [CrossRef]

26. Ganguly, A.K.; Birman, J.L. Theory of Lattice Raman Scattering in Insulators. *Phys. Rev.* **1967**, *162*, 806–816. [CrossRef]

27. Cardona, M.; Güntherodt, G. *Light-Scattering in Solids*; Topics in Applied Physics; Springer-Verlag: Berlin, Germany, 1989; Volume 66, pp. 2–12.

28. Nafie, L.A. Recent advances in linear and non-linear Raman spectroscopy. Part X. *J. Raman Spectrosc.* **2016**, *47*, 1548–1565. [CrossRef]

29. Gouadec, G.; Colomban, P. Raman spectroscopy of nanomaterials: How spectra relate to disorder, particle size and mechanical properties. *Prog. Cryst. Growth Charact. Mater.* **2007**, *53*, 1–56. [CrossRef]

30. Rocard, Y. Role des vibrations des atomes dans les molécules dans le phénomène de diffusion de la lumière. *Compt. Rend.* **1927**, *185*, 1026–1028. (In French)

31. Rocard, Y. Les nouvelles radiations diffusées. *Compt. Rend.* **1928**, *186*, 1107–1109. (In French)

32. Raman, C.V. The scattering of light in crystals and the nature of their vibration spectra. *Proc. Indian Acad. Sci.* **1951**, *34*, 61–71.

33. Poulet, H.; Mathieu, J.P. Détermination des vibrations fondamentales du sulfure de cadmium cristallisé. *Ann. Phys.* **1964**, *13*, 549–552. (In French) [CrossRef]

34. Landsberg, G.; Mandelstam, L. Über die lichtzerstreuung in kristallen. *Z. Phys.* **1928**, *50*, 769–780. (In German) [CrossRef]

35. Boyle, W.S.; Smith, G.E. Charge coupled semiconductors devices. *Bell Syst. Tech. J.* **1970**, *49*, 587–593. [CrossRef]

36. Dierker, S.B.; Murray, C.A.; Legrange, J.D.; Schlotter, N.E. Characterization of order in langmuir-blodgett monolayers by unenhanced raman-spectroscopy. *Chem. Phys. Lett.* **1987**, *137*, 453–457. [CrossRef]

37. Rabolt, J.F.; Santo, R.; Swalen, J.D. Raman measurements on thin polymer-films and organic monolayers. *Appl. Spectrosc.* **1980**, *34*, 517–521. [CrossRef]

38. Adar, F.; Delhaye, M.; DaSilva, E. Evolution of instrumentation for detection of the Raman effect as driven by available technologies and by developing applications. *J. Chem. Educ.* **2007**, *84*, 50–60. [CrossRef]

39. Long, D.A. Early history of the Raman effect. *Int. Rev. Phys. Chem.* **1988**, *7*, 317–349. [CrossRef]

40. Langeluddecke, L.; Singh, P.; Deckert, V. Exploring the nanoscale: Fifteen years of tip-enhanced Raman spectroscopy. *Appl. Spectrosc.* **2015**, *69*, 1357–1371. [CrossRef] [PubMed]

41. Long, D.A. *The Raman Effect: A Unified Treatment of the Theory of Raman Scattering by Molecules*; John Wiley & Sons, Ltd.: Chichester, UK, 2002.

42. Cardona, M. Resonance phenomena. *Top. Appl. Phys.* **1982**, *50*, 19–178.

43. Cantarero, A.; Tralleroginer, C.; Cardona, M. Excitons in one-phonon resonant Raman scattering: Deformation-potential interaction. *Phys. Rev. B* **1989**, *39*, 8388–8397. [CrossRef]

44. Dresselhaus, M.S.; Dresselhaus, G.; Jorio, A. *Applications of Group Theory to the Physics of Solids*; Springer: New York, NY, USA, 2008.

45. Yu, P.Y.; Cardona, M. *Fundamentals of Semiconductors, Physics and Materials Properties*, 4th ed.; Springer: Berlin, Germany, 2010.

46. Ramsteiner, M.; Wild, C.; Wagner, J. Interference effects in the raman-scattering intensity from thin-films. *Appl. Opt.* **1989**, *28*, 4017–4023. [CrossRef] [PubMed]

47. Wang, Y.Y.; Ni, Z.H.; Yu, T.; Shen, Z.X.; Wang, H.M.; Wu, Y.H.; Chen, W.; Wee, A.T.S. Raman studies of monolayer graphene: The substrate effect. *J. Phys. Chem. C* **2008**, *112*, 10637–10640. [CrossRef]

48. Yoon, D.; Moon, H.; Son, Y.W.; Choi, J.S.; Park, B.H.; Cha, Y.H.; Kim, Y.D.; Cheong, H. Interference effect on Raman spectrum of graphene on SiO_2/Si. *Phys. Rev. B* **2009**, *80*, 125422. [CrossRef]

49. Klar, P.; Lidorikis, E.; Eckmann, A.; Verzhbitskiy, I.A.; Ferrari, A.C.; Casiraghi, C. Raman scattering efficiency of graphene. *Phys. Rev. B* **2013**, *87*, 205435. [CrossRef]

50. Hirsch, A. The era of carbon allotropes. *Nat. Mater.* **2010**, *9*, 868–871. [CrossRef] [PubMed]

51. Ostrikov, K.; Neyts, E.C.; Meyyappan, M. Plasma nanoscience: From nano-solids in plasmas to nano-plasmas in solids. *Adv. Phys.* **2013**, *62*, 113–224. [CrossRef]

52. Bernier, P.; Lefrant, S. *Le Carbone Dans Tous Ses Etats*; Gordon and Breach Science Publishers: Philadelphia, PA, USA, 1998.

53. Castro Neto, A.H.; Guinea, F.; Peres, N.M.R.; Novoselov, K.S.; Geim, A.K. The electronic properties of graphene. *Rev. Mod. Phys.* **2009**, *81*, 109–162. [CrossRef]

54. Wallace, P. The band theory of graphite. *Phys. Rev.* **1947**, *71*, 622–634. [CrossRef]

55. Ivanovskaya, V.V.; Zobelli, A.; Teillet-Billy, D.; Rougeau, N.; Sidis, V.; Briddon, P.R. Hydrogen adsorption on graphene: A first principles study. *Eur. Phys. J. B* **2010**, *76*, 481–486. [CrossRef]

56. Wirtz, L.; Rubio, A. The phonon dispersion of graphite revisited. *Solid State Commun.* **2004**, *131*, 141–152. [CrossRef]

57. Reich, S.; Thomsen, C. Raman spectroscopy of graphite. *Philos. Trans. R. Soc. A Math. Phys. Eng. Sci.* **2004**, *362*, 2271–2288. [CrossRef] [PubMed]

58. Lazzeri, M.; Piscanec, S.; Mauri, F.; Ferrari, A.C.; Robertson, J. Phonon linewidths and electron-phonon coupling in graphite and nanotubes. *Phys. Rev. B* **2006**, *73*, 155426. [CrossRef]

59. Cancado, L.G.; Takai, K.; Enoki, T.; Endo, M.; Kim, Y.A.; Mizusaki, H.; Speziali, N.L.; Jorio, A.; Pimenta, M.A. Measuring the degree of stacking order in graphite by Raman spectroscopy. *Carbon* **2008**, *46*, 272–275. [CrossRef]

60. Kawashima, Y.; Katagiri, G. Fundamentals, overtones, and combinations in the raman-spectrum of graphite. *Phys. Rev. B* **1995**, *52*, 10053–10059. [CrossRef]

61. Couzi, M.; Bruneel, J.L.; Talaga, D.; Bokobza, L. A multi wavelength Raman scattering study of defective graphitic carbon materials: The first order Raman spectra revisited. *Carbon* **2016**, *107*, 388–394. [CrossRef]

62. Ferrari, A.C.; Basko, D.M. Raman spectroscopy as a versatile tool for studying the properties of graphene. *Nat. Nanotechnol.* **2013**, *8*, 235–246. [CrossRef] [PubMed]

63. Brar, V.W.; Samsonidze, G.G.; Dresselhaus, M.S.; Dresselhaus, G.; Saito, R.; Swan, A.K.; Unlu, M.S.; Goldberg, B.B.; Souza, A.G.; Jorio, A. Second-order harmonic and combination modes in graphite, single-wall carbon nanotube bundles, and isolated single-wall carbon nanotubes. *Phys. Rev. B* **2002**, *66*, 155418. [CrossRef]

64. Tuinstra, F.; Koenig, J.L. Raman spectrum of graphite. *J. Chem. Phys.* **1970**, *53*, 1126–1130. [CrossRef]

65. Cançado, L.G.; Jorio, A.; Pimenta, M.A. Measuring the absolute Raman cross section of nanographites as a function of laser energy and crystallite size. *Phys. Rev. B* **2007**, *76*, 064304. [CrossRef]

66. Pardanaud, C.; Martin, C.; Roubin, P. Multiwavelength Raman spectroscopy analysis of a large sampling of disordered carbons extracted from the Tore Supra tokamak. *Vib. Spectrosc.* **2014**, *70*, 187–192. [CrossRef]

67. Cancado, L.G.; Takai, K.; Enoki, T.; Endo, M.; Kim, Y.A.; Mizusaki, H.; Jorio, A.; Coelho, L.N.; Magalhaes-Paniago, R.; Pimenta, M.A. General equation for the determination of the crystallite size L_a of nanographite by Raman spectroscopy. *Appl. Phys. Lett.* **2006**, *88*, 163106. [CrossRef]

68. Mernagh, T.P.; Cooney, R.P.; Johnson, R.A. Raman-spectra of graphon carbon-black. *Carbon* **1984**, *22*, 39–42. [CrossRef]

69. Venezuela, P.; Lazzeri, M.; Mauri, F. Theory of double-resonant Raman spectra in graphene: Intensity and line shape of defect-induced and two-phonon bands. *Phys. Rev. B* **2011**, *84*, 035433. [CrossRef]

70. Ferrari, A.C.; Robertson, J. Interpretation of Raman spectra of disordered and amorphous carbon. *Phys. Rev. B* **2000**, *61*, 14095–14107. [CrossRef]

71. Robertson, J.; Oreilly, E.P. Electronic and atomic-structure of amorphous-carbon. *Phys. Rev. B* **1987**, *35*, 2946–2957. [CrossRef]

72. Lajaunie, L.; Pardanaud, C.; Martin, C.; Puech, P.; Hu, C.; Biggs, M.J.; Arenal, R. Advanced spectroscopic analyses on a:C–H materials: Revisiting the EELS characterization and its coupling with multi-wavelength Raman spectroscopy. *Carbon* **2017**, *112*, 149–161. [CrossRef]

73. Ramaswamy, C. Raman effect in diamond. *Nature* **1930**, *125*, 704. [CrossRef]

74. Solin, S.A.; Ramdas, A.K. Raman spectrum of diamond. *Phys. Rev. B* **1970**, *1*, 1687. [CrossRef]

75. Yoshimori, A.; Kitano, Y. Theory of the lattice vibration of graphite. *J. Phys. Soc. Jpn.* **1956**, *2*, 352. [CrossRef]

76. Young, J.A.; Koppel, J.U. Phonon spectrum of graphite. *J. Chem. Phys.* **1965**, *42*, 357. [CrossRef]

77. Tsu, R.; Gonzalez, J.; Hernandez, I. Observation of splitting of E_{2g} mode and two-phonon spectrum in graphites. *Solid State Commun.* **1978**, *27*, 507–510. [CrossRef]

78. Vidano, R.; Fischbach, D.B. New lines in raman-spectra of carbons and graphite. *J. Am. Ceram. Soc.* **1978**, *61*, 13–17. [CrossRef]

79. Lespade, P.; Marchand, A.; Couzi, M.; Cruege, F. Characterization of carbon materials with Raman microspectrometry. *Carbon* **1984**, *22*, 375–385. [CrossRef]

80. Nakamizo, M.; Kammereck, R.; Walker, P.L. Laser Raman studies on carbons. *Carbon* **1974**, *12*, 259–267. [CrossRef]

81. Nemanich, R.J.; Solin, S.A. 1st-order and 2nd-order raman-scattering from finite-size crystals of graphite. *Phys. Rev. B* **1979**, *20*, 392–401. [CrossRef]

82. Vidano, R.P.; Fischbach, D.B.; Willis, L.J.; Loehr, T.M. Observation of raman band shifting with excitation wavelength for carbons and graphites. *Solid State Commun.* **1981**, *39*, 341–344. [CrossRef]

83. Baranov, A.V.; Bekhterev, A.N.; Bobovich, Y.S.; Petrov, V.I. Interpretation of some singularities in raman-spectra of graphite and glass carbon. *Opt. I Spektrosk.* **1987**, *62*, 1036–1042.

84. Wang, Y.; Alsmeyer, D.C.; McCreery, R.L. Raman-spectroscopy of carbon materials—Structural basis of observed spectra. *Chem. Mater.* **1990**, *2*, 557–563. [CrossRef]

85. Ramsteiner, M.; Wagner, J. Resonant raman-scattering of hydrogenated amorphous-carbon—Evidence for pi-bonded carbon clusters. *Appl. Phys. Lett.* **1987**, *51*, 1355–1357. [CrossRef]

86. Kastner, J.; Pichler, T.; Kuzmany, H.; Curran, S.; Blau, W.; Weldon, D.N.; Delamesiere, M.; Draper, S.; Zandbergen, H. Resonance Raman and infrared-spectroscopy of carbon nanotubues. *Chem. Phys. Lett.* **1994**, *221*, 53–58. [CrossRef]

87. Rao, A.M.; Richter, E.; Bandow, S.; Chase, B.; Eklund, P.C.; Williams, K.A.; Fang, S.; Subbaswamy, K.R.; Menon, M.; Thess, A.; et al. Diameter-selective Raman scattering from vibrational modes in carbon nanotubes. *Science* **1997**, *275*, 187–191. [CrossRef] [PubMed]

88. Knight, D.S.; White, W.B. Characterization of diamond films by Raman-spectroscopy. *J. Mater. Res.* **1989**, *4*, 385–393. [CrossRef]

89. Tan, P.H.; Deng, Y.M.; Zhao, Q. Temperature-dependent Raman spectra and anomalous Raman phenomenon of highly oriented pyrolytic graphite. *Phys. Rev. B* **1998**, *58*, 5435–5439. [CrossRef]

90. Pocsik, I.; Hundhausen, M.; Koos, M.; Ley, L. Origin of the D peak in the Raman spectrum of microcrystalline graphite. *J. Non Cryst. Solids* **1998**, *227*, 1083–1086. [CrossRef]

91. Matthews, M.J.; Pimenta, M.A.; Dresselhaus, G.; Dresselhaus, M.S.; Endo, M. Origin of dispersive effects of the Raman D band in carbon materials. *Phys. Rev. B* **1999**, *59*, R6585–R6588. [CrossRef]

92. Thomsen, C.; Reich, S. Double resonant Raman scattering in graphite. *Phys. Rev. Lett.* **2000**, *85*, 5214–5217. [CrossRef] [PubMed]

93. Saito, R.; Jorio, A.; Souza, A.G.; Dresselhaus, G.; Dresselhaus, M.S.; Pimenta, M.A. Probing phonon dispersion relations of graphite by double resonance Raman scattering. *Phys. Rev. Lett.* **2002**, *88*, 027401. [CrossRef] [PubMed]

94. Pimenta, M.A.; Dresselhaus, G.; Dresselhaus, M.S.; Cancado, L.G.; Jorio, A.; Saito, R. Studying disorder in graphite-based systems by Raman spectroscopy. *Phys. Chem. Chem. Phys.* **2007**, *9*, 1276–1290. [CrossRef] [PubMed]

95. Novoselov, K.S.; Geim, A.K.; Morozov, S.V.; Jiang, D.; Zhang, Y.; Dubonos, S.V.; Grigorieva, I.V.; Firsov, A.A. Electric field effect in atomically thin carbon films. *Science* **2004**, *306*, 666–669. [CrossRef] [PubMed]

96. Geim, A.K.; Novoselov, K.S. The rise of graphene. *Nat. Mater.* **2007**, *6*, 183–191. [CrossRef] [PubMed]

97. Cançado, L.G.; da Silva, M.G.; Ferreira, E.H.M.; Hof, F.; Kampioti, K.; Huang, K.; Penicaud, A.; Achete, C.A.; Capaz, R.B.; Jorio, A. Disentangling contributions of point and line defects in the Raman spectra of graphene-related materials. *2D Mater.* **2017**, *4*, 015039.

98. Cançado, L.G.; Jorio, A.; Martins Ferreira, E.H.; Stavale, F.; Achete, C.A.; Capaz, R.B.; Moutinho, M.V.O.; Lombardo, A.; Kulmala, T.S.; Ferrari, A.C. Quantifying defects in graphene via Raman spectroscopy at different excitation energies. *Nano Lett.* **2011**, *11*, 3190–3196. [CrossRef] [PubMed]

99. Ferreira, E.H.M.; Moutinho, M.V.O.; Stavale, F.; Lucchese, M.M.; Capaz, R.B.; Achete, C.A.; Jorio, A. Evolution of the Raman spectra from single-, few-, and many-layer graphene with increasing disorder. *Phys. Rev. B* **2010**, *82*, 125429. [CrossRef]

100. Giro, R.; Archanjo, B.S.; Martins Ferreira, E.H.; Capaz, R.B.; Jorio, A.; Achete, C.A. Quantifying defects in N-layer graphene via a phenomenological model of Raman spectroscopy. *Nucl. Instrum. Methods Phys. Res. Sec. B Beam Interact. Mater. Atoms* **2014**, *319*, 71–74. [CrossRef]

101. Lucchese, M.M.; Stavale, F.; Ferreira, E.H.M.; Vilani, C.; Moutinho, M.V.O.; Capaz, R.B.; Achete, C.A.; Jorio, A. Quantifying ion-induced defects and Raman relaxation length in graphene. *Carbon* **2010**, *48*, 1592–1597. [CrossRef]

102. Ribeiro-Soares, J.; Oliveros, M.E.; Garin, C.; David, M.V.; Martins, L.G.P.; Almeida, C.A.; Martins-Ferreira, E.H.; Takai, K.; Enoki, T.; Magalhaes-Paniago, R.; et al. Structural analysis of polycrystalline graphene systems by Raman spectroscopy. *Carbon* **2015**, *95*, 646–652. [CrossRef]

103. Cong, C.X.; Yu, T.; Saito, R.; Dresselhaus, G.F.; Dresselhaus, M.S. Second-order overtone and combination Raman modes of graphene layers in the range of 1690–2150 cm^{-1}. *Acs Nano* **2011**, *5*, 1600–1605. [CrossRef] [PubMed]

104. Popov, V.N. Two-phonon Raman bands of bilayer graphene: Revisited. *Carbon* **2015**, *91*, 436–444. [CrossRef]

105. Eckmann, A.; Felten, A.; Mishchenko, A.; Britnell, L.; Krupke, R.; Novoselov, K.S.; Casiraghi, C. Probing the nature of defects in graphene by Raman spectroscopy. *Nano Lett.* **2012**, *12*, 3925–3930. [CrossRef] [PubMed]

106. Pimenta, M.A.; del Corro, E.; Carvalho, B.R.; Fantini, C.; Malard, L.M. Comparative study of Raman spectroscopy in graphene and MoS$_2$-type transition metal dichalcogenides. *Accounts Chem. Res.* **2015**, *48*, 41–47. [CrossRef] [PubMed]

107. Carvalho, B.R.; Wang, Y.X.; Mignuzzi, S.; Roy, D.; Terrones, M.; Fantini, C.; Crespi, V.H.; Malard, L.M.; Pimenta, M.A. Intervalley scattering by acoustic phonons in two-dimensional MoS$_2$ revealed by double-resonance Raman spectroscopy. *Nat. Commun.* **2017**, *8*, 14670. [CrossRef] [PubMed]

108. Guo, H.H.; Yang, T.; Yamamoto, M.; Zhou, L.; Ishikawa, R.; Ueno, K.; Tsukagoshi, K.; Zhang, Z.D.; Dresselhaus, M.S.; Saito, R. Double resonance Raman modes in monolayer and few-layer MoTe$_2$. *Phys. Rev. B* **2015**, *91*, 205415. [CrossRef]

109. Castiglioni, C.; Di Donato, E.; Tommasini, M.; Negri, F.; Zerbi, G. Multi-wavelength Raman response of disordered graphitic materials: Models and simulations. *Synth. Met.* **2003**, *139*, 885–888. [CrossRef]

110. Castiglioni, C.; Negri, F.; Tommasini, M.; Di Donato, E.; Zerbi, G. Raman spectra and structure of sp^2 carbon-based materials: Electron-phonon coupling, vibrational dynamics and Raman activity. *Carbon* **2006**, *100*, 381–402.

111. Castiglioni, C.; Tommasini, M.; Zerbi, G. Raman spectroscopy of polyconjugated molecules and materials: Confinement effect in one and two dimensions. *Philos. Trans. R. Soc.* **2004**, *362*, 2425–2459. [CrossRef] [PubMed]

112. Di Donato, E.; Tommasini, M.; Fustella, G.; Brambilla, L.; Castiglioni, C.; Zerbi, G.; Simpson, C.D.; Mullen, K.; Negri, F. Wavelength-dependent Raman activity of D_{2h} symmetry polycyclic aromatic hydrocarbons in the D-band and acoustic phonon regions. *Chem. Phys.* **2004**, *301*, 81–93. [CrossRef]

113. Negri, F.; Castiglioni, C.; Tommasini, M.; Zerbi, G. A computational study of the Raman spectra of large polycyclic aromatic hydrocarbons: Toward molecularly defined subunits of graphite. *J. Phys. Chem. A* **2002**, *106*, 3306–3317. [CrossRef]

114. Negri, F.; di Donato, E.; Tommasini, M.; Castiglioni, C.; Zerbi, G.; Mullen, K. Resonance Raman contribution to the D band of carbon materials: Modeling defects with quantum chemistry. *J. Chem. Phys.* **2004**, *120*, 11889–11900. [CrossRef] [PubMed]

115. Tommasini, M.; Di Donato, E.; Castiglioni, C.; Zerbi, G.; Severin, N.; Bohme, T.; Rabe, J.P. Resonant Raman spectroscopy of nanostructured carbon-based materials: The molecular approach. In *Electronic Properties of Synthetic Nanostructures*; Kuzmany, H., Fink, J., Mehring, M., Roth, S., Eds.; American Institute of Physics: College Park, MD, USA, 2004; Volume 723, pp. 334–338.

116. Tommasini, M.; Castiglioni, C.; Zerbi, G. Raman scattering of molecular graphenes. *Phys. Chem. Chem. Phys.* **2009**, *11*, 10185–10194. [CrossRef] [PubMed]

117. Maghsoumi, A.; Brambilla, L.; Castiglioni, C.; Mullen, K.; Tommasini, M. Overtone and combination features of G and D peaks in resonance Raman spectroscopy of the $C_{78}H_{26}$ polycyclic aromatic hydrocarbon. *J. Raman Spectrosc.* **2015**, *46*, 757–764. [CrossRef]

118. Heller, E.J.; Yang, Y.; Kocia, L.; Chen, W.; Fang, S.A.; Borunda, M.; Kaxiras, E. Theory of graphene Raman scattering. *Acs Nano* **2016**, *10*, 2803–2818. [CrossRef] [PubMed]

119. Luo, X.; Lu, X.; Cong, C.X.; Yu, T.; Xiong, Q.H.; Quek, S.Y. Stacking sequence determines Raman intensities of observed interlayer shear modes in 2D layered materials—A general bond polarizability model. *Sci. Rep.* **2015**, *5*, 14565. [CrossRef] [PubMed]

120. Benybassez, C.; Rouzaud, J.N. Characterization of carbonaceous materials by correlated electron and optical microscopy and raman microspectroscopy. *Scanning Electron Microsc.* **1985**, *1*, 119–132.

121. Jawhari, T.; Roig, A.; Casado, J. Raman-spectroscopic characterization of some commercially available carbon-black materials. *Carbon* **1995**, *33*, 1561–1565. [CrossRef]

122. Sadezky, A.; Muckenhuber, H.; Grothe, H.; Niessner, R.; Poschl, U. Raman micro spectroscopy of soot and related carbonaceous materials: Spectral analysis and structural information. *Carbon* **2005**, *43*, 1731–1742. [CrossRef]

123. Vallerot, J.M.; Bourrat, X.; Mouchon, A.; Chollon, G. Quantitative structural and textural assessment of laminar pyrocarbons through Raman spectroscopy, electron diffraction and few other techniques. *Carbon* **2006**, *44*, 1833–1844. [CrossRef]

124. Jacob, W.; Moller, W. On the structure of thin hydrocarbon films. *Appl. Phys. Lett.* **1993**, *63*, 1771–1773. [CrossRef]

125. Hopf, C.; Angot, T.; Areou, E.; Duerbeck, T.; Jacob, W.; Martin, C.; Pardanaud, C.; Roubin, P.; Schwarz-Selinger, T. Characterization of temperature-induced changes in amorphous hydrogenated carbon thin films. *Diam. Relat. Mater.* **2013**, *37*, 97–103. [CrossRef]

126. Schwarz-Selinger, T.; von Keudell, A.; Jacob, W. Plasma chemical vapor deposition of hydrocarbon films: The influence of hydrocarbon source gas on the film properties. *J. Appl. Phys.* **1999**, *86*, 3988–3996. [CrossRef]

127. Wagner, J.; Ramsteiner, M.; Wild, C.; Koidl, P. Resonant raman-scattering of amorphous-carbon and polycrystalline diamond films. *Phys. Rev. B* **1989**, *40*, 1817–1824. [CrossRef]

128. Ferrari, A.C.; Li Bassi, A.; Tanner, B.K.; Stolojan, V.; Yuan, J.; Brown, L.M.; Rodil, S.E.; Kleinsorge, B.; Robertson, J. Density, sp^3 fraction, and cross-sectional structure of amorphous carbon films determined by X-ray reflectivity and electron energy-loss spectroscopy. *Phys. Rev. B* **2000**, *62*, 11089–11103. [CrossRef]

129. Kroto, H.W.; Heath, J.R.; O'Brien, S.C.; Curl, R.F.; Smalley, R.E. C_{60}: Buckminsterfullerene. *Nature* **1985**, *318*, 162–163. [CrossRef]

130. Matus, M.; Kuzmany, H.; Kratschmer, W. Resonance raman-scattering and electronic-transitions in C_{60}. *Solid State Commun.* **1991**, *80*, 839–842. [CrossRef]

131. Sinha, K.; Menendez, J.; Hanson, R.C.; Adams, G.B.; Page, J.B.; Sankey, O.F.; Lamb, L.D.; Huffman, D.R. Evidence for solid-state effects in the electronic-structure of C_{60} films—A resonance-Raman study. *Chem. Phys. Lett.* **1991**, *186*, 287–290. [CrossRef]

132. Vanloosdrecht, P.H.M.; Vanbentum, P.J.M.; Verheijen, M.A.; Meijer, G. Raman-scattering in single-crystal C_{60}. *Chem. Phys. Lett.* **1992**, *198*, 587–595. [CrossRef]

133. Monthioux, M.; Kuznetsov, V.L. Who should be given the credit for the discovery of carbon nanotubes? *Carbon* **2006**, *44*, 1621–1623. [CrossRef]

134. Iijima, S. Helical microtubules of graphitic carbon. *Nature* **1991**, *354*, 56–58. [CrossRef]

135. Kataura, H.; Kumazawa, Y.; Maniwa, Y.; Umezu, I.; Suzuki, S.; Ohtsuka, Y.; Achiba, Y. Optical properties of single-wall carbon nanotubes. *Synth. Met.* **1999**, *103*, 2555–2558. [CrossRef]

136. Barros, E.B.; Jorio, A.; Samsonidze, G.G.; Capaz, R.B.; Souza, A.G.; Mendes, J.; Dresselhaus, G.; Dresselhaus, M.S. Review on the symmetry-related properties of carbon nanotubes. *Phys. Rep. Rev. Sec. Phys. Lett.* **2006**, *431*, 261–302. [CrossRef]

137. Ferrari, A.C. Raman spectroscopy of graphene and graphite: Disorder, electron-phonon coupling, doping and nonadiabatic effects. *Solid State Commun.* **2007**, *143*, 47–57. [CrossRef]

138. Dresselhaus, M.S.; Jorio, A.; Souza, A.G.; Saito, R. Defect characterization in graphene and carbon nanotubes using Raman spectroscopy. *Philos. Trans. R. Soc. A Math. Phys. Eng. Sci.* **2010**, *368*, 5355–5377. [CrossRef] [PubMed]

139. Malard, L.M.; Pimenta, M.A.; Dresselhaus, G.; Dresselhaus, M.S. Raman spectroscopy in graphene. *Phys. Rep. Rev. Sec. Phys. Lett.* **2009**, *473*, 51–87. [CrossRef]

140. Mohr, M.; Maultzsch, J.; Thomsen, C. Splitting of the Raman 2D band of graphene subjected to strain. *Phys. Rev. B* **2010**, *82*, 201409. [CrossRef]

141. Bonini, N.; Lazzeri, M.; Marzari, N.; Mauri, F. Phonon anharmonicities in graphite and graphene. *Phys. Rev. Lett.* **2007**, *99*, 176802. [CrossRef] [PubMed]

142. Schwan, J.; Ulrich, S.; Batori, V.; Ehrhardt, H.; Silva, S.R.P. Raman spectroscopy on amorphous carbon films. *J. Appl. Phys.* **1996**, *80*, 440–447. [CrossRef]

143. Chu, P.K.; Li, L.H. Characterization of amorphous and nanocrystalline carbon films. *Mater. Chem. Phys.* **2006**, *96*, 253–277. [CrossRef]

144. Wang, Q.; Allred, D.D.; Knight, L.V. Deconvolution of the Raman spectrum of amorphous carbon. *J. Raman Spectrosc.* **1995**, *26*, 1039–1043. [CrossRef]

145. Richter, H.; Wang, Z.P.; Ley, L. The one phonon raman-spectrum in microcrystalline silicon. *Solid State Commun.* **1981**, *39*, 625–629. [CrossRef]

146. Puech, P.; Plewa, J.M.; Mallet-Ladeira, P.; Monthioux, M. Spatial confinement model applied to phonons in disordered graphene-based carbons. *Carbon* **2016**, *105*, 275–281. [CrossRef]

147. Jorio, A.; Ferreira, E.H.M.; Moutinho, M.V.O.; Stavale, F.; Achete, C.A.; Capaz, R.B. Measuring disorder in graphene with the G and D bands. *Phys. Status Solidi B* **2010**, *247*, 2980–2982. [CrossRef]

148. Baroni, S.; de Gironcoli, S.; Dal Corso, A.; Giannozzi, P. Phonons and related crystal properties from density-functional perturbation theory. *Rev. Mod. Phys.* **2001**, *73*, 515–562. [CrossRef]

149. Giannozzi, P.; Baroni, S.; Bonini, N.; Calandra, M.; Car, R.; Cavazzoni, C.; Ceresoli, D.; Chiarotti, G.L.; Cococcioni, M.; Dabo, I.; et al. QUANTUM ESPRESSO: A modular and open-source software project for quantum simulations of materials. *J. Phys. Condens. Matter* **2009**, *21*, 395502. [CrossRef] [PubMed]

150. Saidi, W.A. Effects of topological defects and diatom vacancies on characteristic vibration modes and Raman intensities of zigzag single-walled carbon nanotubes. *J. Phys. Chem. A* **2014**, *118*, 7235–7241. [CrossRef] [PubMed]

151. Saidi, W.A.; Norman, P. Probing single-walled carbon nanotube defect chemistry using resonance Raman spectroscopy. *Carbon* **2014**, *67*, 17–26. [CrossRef]

152. Saidi, W.A.; Norman, P. Spectroscopic signatures of topological and diatom-vacancy defects in single-walled carbon nanotubes. *Phys. Chem. Chem. Phys.* **2014**, *16*, 1479–1486. [CrossRef] [PubMed]

153. Kudin, K.N.; Ozbas, B.; Schniepp, H.C.; Prud'homme, R.K.; Aksay, I.A.; Car, R. Raman spectra of graphite oxide and functionalized graphene sheets. *Nano Lett.* **2008**, *8*, 36–41. [CrossRef] [PubMed]

154. Aggarwal, R.L.; Farrar, L.W.; Saikin, S.K.; Andrade, X.; Aspuru-Guzik, A.; Polla, D.L. Measurement of the absolute Raman cross section of the optical phonons in type Ia natural diamond. *Solid State Commun.* **2012**, *152*, 204–209. [CrossRef]

155. Skinner, J.G.; Nilsen, W.G. Absolute Raman scattering cross-section measurement of the 992 cm^{-1} line of benzene. *J. Opt. Soc. Am.* **1968**, *58*, 113–119. [CrossRef]

156. Aggarwal, R.L.; Farrar, L.W.; Saikin, S.K.; Aspuru-Guzik, A.; Stopa, M.; Polla, D.L. Measurement of the absolute Raman cross section of the optical phonon in silicon. *Solid State Commun.* **2011**, *151*, 553–556. [CrossRef]

157. Pettinger, B.; Picardi, G.; Schuster, R.; Ertl, G. Surface-enhanced and STM-tip-enhanced Raman spectroscopy at metal surfaces. *Single Mol.* **2002**, *3*, 285–294. [CrossRef]

158. Wang, Y.Y.; Ni, Z.H.; Shen, Z.X.; Wang, H.M.; Wu, Y.H. Interference enhancement of Raman signal of graphene. *Appl. Phys. Lett.* **2008**, *92*, 043121. [CrossRef]

159. Peres, N.M.R. Colloquium: The transport properties of graphene: An introduction. *Rev. Mod. Phys.* **2010**, *82*, 2673–2700. [CrossRef]

160. Beams, R.; Cancado, L.G.; Novotny, L. Raman characterization of defects and dopants in graphene. *J. Phys. Condens. Matter* **2015**, *27*, 083002. [CrossRef] [PubMed]

161. Bayle, M.; Reckinger, N.; Huntzinger, J.R.; Felten, A.; Bakaraki, A.; Landois, P.; Colomer, J.F.; Henrard, L.; Zahab, A.A.; Sauvajol, J.L.; et al. Dependence of the Raman spectrum characteristics on the number of layers and stacking orientation in few-layer graphene. *Phys. Status Solidi B* **2015**, *252*, 2375–2379. [CrossRef]

162. Poncharal, P.; Ayari, A.; Michel, T.; Sauvajol, J.L. Raman spectra of misoriented bilayer graphene. *Phys. Rev. B* **2008**, *78*, 113407. [CrossRef]

163. Poncharal, P.; Ayari, A.; Michel, T.; Sauvajol, J.L. Effect of rotational stacking faults on the Raman spectra of folded graphene. *Phys. Rev. B* **2009**, *79*, 195417. [CrossRef]

164. Das, A.; Chakraborty, B.; Sood, A.K. Raman spectroscopy of graphene on different substrates and influence of defects. *Bull. Mater. Sci.* **2008**, *31*, 579–584. [CrossRef]

165. Tan, P.H.; Han, W.P.; Zhao, W.J.; Wu, Z.H.; Chang, K.; Wang, H.; Wang, Y.F.; Bonini, N.; Marzari, N.; Pugno, N.; et al. The shear mode of multilayer graphene. *Nat. Mater.* **2012**, *11*, 294–300. [CrossRef] [PubMed]

166. Araujo, P.T.; Terrones, M.; Dresselhaus, M.S. Defects and impurities in graphene-like materials. *Mater. Today* **2012**, *15*, 98–109. [CrossRef]

167. Casiraghi, C. Doping dependence of the Raman peaks intensity of graphene close to the Dirac point. *Phys. Rev. B* **2009**, *80*, 233407. [CrossRef]

168. Casiraghi, C.; Pisana, S.; Novoselov, K.S.; Geim, A.K.; Ferrari, A.C. Raman fingerprint of charged impurities in graphene. *Appl. Phys. Lett.* **2007**, *91*, 233108. [CrossRef]

169. Kalbac, M.; Reina-Cecco, A.; Farhat, H.; Kong, J.; Kavan, L.; Dresselhaus, M.S. The influence of strong electron and hole doping on the Raman intensity of chemical vapor-deposition graphene. *ACS Nano* **2010**, *4*, 6055–6063. [CrossRef] [PubMed]

170. Liu, J.K.; Li, Q.Q.; Zou, Y.; Qian, Q.K.; Jin, Y.H.; Li, G.H.; Jiang, K.L.; Fan, S.S. The dependence of graphene Raman D-band on carrier density. *Nano Lett.* **2013**, *13*, 6170–6175. [CrossRef] [PubMed]

171. Ni, Z.H.; Ponomarenko, L.A.; Nair, R.R.; Yang, R.; Anissimova, S.; Grigorieva, I.V.; Schedin, F.; Blake, P.; Shen, Z.X.; Hill, E.H.; et al. On resonant scatterers as a factor limiting carrier mobility in graphene. *Nano Lett.* **2010**, *10*, 3868–3872. [CrossRef] [PubMed]

172. Calizo, I.; Ghosh, S.; Bao, W.Z.; Miao, F.; Lau, C.N.; Balandin, A.A. Raman nanometrology of graphene: Temperature and substrate effects. *Solid State Commun.* **2009**, *149*, 1132–1135. [CrossRef]

173. Calizo, I.; Bao, W.Z.; Miao, F.; Lau, C.N.; Balandin, A.A. The effect of substrates on the Raman spectrum of graphene: Graphene-on-sapphire and graphene-on-glass. *Appl. Phys. Lett.* **2007**, *91*, 201904. [CrossRef]

174. Frank, O.; Mohr, M.; Maultzsch, J.; Thomsen, C.; Riaz, I.; Jalil, R.; Novoselov, K.S.; Tsoukleri, G.; Parthenios, J.; Papagelis, K.; et al. Raman 2D-band splitting in graphene: Theory and experiment. *Acs Nano* **2011**, *5*, 2231–2239. [CrossRef] [PubMed]

175. Yang, R.; Huang, Q.S.; Chen, X.L.; Zhang, G.Y.; Gao, H.J. Substrate doping effects on Raman spectrum of epitaxial graphene on SiC. *J. Appl. Phys.* **2010**, *107*, 034305. [CrossRef]

176. Das, A.; Pisana, S.; Chakraborty, B.; Piscanec, S.; Saha, S.K.; Waghmare, U.V.; Novoselov, K.S.; Krishnamurthy, H.R.; Geim, A.K.; Ferrari, A.C.; et al. Monitoring dopants by Raman scattering in an electrochemically top-gated graphene transistor. *Nat. Nanotechnol.* **2008**, *3*, 210–215. [CrossRef] [PubMed]

177. Lee, J.; Novoselov, K.S.; Shin, H.S. Interaction between metal and graphene: Dependence on the Layer number of graphene. *Acs Nano* **2011**, *5*, 608–612. [CrossRef] [PubMed]

178. Xu, W.G.; Mao, N.N.; Zhang, J. Graphene: A platform for surface-enhanced Raman spectroscopy. *Small* **2013**, *9*, 1206–1224. [CrossRef] [PubMed]

179. Bronsgeest, M.S.; Bendiab, N.; Mathur, S.; Kimouche, A.; Johnson, H.T.; Coraux, J.; Pochet, P. Strain relaxation in CVD graphene: Wrinkling with shear lag. *Nano Lett.* **2015**, *15*, 5098–5104. [CrossRef] [PubMed]

180. Liu, H.L.; Siregar, S.; Hasdeo, E.H.; Kumamoto, Y.; Shen, C.C.; Cheng, C.C.; Li, L.J.; Saito, R.; Kawata, S. Deep-ultraviolet Raman scattering studies of monolayer graphene thin films. *Carbon* **2015**, *81*, 807–813. [CrossRef]

181. Tyborski, C.; Herziger, F.; Gillen, R.; Maultzsch, J. Beyond double-resonant Raman scattering: Ultraviolet Raman spectroscopy on graphene, graphite, and carbon nanotubes. *Phys. Rev. B* **2015**, *92*, 041401. [CrossRef]

182. Saito, R.; Nugraha, A.R.T.; Hasdeo, E.H.; Siregar, S.; Guo, H.H.; Yang, T. Ultraviolet Raman spectroscopy of graphene and transition-metal dichalcogenides. *Phys. Status Solidi B* **2015**, *252*, 2363–2374. [CrossRef]

183. Zhou, W.; Zeng, J.W.; Li, X.F.; Xu, J.; Shi, Y.; Ren, W.; Miao, F.; Wang, B.G.; Xing, D.Y. Ultraviolet Raman spectra of double-resonant modes of graphene. *Carbon* **2016**, *101*, 235–238. [CrossRef]

184. Herziger, F.; Calandra, M.; Gava, P.; May, P.; Lazzeri, M.; Mauri, F.; Maultzsch, J. Two-dimensional analysis of the double-resonant 2D Raman mode in bilayer graphene. *Phys. Rev. Lett.* **2014**, *113*, 187401. [CrossRef] [PubMed]

185. Zandiatashbar, A.; Lee, G.-H.; An, S.J.; Lee, S.; Mathew, N.; Terrones, M.; Hayashi, T.; Picu, C.R.; Hone, J.; Koratkar, N. Effect of defects on the intrinsic strength and stiffness of graphene. *Nat. Commun.* **2014**, *5*, 3186. [CrossRef] [PubMed]

186. Ferralis, N. Probing mechanical properties of graphene with Raman spectroscopy. *J. Mater. Sci.* **2010**, *45*, 5135–5149. [CrossRef]

187. Yu, Q.K.; Jauregui, L.A.; Wu, W.; Colby, R.; Tian, J.F.; Su, Z.H.; Cao, H.L.; Liu, Z.H.; Pandey, D.; Wei, D.G.; et al. Control and characterization of individual grains and grain boundaries in graphene grown by chemical vapour deposition. *Nat. Mater.* **2011**, *10*, 443–449. [CrossRef] [PubMed]

188. Lee, J.Y.; Lee, J.H.; Kim, M.J.; Dash, J.K.; Lee, C.H.; Joshi, R.; Lee, S.; Hone, J.; Soon, A.; Lee, G.H. Direct observation of grain boundaries in chemical vapor deposited graphene. *Carbon* **2017**, *115*, 147–153. [CrossRef]

189. Chen, S.S.; Moore, A.L.; Cai, W.W.; Suk, J.W.; An, J.H.; Mishra, C.; Amos, C.; Magnuson, C.W.; Kang, J.Y.; Shi, L.; et al. Raman measurements of thermal transport in suspended monolayer graphene of variable sizes in vacuum and gaseous environments. *Acs Nano* **2011**, *5*, 321–328. [CrossRef] [PubMed]

190. Metten, D.; Froehlicher, G.; Berciaud, S. Monitoring electrostatically-induced deflection, strain and doping in suspended graphene using Raman spectroscopy. *2D Mater.* **2017**, *4*. [CrossRef]

191. Suarez-Martinez, I.; Grobert, N.; Ewels, C.P. Nomenclature of sp^2 carbon nanoforms. *Carbon* **2012**, *50*, 741–747. [CrossRef]

192. Bianco, A.; Cheng, H.M.; Enoki, T.; Gogotsi, Y.; Hurt, R.H.; Koratkar, N.; Kyotani, T.; Monthioux, M.; Park, C.R.; Tascon, J.M.D.; et al. All in the graphene family A recommended nomenclature for two-dimensional carbon materials. *Carbon* **2013**, *65*, 1–6. [CrossRef]

193. Wick, P.; Louw Gaume, A.F.; Kucki, M.; Krug, H.F.; Kostarelos, K.; Fadeel, B.; Dawson, K.A.; Salvati, A.; Vazquez, E.; Ballerini, L.; et al. Classification framework for graphene-based materials. *Angew. Chem. Int. Ed.* **2014**, *53*, 7714–7718. [CrossRef] [PubMed]

194. Cancado, L.G.; Pimenta, M.A.; Neves, B.R.A.; Dantas, M.S.S.; Jorio, A. Influence of the atomic structure on the Raman spectra of graphite edges. *Phys. Rev. Lett.* **2004**, *93*, 247401. [CrossRef] [PubMed]

195. You, Y.M.; Ni, Z.H.; Yu, T.; Shen, Z.X. Edge chirality determination of graphene by Raman spectroscopy. *Appl. Phys. Lett.* **2008**, *93*, 163112. [CrossRef]

196. Casiraghi, C.; Hartschuh, A.; Qian, H.; Piscanec, S.; Georgi, C.; Fasoli, A.; Novoselov, K.S.; Basko, D.M.; Ferrari, A.C. Raman spectroscopy of graphene edges. *Nano Lett.* **2009**, *9*, 1433–1441. [CrossRef] [PubMed]

197. Islam, M.S.; Tamakawa, D.; Tanaka, S.; Makino, T.; Hashimoto, A. Polarized microscopic laser Raman scattering spectroscopy for edge structure of epitaxial graphene and localized vibrational mode. *Carbon* **2014**, *77*, 1073–1081. [CrossRef]

198. Ren, W.C.; Saito, R.; Gao, L.B.; Zheng, F.W.; Wu, Z.S.; Liu, B.L.; Furukawa, M.; Zhao, J.P.; Chen, Z.P.; Cheng, H.M. Edge phonon state of mono- and few-layer graphene nanoribbons observed by surface and interference co-enhanced Raman spectroscopy. *Phys. Rev. B* **2010**, *81*, 035412. [CrossRef]

199. Mazzamuto, F.; Saint-Martin, J.; Valentin, A.; Chassat, C.; Dollfus, P. Edge shape effect on vibrational modes in graphene nanoribbons: A numerical study. *J. Appl. Phys.* **2011**, *109*, 064516. [CrossRef]

200. Saito, R.; Furukawa, M.; Dresselhaus, G.; Dresselhaus, M.S. Raman spectra of graphene ribbons. *J. Phys. Condens. Matter* **2010**, *22*, 334203. [CrossRef] [PubMed]

201. Yu, F.; Zhou, H.Q.; Zhang, Z.X.; Tang, D.S.; Chen, M.J.; Yang, H.C.; Wang, G.; Yang, H.F.; Gu, C.Z.; Sun, L.F. Experimental observation of radial breathing-like mode of graphene nanoribbons. *Appl. Phys. Lett.* **2012**, *100*, 101904. [CrossRef]

202. Zhou, J.; Dong, J. Vibrational property and Raman spectrum of carbon nanoribbon. *Appl. Phys. Lett.* **2007**, *91*, 173108. [CrossRef]

203. Verzhbitskiy, I.A.; De Corato, M.; Ruini, A.; Molinari, E.; Narita, A.; Hu, Y.; Schwab, M.G.; Bruna, M.; Yoon, D.; Milana, S.; et al. Raman fingerprints of atomically precise graphene nanoribbons. *Nano Lett.* **2016**, *16*, 3442–3447. [CrossRef] [PubMed]

204. Casiraghi, C.; Prezzi, D. Raman spectroscopy of graphene nanoribbons: A review. In *GraphITA: Selected papers from the Workshop on Synthesis, Characterization and Technological Exploitation of Graphene and 2D Materials Beyond Graphene*; Carbon Nanostructures; Morandi, V., Ottaviano, L., Eds.; Springer International Publishing: Cham, Switzerland, 2017.

205. Reich, S.; Thomsen, C.; Maultzsch, J. *Carbon Nanotubes: Basic Concepts and Physical Properties*; Wiley-VCH: Weinheim, Germany, 2004.

206. Saito, R.; Dresselhaus, G.; Dresselhaus, M.S. *Physical Properties of Carbon Nanotubes*; Imperial College Press: London, UK, 1998.

207. Bohn, J.E.; Etchegoin, P.G.; le Ru, E.C.; Xiang, R.; Chiashi, S.; Maruyama, S. Estimating the Raman cross sections of single carbon nanotubes. *Acs Nano* **2010**, *4*, 3466–3470. [CrossRef] [PubMed]

208. Sauvajol, J.-L.; Anglaret, E.; Rols, S.; Stephan, O. Spectroscopies on carbon nanotubes. In *Understanding Carbon Nanotubes: From Basics to Applications*; Loiseau, A., Launois, P., Petit, P., Roche, S., Salvetat, J.-P., Eds.; Springer: Berlin, Germany, 2006; Volume 677, pp. 277–334.

209. Ghavanloo, E.; Fazelzadeh, S.A.; Rafii-Tabar, H. Analysis of radial breathing-mode of nanostructures with various morphologies: A critical review. *Int. Mater. Rev.* **2015**, *60*, 312–329. [CrossRef]

210. Jorio, A.; Pimenta, M.A.; Souza, A.G.; Saito, R.; Dresselhaus, G.; Dresselhaus, M.S. Characterizing carbon nanotube samples with resonance Raman scattering. *New J. Phys.* **2003**, *5*, 139. [CrossRef]

211. Brown, S.D.M.; Jorio, A.; Corio, P.; Dresselhaus, M.S.; Dresselhaus, G.; Saito, R.; Kneipp, K. Origin of the Breit-Wigner-Fano lineshape of the tangential G-band feature of metallic carbon nanotubes. *Phys. Rev. B* **2001**, *63*, 155414. [CrossRef]

212. Song, L.; Ci, L.J.; Sun, L.F.; Jin, C.H.; Liu, L.F.; Ma, W.J.; Liu, D.F.; Zhao, X.W.; Luo, S.D.; Zhang, Z.X.; et al. Large-scale synthesis of rings of bundled single-walled carbon nanotubes by floating chemical vapor deposition. *Adv. Mater.* **2006**, *18*, 1817–1821. [CrossRef]

213. Ren, Y.; Song, L.; Ma, W.J.; Zhao, Y.C.; Sun, L.F.; Gu, C.Z.; Zhou, W.Y.; Xie, S.S. Additional curvature-induced Raman splitting in carbon nanotube ring structures. *Phys. Rev. B* **2009**, *80*, 113412. [CrossRef]

214. Dunk, P.W.; Niwa, H.; Shinohara, H.; Marshall, A.G.; Kroto, H.W. Large fullerenes in mass spectra. *Mol. Phys.* **2015**, *113*, 2359–2361. [CrossRef]

215. Dresselhaus, M.S.; Dresselhaus, G.; Eklund, P.C. Raman scattering in fullerenes. *J. Raman Spectrosc.* **1996**, *27*, 351–371. [CrossRef]

216. Kuzmany, H.; Pfeiffer, R.; Hulman, M.; Kramberger, C. Raman spectroscopy of fullerenes and fullerene-nanotube composites. *Philos. Trans. R. Soc. A Math. Phys. Eng. Sci.* **2004**, *362*, 2375–2406. [CrossRef] [PubMed]

217. Bardelang, D.; Giorgi, M.; Pardanaud, C.; Hornebecq, V.; Rizzato, E.; Tordo, P.; Ouari, O. Organic multishell isostructural host-guest crystals: Fullerenes C_{60} inside a nitroxide open framework. *Chem. Commun.* **2013**, *49*, 3519–3521. [CrossRef] [PubMed]

218. Jishi, R.A.; Dresselhaus, M.S.; Dresselhaus, G.; Wang, K.A.; Zhou, P.; Rao, A.M.; Eklund, P.C. Vibrational-mode frequencies in C_{70}. *Chem. Phys. Lett.* **1993**, *206*, 187–192. [CrossRef]

219. Eklund, P.C.; Rao, A.M.; Zhou, P.; Wang, Y. Photochemical transformation of C_{60} and C_{70} films. *Thin Solid Films* **1995**, *257*, 185–203. [CrossRef]

220. Brazhkin, V.V.; Lyapin, A.G.; Popova, S.V.; Voloshin, R.N.; Antonov, Y.V.; Lyapin, S.G.; Kluev, Y.A.; Naletov, A.M.; Melnik, N.N. Metastable crystalline and amorphous carbon phases obtained from fullerite C_{60} by high-pressuse-high-temperature treatment. *Phys. Rev. B* **1997**, *56*, 11465–11472. [CrossRef]

221. Karousis, N.; Suarez-Martinez, I.; Ewels, C.P.; Tagmatarchis, N. Structure, properties, functionalization, and applications of carbon nanohorns. *Chem. Rev.* **2016**, *116*, 4850–4883. [CrossRef] [PubMed]
222. Iijima, S.; Yudasaka, M.; Yamada, R.; Bandow, S.; Suenaga, K.; Kokai, F.; Takahashi, K. Nano-aggregates of single-walled graphitic carbon nano-horns. *Chem. Phys. Lett.* **1999**, *309*, 165–170. [CrossRef]
223. Pena-Alvarez, M.; del Corro, E.; Langa, F.; Baonza, V.G.; Taravillo, M. Morphological changes in carbon nanohorns under stress: A combined Raman spectroscopy and TEM study. *RSC Adv.* **2016**, *6*, 49543–49550. [CrossRef]
224. Sasaki, K.; Sekine, Y.; Tateno, K.; Gotoh, H. Topological Raman band in the carbon nanohorn. *Phys. Rev. Lett.* **2013**, *111*, 116801. [CrossRef] [PubMed]
225. Pardanaud, C.; Martin, C.; Giacometti, G.; Mellet, N.; Pegourie, B.; Roubin, P. Thermal stability and long term hydrogen/deuterium release from soft to hard amorphous carbon layers analyzed using in-situ Raman spectroscopy. Comparison with Tore Supra deposits. *Thin Solid Films* **2015**, *581*, 92–98. [CrossRef]
226. Pardanaud, C.; Martin, C.; Giacometti, G.; Roubin, P.; Pegourie, B.; Hopf, C.; Schwarz-Selinger, T.; Jacob, W.; Buijnsters, J.G. Long-term H-release of hard and intermediate between hard and soft amorphous carbon evidenced by in situ Raman microscopy under isothermal heating. *Diam. Relat. Mater.* **2013**, *37*, 92–96. [CrossRef]
227. Niwase, K.; Tanabe, T.; Sugimoto, M.; Fujita, F.E. Modification of graphite structure by D^+ and He^+ bombardment. *J. Nucl. Mater.* **1989**, *162*, 856–860. [CrossRef]
228. Oschatz, M.; Pre, P.; Dorfler, S.; Nickel, W.; Beaunier, P.; Rouzaud, J.N.; Fischer, C.; Brunner, E.; Kaskel, S. Nanostructure characterization of carbide-derived carbons by morphological analysis of transmission electron microscopy images combined with physisorption and Raman spectroscopy. *Carbon* **2016**, *105*, 314–322. [CrossRef]
229. Da Costa, J.P.; Weisbecker, P.; Farbos, B.; Leyssale, J.M.; Vignoles, G.L.; Germain, C. Investigating carbon materials nanostructure using image orientation statistics. *Carbon* **2015**, *84*, 160–173. [CrossRef]
230. Niwase, K. Irradiation-induced amorphization of graphite. *Phys. Rev. B* **1995**, *52*, 15785–15798. [CrossRef]
231. Pardanaud, C.; Martin, C.; Cartry, G.; Ahmad, A.; Schiesko, L.; Giacometti, G.; Carrere, M.; Roubin, P. In-plane and out-of-plane defects of graphite bombarded by H, D and He investigated by atomic force and Raman microscopies. *J. Raman Spectrosc.* **2015**, *46*, 256–265. [CrossRef]
232. Banhart, F.; Kotakoski, J.; Krasheninnikov, A.V. Structural defects in graphene. *Acs Nano* **2011**, *5*, 26–41. [CrossRef] [PubMed]
233. Kotakoski, J.; Krasheninnikov, A.V.; Kaiser, U.; Meyer, J.C. From point defects in graphene to two-dimensional amorphous carbon. *Phys. Rev. Lett.* **2011**, *106*, 105505. [CrossRef] [PubMed]
234. Elman, B.S.; Dresselhaus, M.S.; Dresselhaus, G.; Maby, E.W.; Mazurek, H. Raman-scattering from ion-implanted graphite. *Phys. Rev. B* **1981**, *24*, 1027–1034. [CrossRef]
235. Compagnini, G.; Puglisi, O.; Foti, G. Raman spectra of virgin and damaged graphite edge planes. *Carbon* **1997**, *35*, 1793–1797. [CrossRef]
236. Nakamura, K.; Fujitsuka, M.; Kitajima, M. Finite size effect on raman-scattering of graphite microcrystals. *Chem. Phys. Lett.* **1990**, *172*, 205–208. [CrossRef]
237. Nakamura, K.; Kitajima, M. Real-time raman measurements of graphite under Ar^+ irradiation. *Appl. Phys. Lett.* **1991**, *59*, 1550–1552. [CrossRef]
238. Nakamura, K.; Kitajima, M. Ion-irradiation effects on the phonon correlation length of graphite studies by raman-spectroscopy. *Phys. Rev. B* **1992**, *45*, 78–82. [CrossRef]
239. Eckmann, A.; Felten, A.; Verzhbitskiy, I.; Davey, R.; Casiraghi, C. Raman study on defective graphene: Effect of the excitation energy, type, and amount of defects. *Phys. Rev. B* **2013**, *88*, 035426. [CrossRef]
240. Sato, K.; Saito, R.; Oyama, Y.; Jiang, J.; Cancado, L.G.; Pimenta, M.A.; Jorio, A.; Samsonidze, G.G.; Dresselhaus, G.; Dresselhaus, M.S. D-band Raman intensity of graphitic materials as a function of laser energy and crystallite size. *Chem. Phys. Lett.* **2006**, *427*, 117–121. [CrossRef]
241. Meyer, J.C.; Kisielowski, C.; Erni, R.; Rossell, M.D.; Crommie, M.F.; Zettl, A. Direct imaging of lattice atoms and topological defects in graphene membranes. *Nano Lett.* **2008**, *8*, 3582–3586. [CrossRef] [PubMed]
242. Ng, T.Y.; Yeo, J.J.; Liu, Z.S. A molecular dynamics study of the thermal conductivity of graphene nanoribbons containing dispersed Stone-Thrower-Wales defects. *Carbon* **2012**, *50*, 4887–4893. [CrossRef]
243. Thrower, P.A. The study of defects in graphite by transmission electron microscopy. In *Chemistry and Physics of Carbon*; Walker, P.L., Jr., Ed.; Marcel Dekker: New York, NY, USA, 1969.

244. Wu, G.; Dong, J.M. Raman characteristic peaks induced by the topological defects of carbon nanotube intramolecular junctions. *Phys. Rev. B* **2006**, *73*, 245414. [CrossRef]

245. Fujimori, T.; Radovic, L.R.; Silva-Tapia, A.B.; Endo, M.; Kaneko, K. Structural importance of Stone-Thrower-Wales defects in rolled and flat graphenes from surface-enhanced Raman scattering. *Carbon* **2012**, *50*, 3274–3279. [CrossRef]

246. Shirodkar, S.N.; Waghmare, U.V. Electronic and vibrational signatures of Stone-Wales defects in graphene: First-principles analysis. *Phys. Rev. B* **2012**, *86*, 165401. [CrossRef]

247. Itoh, T.; Yamamoto, Y.S.; Biju, V.; Tamaru, H.; Wakida, S. Fluctuating single sp^2 carbon clusters at single hotspots of silver nanoparticle dimers investigated by surface-enhanced resonance Raman scattering. *AIP Adv.* **2015**, *5*, 127113. [CrossRef]

248. Podila, R.; Rao, R.; Tsuchikawa, R.; Ishigami, M.; Rao, A.M. Raman spectroscopy of folded and scrolled graphene. *Acs Nano* **2012**, *6*, 5784–5790. [CrossRef] [PubMed]

249. Larouche, N.; Stansfield, B.L. Classifying nanostructured carbons using graphitic indices derived from Raman spectra. *Carbon* **2010**, *48*, 620–629. [CrossRef]

250. Yehliu, K.; Vander Wal, R.L.; Boehman, A.L. Development of an HRTEM image analysis method to quantify carbon nanostructure. *Combust. Flame* **2011**, *158*, 1837–1851. [CrossRef]

251. Pre, P.; Huchet, G.; Jeulin, D.; Rouzaud, J.N.; Sennour, M.; Thorel, A. A new approach to characterize the nanostructure of activated carbons from mathematical morphology applied to high resolution transmission electron microscopy images. *Carbon* **2013**, *52*, 239–258. [CrossRef]

252. Bourrat, X.; Langlais, F.; Chollon, G.; Vignoles, G.L. Low temperature pyrocarbons: A review. *J. Braz. Chem. Soc.* **2006**, *17*, 1090–1095. [CrossRef]

253. Rouzaud, J.N.; Oberlin, A.; Benybassez, C. Carbon-films—Structure and microtexture (optical and electron-microscopy, raman-spectroscopy). *Thin Solid Films* **1983**, *105*, 75–96. [CrossRef]

254. Brunetto, R.; Pino, T.; Dartois, E.; Cao, A.T.; d'Hendecourt, L.; Strazzulla, G.; Brechignac, P. Comparison of the Raman spectra of ion irradiated soot and collected extraterrestrial carbon. *Icarus* **2009**, *200*, 323–337. [CrossRef]

255. Schmid, J.; Grob, B.; Niessner, R.; Ivleva, N.P. Multiwavelength Raman microspectroscopy for rapid prediction of soot oxidation reactivity. *Anal. Chem.* **2011**, *83*, 1173–1179. [CrossRef] [PubMed]

256. Russo, C.; Ciajolo, A. Effect of the flame environment on soot nanostructure inferred by Raman spectroscopy at different excitation wavelengths. *Combust. Flame* **2015**, *162*, 2431–2441. [CrossRef]

257. Ess, M.N.; Ferry, D.; Kireeva, E.D.; Niessner, R.; Ouf, F.X.; Ivleva, N.P. In situ Raman microspectroscopic analysis of soot samples with different organic carbon content: Structural changes during heating. *Carbon* **2016**, *105*, 572–585. [CrossRef]

258. Bourrat, X.; Fillion, A.; Naslain, R.; Chollon, G.; Brendle, M. Regenerative laminar pyrocarbon. *Carbon* **2002**, *40*, 2931–2945. [CrossRef]

259. Bourrat, X.; Lavenac, J.; Langlais, F.; Naslain, R. The role of pentagons in the growth of laminar pyrocarbon. *Carbon* **2001**, *39*, 2376–2380. [CrossRef]

260. Vallerot, J.M. Matrice de Pyrocarbone: Propriétés, Structure et Anisotropie Optique. Ph.D. Thesis, Université de Bordeaux 1, Bordeaux, France, 2004. (In French)

261. Prawer, S.; Nemanich, R.J. Raman spectroscopy of diamond and doped diamond. *Philos. Trans. R. Soc.* **2004**, *362*, 2537–2565. [CrossRef] [PubMed]

262. Hu, C.; Sedghi, S.; Silvestre-Albero, A.; Andersson, G.G.; Sharma, A.; Pendleton, P.; Rodriguez-Reinoso, F.; Kaneko, K.; Biggs, M.J. Raman spectroscopy study of the transformation of the carbonaceous skeleton of a polymer-based nanoporous carbon along the thermal annealing pathway. *Carbon* **2015**, *85*, 147–158. [CrossRef]

263. May, P.; Lazzeri, M.; Venezuela, P.; Herziger, F.; Callsen, G.; Reparaz, J.S.; Hoffmann, A.; Mauri, F.; Maultzsch, J. Signature of the two-dimensional phonon dispersion in graphene probed by double-resonant Raman scattering. *Phys. Rev. B* **2013**, *87*, 075402. [CrossRef]

264. Herziger, F.; Tyborski, C.; Ochedowski, O.; Schleberger, M.; Maultzsch, J. Double-resonant LA phonon scattering in defective graphene and carbon nanotubes. *Phys. Rev. B* **2014**, *90*, 45431. [CrossRef]

265. Chernyak, S.A.; Ivanov, A.S.; Maslakov, K.I.; Egorov, A.V.; Shen, Z.X.; Savilov, S.S.; Lunin, V.V. Oxidation, defunctionalization and catalyst life cycle of carbon nanotubes: A Raman spectroscopy view. *Phys. Chem. Chem. Phys.* **2017**, *19*, 2276–2285. [CrossRef] [PubMed]

266. Chacon-Torres, J.C.; Wirtz, L.; Pichler, T. Raman spectroscopy of graphite intercalation compounds: Charge transfer, strain, and electron-phonon coupling in graphene layers. *Phys. Status Solidi B* **2014**, *251*, 2337–2355. [CrossRef]

267. Allen, M.J.; Tung, V.C.; Kaner, R.B. Honeycomb carbon: A review of graphene. *Chem. Rev.* **2010**, *110*, 132–145. [CrossRef] [PubMed]

268. Dresselhaus, M.S.; Dresselhaus, G. Intercalation compounds of graphite. *Adv. Phys.* **2002**, *51*, 1–186. [CrossRef]

269. Abdelkader, A.M.; Cooper, A.J.; Dryfe, R.A.W.; Kinloch, I.A. How to get between the sheets: A review of recent works on the electrochemical exfoliation of graphene materials from bulk graphite. *Nanoscale* **2015**, *7*, 6944–6956. [CrossRef] [PubMed]

270. Gupta, V.; Scharff, P.; Risch, K.; Romanus, H.; Muller, R. Synthesis of C_{60} intercalated graphite. *Solid State Commun.* **2004**, *131*, 153–155. [CrossRef]

271. Zhao, W.J.; Tan, P.H.; Liu, J.; Ferrari, A.C. Intercalation of few-layer graphite flakes with $FeCl_3$: Raman determination of fermi level, layer by layer decoupling, and stability. *J. Am. Chem. Soc.* **2011**, *133*, 5941–5946. [CrossRef] [PubMed]

272. Salvatore, M.; Carotenuto, G.; de Nicola, S.; Camerlingo, C.; Ambrogi, V.; Carfagna, C. Synthesis and characterization of highly intercalated graphite bisulfate. *Nanoscale Res. Lett.* **2017**, *12*, 167. [CrossRef] [PubMed]

273. Eklund, P.C.; Falardeau, E.R.; Fischer, J.E. Raman-scattering in low stage compounds of graphite intercalated with AsF_5, HNO_3 and $SbCl_5$. *Solid State Commun.* **1979**, *32*, 631–634. [CrossRef]

274. Chacon-Torres, J.C.; Wirtz, L.; Pichler, T. Manifestation of charged and strained graphene layers in the Raman response of graphite intercalation compounds. *ACS Nano* **2013**, *7*, 9249–9259. [CrossRef] [PubMed]

275. Solin, S.A. Raman and IR studies of graphite intercalates. *Phys. B+C* **1980**, *99*, 443–452. [CrossRef]

276. Calandra, M.; Mauri, F. Theoretical explanation of superconductivity in C_6Ca. *Phys. Rev. Lett.* **2005**, *95*, 237002. [CrossRef] [PubMed]

277. Robertson, J. Diamond-like amorphous carbon. *Mater. Sci. Eng. R Rep.* **2002**, *37*, 129–281. [CrossRef]

278. Zhang, L.; Wei, X.; Lin, Y.; Wang, F. A ternary phase diagram for amorphous carbon. *Carbon* **2015**, *94*, 202–213. [CrossRef]

279. Dillon, R.O.; Woollam, J.A.; Katkanant, V. Use of raman-scattering to investigate disorder and crystallite formation in as-deposited and annealed carbon-films. *Phys. Rev. B* **1984**, *29*, 3482–3489. [CrossRef]

280. Peter, S.; Guenther, M.; Gordan, O.; Berg, S.; Zahn, D.R.T.; Seyller, T. Experimental analysis of the thermal annealing of hard a-C:H films. *Diam. Relat. Mater.* **2014**, *45*, 43–57. [CrossRef]

281. Mangolini, F.; Rose, F.; Hilbert, J.; Carpick, R.W. Thermally induced evolution of hydrogenated amorphous carbon. *Appl. Phys. Lett.* **2013**, *103*, 161605. [CrossRef]

282. Rose, F.; Wang, N.; Smith, R.; Xiao, Q.-F.; Inaba, H.; Matsumura, T.; Saito, Y.; Matsumoto, H.; Dai, Q.; Marchon, B.; et al. Complete characterization by Raman spectroscopy of the structural properties of thin hydrogenated diamond-like carbon films exposed to rapid thermal annealing. *J. Appl. Phys.* **2014**, *116*, 123516. [CrossRef]

283. Casiraghi, C. Effect of hydrogen on the UV Raman intensities of diamond-like carbon. *Diam. Relat. Mater.* **2011**, *20*, 120–122. [CrossRef]

284. Wagner, J.; Wild, C.; Koidl, P. Resonance effects in raman-scattering from polycrystalline diamond films. *Appl. Phys. Lett.* **1991**, *59*, 779–781. [CrossRef]

285. Buijnsters, J.G.; Gago, R.; Jimenez, I.; Camero, M.; Agullo-Rueda, F.; Gomez-Aleixandre, C. Hydrogen quantification in hydrogenated amorphous carbon films by infrared, Raman, and X-ray absorption near edge spectroscopies. *J. Appl. Phys.* **2009**, *105*, 093510. [CrossRef]

286. Cui, W.G.; Lai, Q.B.; Zhang, L.; Wang, F.M. Quantitative measurements of sp^3 content in DLC films with Raman spectroscopy. *Surf. Coat. Technol.* **2010**, *205*, 1995–1999. [CrossRef]

287. Wu, A.; Cremer, D. Correlation of the vibrational spectra of isotopomers: Theory and application. *J. Phys. Chem. A* **2003**, *107*, 10272–10279. [CrossRef]

288. Mallet-Ladeira, P.; Puech, P.; Toulouse, C.; Cazayous, M.; Ratel-Ramond, N.; Weisbecker, P.; Vignoles, G.L.; Monthioux, M. A Raman study to obtain crystallite size of carbon materials: A better alternative to the Tuinstra-Koenig law. *Carbon* **2014**, *80*, 629–639. [CrossRef]

289. Martin, C.; Pegourie, B.; Ruffe, R.; Marandet, Y.; Giacometti, G.; Pardanaud, C.; Languille, P.; Panayotis, S.; Tsitrone, E.; Roubin, P. Structural analysis of eroded carbon fiber composite tiles of Tore Supra: Insights on ion transport and erosion parameters. *Phys. Scr.* **2011**, *2011*, 01024. [CrossRef]

290. Ruiz, M.P.; de Villoria, R.G.; Millera, A.; Alzueta, M.U.; Bilbao, R. Influence of the temperature on the properties of the soot formed from C_2H_2 pyrolysis. *Chem. Eng. J.* **2007**, *127*, 1–9. [CrossRef]

291. Chen, P.W.; Huang, F.L.; Yun, S.R. Optical characterization of nanocarbon phases in detonation soot and shocked graphite. *Diam. Relat. Mater.* **2006**, *15*, 1400–1404. [CrossRef]

292. Beyssac, O.; Goffe, B.; Petitet, J.P.; Froigneux, E.; Moreau, M.; Rouzaud, J.N. On the characterization of disordered and heterogeneous carbonaceous materials by Raman spectroscopy. *Spectrochim. Acta Part A Mol. Biomol. Spectrosc.* **2003**, *59*, 2267–2276. [CrossRef]

293. Farbos, B.; Weisbecker, P.; Fischer, H.E.; da Costa, J.P.; Lalanne, M.; Chollon, G.; Germain, C.; Vignoles, G.L.; Leyssale, J.M. Nanoscale structure and texture of highly anisotropic pyrocarbons revisited with transmission electron microscopy, image processing, neutron diffraction and atomistic modeling. *Carbon* **2014**, *80*, 472–489. [CrossRef]

294. Shin, Y.Y.; Lozada-Hidalgo, M.; Sambricio, J.L.; Grigorieva, I.V.; Geim, A.K.; Casiraghi, C. Raman spectroscopy of highly pressurized graphene membranes. *Appl. Phys. Lett.* **2016**, *108*, 221907. [CrossRef]

Article

Corrosion Resistance and Durability of Superhydrophobic Copper Surface in Corrosive NaCl Aqueous Solution

Chun-Wei Yao [1,*], Divine Sebastian [1], Ian Lian [2], Özge Günaydın-Şen [3], Robbie Clarke [1], Kirby Clayton [1], Chiou-Yun Chen [1], Krishna Kharel [3], Yanyu Chen [4] and Qibo Li [4]

[1]   Department of Mechanical Engineering, Lamar University, Beaumont, TX 77710, USA;
      dsebastian1@lamar.edu (D.S.); rclarke@lamar.edu (R.C.); kirbyelizabethclayton@gmail.com (K.C.);
      cchen4@lamar.edu (C.-Y.C.)
[2]   Department of Biology, Lamar University, Beaumont, TX 77710, USA; ilian@lamar.edu
[3]   Department of Chemistry and Biochemistry, Lamar University, Beaumont, TX 77710, USA;
      osen@lamar.edu (Ö.G.-Ş.); kkharel@lamar.edu (K.K.)
[4]   Transportation and Hydrogen Systems Center, National Renewable Energy Laboratory, Golden, CO 80401,
      USA; yanyu.chen@alumni.stonybrook.edu (Y.C.); li3902@tamu.edu (Q.L.)
*   Correspondence: cyao@lamar.edu; Tel.: +1-409-880-7008

Received: 20 December 2017; Accepted: 9 February 2018; Published: 11 February 2018

Abstract: Artificial superhydrophobic copper surfaces play an important role in modern applications such as self-cleaning and dropwise condensation; however, corrosion resistance and durability often present as major concerns in such applications. In this study, the anti-corrosion properties and mechanical durability of superhydrophobic copper surface have been investigated. The superhydrophobic copper surfaces were achieved with wet chemical etching and an immersion method to reduce the complexity of the fabrication process. The surface structures and materials were characterized using scanning electron microscope (SEM), energy dispersive X-ray spectroscopy (EDX), and Fourier transform infrared spectrometer (FTIR). The corrosion resistance and mechanical properties of the superhydrophobic copper surface were characterized after immersing surfaces in a 3.5 wt % NaCl solution. The chemical stability of the superhydrophobic copper surface in the NaCl solution for a short period of time was also evaluated. An abrasion test and an ultrasound oscillation were conducted to confirm that the copper surface contained durable superhydrophobic properties. In addition, an atomic force microscope was employed to study the surface mechanical property in the corrosion conditions. The present study shows that the resulting superhydrophobic copper surface exhibit enhanced corrosion resistance and durability.

Keywords: superhydrophobic copper surface; anti-corrosion; durability; AFM

---

1. Introduction

Every natural phenomenon bears special features and properties. For instance, a feature of the lotus leaf is the behavior by virtue of which its surface repels water. The action in which a surface repels water droplets is called hydrophobic behavior and has vast implication in engineering applications. A hydrophobic surface requires a suitable type of morphology, specifically roughness, on the right materials exhibiting low surface free energy [1–4]. Scientists currently employ various methods for the fabrication and study of superhydrophobic surfaces, including chemical etching [5], chemical vapor deposition [6], solution immersion method [7], sol-gel method [8], and laser fabrication [9].

Notably, superhydrophobic surfaces that possess apparent contact angles greater than 150° have been investigated for applications in areas of self-cleaning [10], anti-icing [11], electronic cooling [12,13], and dropwise condensation [14]. In such applications, copper, which has excellent conductive properties, is widely used as substrate material and turned into superhydrophobic surfaces. However, previous literature has primarily focused on thermal performance, such as heat transfer rate, in applications. To date, few methods have been presented for the fabrication of superhydrophobic copper surfaces for corrosion protection [15–18]. Durability and corrosion resistance of a superhydrophobic copper surface are still questions in implementation.

The main goal of the present study is to investigate the corrosion resistance and durability of superhydrophobic copper surfaces. In this research, superhydrophobic copper surfaces were fabricated by taking advantage of the nanostructures and the alteration of surface chemistry. The chemical stability of the superhydrophobic copper surfaces in the corrosive NaCl aqueous solution (3.5 wt %) for a short period of time were evaluated. The corrosion resistances were characterized by electrochemical methods after immersing surfaces in the NaCl solution. To test the durability of the superhydrophobic copper surfaces, mild abrasion testing and ultrasound oscillation were conducted. In addition, the degradation of the mechanical property after immersing the superhydrophobic copper surfaces in the NaCl solution was also investigated by using an atomic force microscope (AFM). The results of the morphology, mechanical property and non-wetting performance of the superhydrophobic copper surface exposed to corrosive medium highlighted the superior anti-corrosive properties of the surface.

2. Materials and Methods

C101 copper plates were used as test samples throughout the experiments performed. Sodium chlorite ($NaClO_2$) (Cole-Parmer, Vernon Hills, IL, USA), sodium hydroxide pellets (NaOH) (Cole-Parmer, Vernon Hills, IL, USA), and sodium phosphate tribasic dodecahydrate ($Na_3PO_4 \cdot 12H_2O$) (Sigma-Aldrich, St. Louis, MO, USA) were used along with deionized water to prepare an alkaline solution. Trichloro(1H,1H,2H,2H-perfluorooctyl) silane (Sigma-Aldrich, St. Louis, MO, USA) was used for the functionalization of the hierarchical copper oxide structure. Toluene (Sigma-Aldrich, St. Louis, MO, USA) was used as a solvent to dilute the solution of Trichloro(1H,1H,2H,2H-perfluorooctyl) silane. Experiments were conducted at room temperature of 23 °C.

The flat copper substrates, dimensions 5 cm by 5 cm by 0.5 mm, were initially cleaned using ultrasonication in acetone followed by rinsing with isopropyl alcohol, ethanol, and deionized water to clean off surface contaminants and grease. After the initial cleaning, the substrates were immersed in 2 M HCl to etch away the native oxide film present on the surfaces and named as pristine copper surface. The obtained pristine copper surfaces were then immersed in a hot alkaline solution composed of $NaClO_2$, NaOH, $Na_3PO_4 \cdot 12H_2O$ and deionized water in a weight proportion of 3.5:5:10:100 [19–21] maintained at 93 °C for a duration of 10 min. During this hot immersion, it was observed that the color of copper substrates gradually darkened to form a greyish texture on the surfaces. The substrates were then functionalized using Trichloro(1H,1H,2H,2H-perfluorooctyl) silane in Toluene solution with 1 mM concentration for a duration of 25 min [22]. After the functionalization process, the processed substrates were oven dried at 70 °C for 1 h and designated as a superhydrophobic copper surface. For preparing pristine copper substrates coated with Trichlorosilane, designated the reference surface, the processes remained the same except for immersing the substrates in the alkaline solution.

The morphologies and chemical compositions of the superhydrophobic copper surfaces were assessed using a scanning electron microscope (SEM, Hitachi S-3400N, Hitachi High Technologies, Tokyo, Japan) equipped with energy-dispersive X-ray spectroscopy, an atomic force microscope (AFM, Park NX10, Park System Co., Suwon, Korea), and a Fourier-transform infrared spectrometer (Nicolet iS50 FT-IR Spectrometer, Thermo Fisher Scientific, Waltham, MA, USA). The water contact angles were measured using a contact angle goniometer (ramé-hart instruments co., Succasunna, NJ, USA). Water droplets (5 µL) were carefully dropped onto the copper surfaces under ambient temperature and atmosphere. Electrochemical measurements were performed in a 3.5 wt % aqueous solution of NaCl at room temperature using a potentiostat system (Autolab PGSTAT204, Metrohm, Riverview, FL, USA). A stainless steel electrode, the superhydrophobic copper surface, and a silver/silver chloride (Ag/AgCl) reference electrode were used as the counter electrode, working electrode, and reference electrode, respectively. The exposed area of the working electrode was 16.9 cm^2 for the calculation of corrosion rate. The current density was based primarily on the exposed geometrical area. The potentiodynamic anodic polarization curves were recorded at a sweep rate of 1 mV/s from -250 to 250 mV versus the open circuit potential.

3. Results and Discussion

3.1. Surface Morphologies and Chemical Compositions

Figure 1 shows the SEM image of the superhydrophobic copper surfaces and pristine copper surfaces. Micro-scratches/grooves can be seen on the pristine copper surfaces. At nano-scale, the superhydrophobic copper surfaces appear as leaf-like structures. These sharp and dense structures contribute to the high roughness associated with the substrates after processing. The energy-dispersive X-ray spectroscope (EDX) was used to verify the chemical composition of the surface, as Figure 2 shows. EDX data indicated that the surface includes Cu and O, suggesting CuO leaf-like nanostructures on the surface. Figure 3 shows the three-dimensional AFM image of the superhydrophobic copper surface. The average surface roughness of the surfaces was found to be 428 ± 12 nm, a relatively high number in terms of meeting the roughness requirements for a highly water repellent nature. The FTIR spectrum was recorded in the range from 400 to 4000 cm^{-1} with 320 scans and 2 cm^{-1} resolution to characterize chemical nature of the superhydrophobic surface. The few micrometers-thick top layer on the fabricated superhydrophobic surface was scraped off using a razor blade. The sample in powdered form was mixed with KCl to form isotropic pellets with 1% loading in mid IR region. Figure 4 shows the FTIR spectrum obtained for the superhydrophobic copper surface. A broad peak around 3411 cm^{-1} along with the peak around 1606 cm^{-1} accounts for OH stretching vibration [23–26]. A sharp peak at 614 cm^{-1} is accounted by the presence of halogens such as chlorine and fluorine [23,26]. The Si-O-Si stretching is evident at 1091 cm^{-1} [23,25], whereas Si–O rocking is characterized around 419 cm^{-1} [27]. Overlapping of PO_4^{3-} with SiO_4^{4-} can be observed at 504 cm^{-1} [28]. The weak band around 953 cm^{-1} is due to the presence of Si–O–Cu vibration [25] which suggests that the coating film is covalently attached to the copper substrate. The bands at wavenumbers 419 cm^{-1}, 504 cm^{-1}, 953 cm^{-1} and 1091 cm^{-1} indicate the covalent bond between silicon and oxygen that may have formed during the functionalization process. The small peak near 1317 cm^{-1} is assigned to the CF$_2$ functional group [29,30]. The peak around 1395 cm^{-1} is due to the presence of the aryl group. The presence of the C–H bond accounts for the peak at 2921 cm^{-1} [24,31]. In terms of the water contact angle, the measured average angle was 83° for pristine copper surface, 110° for the copper surface coated with Trichlorosilane (reference surface), and 169° for the superhydrophobic copper surface.

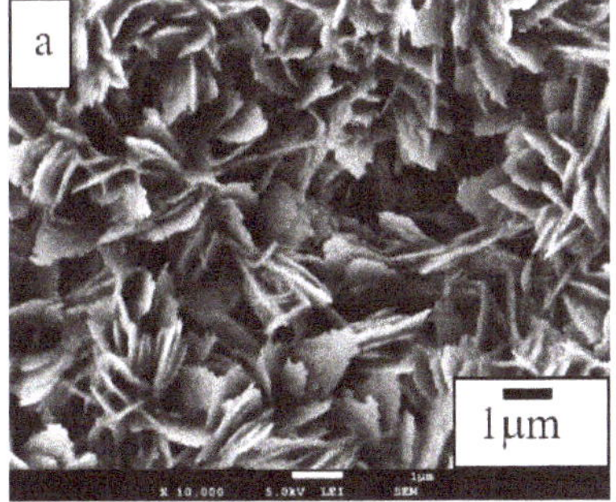

Figure 1. *Cont.*

Figure 1. SEM images: (**a**) the superhydrophobic copper surface; (**b**) the superhydrophobic copper surface at higher magnification; (**c**) the pristine copper surface.

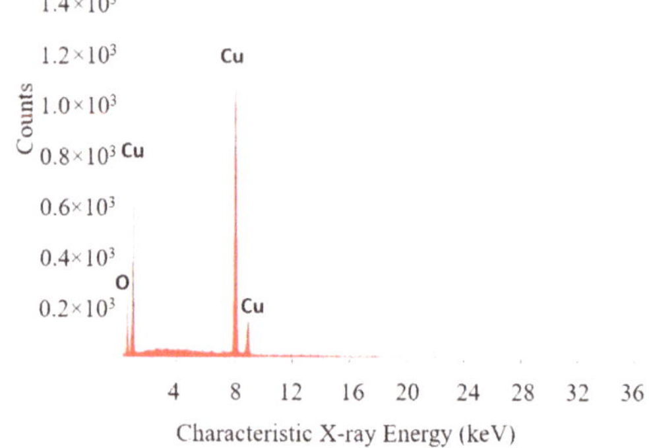

Figure 2. EDX spectrum analysis of the superhydrophobic copper surface.

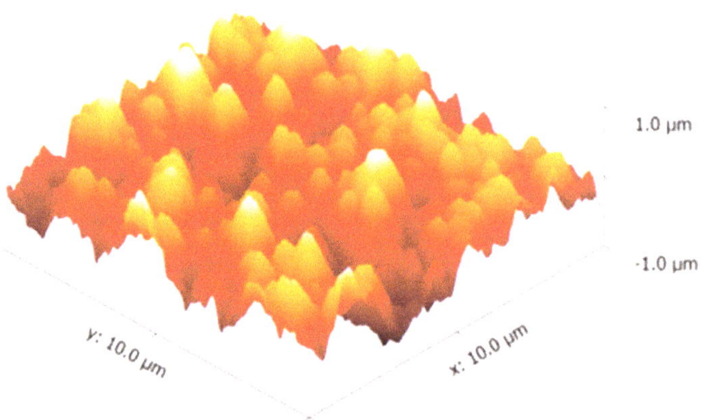

Figure 3. 3-Dimensional AFM image of the superhydrophobic copper surface.

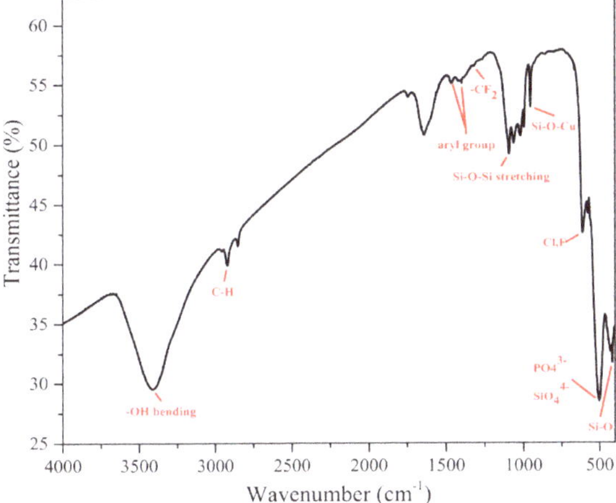

Figure 4. FTIR spectrum of the superhydrophobic copper surface.

3.2. Chemical Stability and Corrosion Resistance

Figure 5 shows the variation in the water contact angles of the superhydrophobic copper surfaces that were immersed in the 3.5 wt % NaCl aqueous solution. The static contact angles did not deviate considerably with respect to an increase in immersion time. Even after 24 h of immersion time, the static contact angle remained around 169°, which indicates that the surfaces maintain superhydrophobicity.

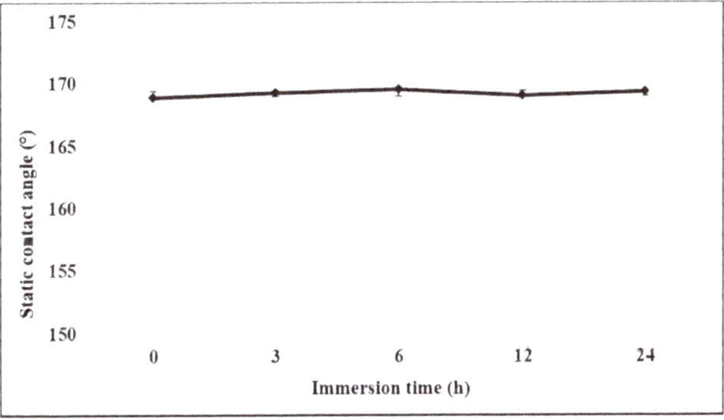

Figure 5. Effect of immersion time on the static water contact angle of the superhydrophobic copper surface.

Corrosion rate analysis was performed on the superhydrophobic copper surface using the potentiodynamic polarization method. All samples were immersed in a 3.5 wt % NaCl solution for one hour before each measurement was taken. The potentiodynamic polarization curve of the superhydrophobic copper surface was compared to the curves of pristine copper surface and copper surface coated with Trichlorosilane (reference surface), as shown in the Figure 6. Table 1 summarizes the corrosion potential (E_{corr}) and corrosion current density (j_{corr}). The corrosion potential for the pristine copper surface was less than -219 mV; whereas, the corrosion potential for the superhydrophobic copper surface exceeded 105 mV. Moreover, the corrosion current density of the superhydrophobic surface was less than 1% that of the pristine copper surface. The corrosion rate can be calculated from

$$R = EW \times J \times K/\rho \tag{1}$$

where R is the corrosion rate (mm/year), EW is the equivalent weight of metal, J represents corrosion current density, K is a constant equal to 3272 [mm/(A·cm·year)]. ρ represents the density of the metal [32]. The exposed geometrical area of the surfaces was used to calculate the current density. The superhydrophobic surface exhibited a very low corrosion rate, which is a reduction of two orders of magnitude from that of the pristine copper surface or the copper surface coated with Trichlorosilane (reference surface), as shown in Table 1. This high-degree reduction in corrosion behavior can be accounted for by the interaction between the electrolyte solution and the surface, due to its liquid-repelling behavior. According to the Cassie-Baxter model, since air can be trapped within the leaf-like nanostructures, the liquid solution has difficulty penetrating the surface structures. Therefore, the trapped air can also inhibit a corrosive solution's ability to penetrate into surface structures [33]. Based on all the results discussed above, it can be concluded that the effect of the trapped air in the nanostructures is more significant than the effect of the silane layer in accounting for the superior corrosion resistance. Therefore the superhydrophobic copper surface possesses favorable corrosion resistance for the Cu substrate.

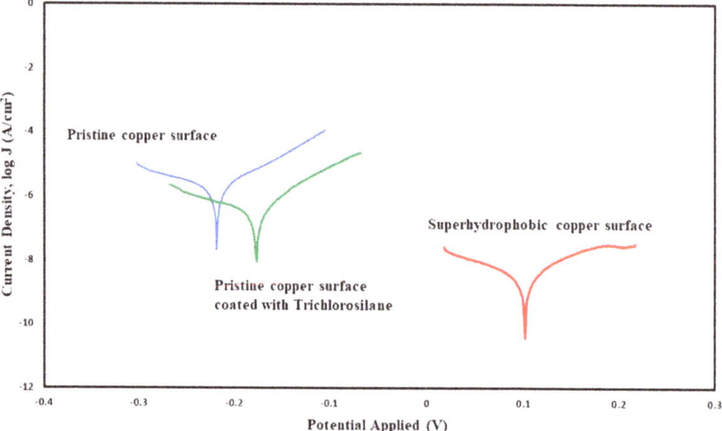

Figure 6. Potentiodynamic polarization curve obtained by using Ag/AgCl electrode as the reference electrode and 3.5 wt % NaCl solution as the electrolyte for pristine copper surface, copper surface coated with Trichlorosilane (reference surface), and superhydrophobic copper surface.

Table 1. Corrosion Potential (E_{corr}), Corrosion Current Density (J_{corr}), and Corrosion Rate of the pristine copper surface, copper surface coated with Trichlorosilane (reference surface), and superhydrophobic copper surface.

Sample	E_{corr} (V)	J_{corr} (A/cm²)	Corrosion Rate (mm/year)
Pristine copper surface	$-2.20 \times 10^{-1} \pm 1.00 \times 10^{-3}$	$2.12 \times 10^{-6} \pm 4.02 \times 10^{-7}$	$4.89 \times 10^{-2} \pm 9.63 \times 10^{-3}$
Copper surface coated with Trichlorosilane	$-1.7 \times 10^{-1} \pm 8.91 \times 10^{-3}$	$6.6 \times 10^{-7} \pm 5.83 \times 10^{-7}$	$1.53 \times 10^{-2} \pm 1.35 \times 10^{-2}$
Superhydrophobic copper	$10.6 \times 10^{-2} \pm 3.61 \times 10^{-2}$	$4.62 \times 10^{-9} \pm 2.99 \times 10^{-9}$	$1.07 \times 10^{-4} \pm 6.95 \times 10^{-5}$

3.3. Mechanical Stability and Property

The mechanical stability and wear resistance of the superhydrophobic coating is significant in view of its usage in long-term applications. This work first used a simple method to quantify the wear resistance of the coating. The superhydrophobic surface was dragged against a 5000 grit sand paper with a 7.75 kPa normal pressure. The specimen, with exerted pressure, was moved in 40 cm cycles, rotating the specimen by 90° at the halfway point of each cycle to manifest even abrasions on the surface. After a total abrasion length of 800 cm was reached, the contact angle was measured as 138°, which indicated that the contact angle of the surface decreased as the distances of abrasion increased due to the disruption of the surface morphology. However, the surface maintained its hydrophobicity fairly well in the given distance range, as shown in Figure 7, due to the persistent nanostructures with covalently attached silane coating. The mechanical stability of the superhydrophobic surface was also characterized through ultrasonication testing in water and in a 3.5 wt % NaCl solution for 5 min. After ultrasonication in water or the 3.5 wt % NaCl solution, the surface retained its high contact angles and its superhydrophobicity as Figure 8 shows. The reduction in contact angle in case of using the NaCl solution is more than that in case of using water because of the effect of corrosive behavior of saline medium during ultrasonication. Furthermore, the influence of the 3.5 wt % NaCl solution on the mechanical property of superhydrophobic surfaces was characterized by AFM. Table 2 shows AFM images for the superhydrophobic surface, the surface after corrosion testing (after immersion in the 3.5 wt % NaCl solution for one hour), and the surface after immersion in the 3.5 wt % NaCl solution for 24 h. From three-dimensional images and roughness values (Table 2), it can be observed that the nanoscale structures on the surfaces fairly retained their sharp profile. The process of induced corrosion did not affect surface stiffness values. In other words, the surface remained firm under varying conditions. These results indicate that the superhydrophobic surface possesses favorable mechanical stability and property.

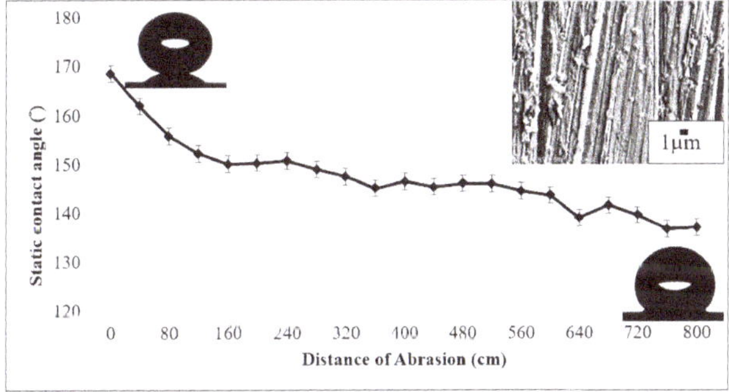

Figure 7. Water contact angles as a result of the abrasion against a 5000 grit sand paper with 7.75 kPa normal load. The inset SEM image (upper right) shows surface morphology after reaching 800 cm of abrasion length.

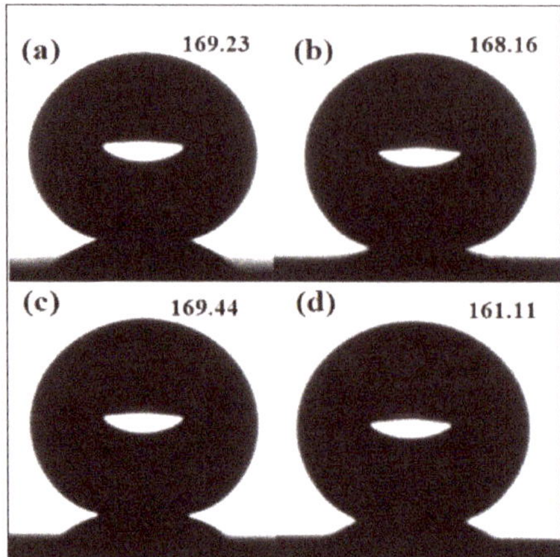

Figure 8. Static contact angle: (**a**) before ultrasonication and (**b**) after ultrasonication in water; Static contact angle (**c**) before ultrasonication and (**d**) after ultrasonication in 3.5 wt % NaCl solution.

Table 2. AFM image, roughness, and stiffness of the superhydrophobic copper surfaces.

Property	Surface		
	Superhydrophobic Copper Surface	Superhydrophoic Copper Surface after Corrosion Test	Superhydrophobic Copper Surface after Immersion in 3.5 wt % NaCl for 24 h
3-Dimensional image			
Roughness R_q (nm)	333.86 ± 5	322.54 ± 16	326.82 ± 5
Stiffness value (N/m)	4.25 ± 0.05	4.24 ± 0.26	4.23 ± 0.20

4. Conclusions

In this work, a stable and durable superhydrophobic copper surface was prepared and investigated. The superhydrophobicity of the surface was the result of the unique nanostructures on microstructures and the assembly of low surface energy components, generating a typical contact angle was around 169°. Short term immersion measurements and potentiodynamic polarization measurements revealed that the superhydrophobic copper surface exhibited an excellent corrosion resistance for the Cu substrate. Specifically, the surface retained its morphologies and surface stiffness after immersion in the 3.5 wt % NaCl solution as revealed by AFM characterization; proven to be a highly lasting treatment method. Due to its corrosion resistance and highly durable features, the superhydrophobic surface demonstrated in the current study may be presented as an effective solution for protecting copper substrates in various engineering applications.

Acknowledgments: This work was supported by Research Enhancement Grants, Lamar University. This work was also supported by grants from Center for Advances in Port Management (CAPM) of Lamar University and Center for Advances in Water and Air Quality (CAWAQ) of Lamar University to Chun-Wei Yao and Ian Lian. Özge Günaydın-Şen and Krishna Kharel were supported by Welch Foundation (V-0004).

Author Contributions: Chun-Wei Yao designed and supervised the study; Divine Sebastian, Robbie Clarke, Kirby Clayton, Chiou-Yun Chen, and Krishna Kharel performed the experiments; Divine Sebastian,

Krishna Kharel, Özge Günaydın-Şen, and Chun-Wei Yao analyzed the data; Ian Lian, Yanyu Chen and Qibo Li contributed to the research guidance; Divine Sebastian and Chun-Wei Yao wrote the paper.

Conflicts of Interest: The authors declare no conflict of interest.

References

1. She, Z.; Li, Q.; Wang, Z.; Li, L.; Chen, F.; Zhou, J. Novel method for controllable fabrication of a superhydrophobic CuO surface on AZ91D magnesium alloy. *ACS Appl. Mater. Interface* **2012**, *4*, 4348–4356. [CrossRef] [PubMed]

2. Zhu, X.; Zhang, Z.; Xu, X.; Men, X.; Yang, J.; Zhou, X.; Xue, Q. Facile fabrication of a superamphiphobic surface on the copper substrate. *J. Colloid Interface Sci.* **2012**, *367*, 443–449. [CrossRef] [PubMed]

3. Chen, Z.; Hao, L.; Chen, A.; Song, Q.; Chen, C. A rapid one-step process for fabrication of superhydrophobic surface by electrodeposition method. *Electrochim. Acta* **2012**, *59*, 168–171. [CrossRef]

4. Yao, C.-W.; Alvarado, J.L.; Marsh, C.P.; Jones, B.G.; Collins, M.K. Wetting behavior on hybrid surfaces with hydrophobic and hydrophilic properties. *Appl. Surf. Sci.* **2014**, *290*, 59–65. [CrossRef]

5. Qian, B.; Shen, Z. Fabrication of superhydrophobic surfaces by dislocation-selective chemical etching on aluminum, copper, and zinc substrates. *Langmuir* **2005**, *21*, 9007–9009. [CrossRef] [PubMed]

6. Yu, J.; Qin, L.; Hao, Y.; Kuang, S.; Bai, X.; Chong, Y.-M.; Zhang, W.; Wang, E. Vertically aligned boron nitride nanosheets: Chemical vapor synthesis, ultraviolet light emission, and superhydrophobicity. *ACS Nano* **2010**, *4*, 414–422. [CrossRef] [PubMed]

7. Chaudhary, A.; Barshilia, H.C. Nanometric multiscale rough CuO/Cu(OH)$_2$ superhydrophobic surfaces prepared by a facile one-step solution-immersion process: Transition to superhydrophilicity with oxygen plasma treatment. *J. Phys. Chem. C* **2011**, *115*, 18213–18220. [CrossRef]

8. Rao, A.V.; Latthe, S.S.; Mahadik, S.A.; Kappenstein, C. Mechanically stable and corrosion resistant superhydrophobic sol–gel coatings on copper substrate. *Appl. Surf. Sci.* **2011**, *257*, 5772–5776. [CrossRef]

9. Dong, C.; Gu, Y.; Zhong, M.; Li, L.; Sezer, K.; Ma, M.; Liu, W. Fabrication of superhydrophobic Cu surfaces with tunable regular micro and random nano-scale structures by hybrid laser texture and chemical etching. *J. Mater. Proc. Technol.* **2011**, *211*, 1234–1240. [CrossRef]

10. Wang, Z.; Li, Q.; She, Z.; Chen, F.; Li, L. Low-cost and large-scale fabrication method for an environmentally-friendly superhydrophobic coating on magnesium alloy. *J. Mater. Chem.* **2012**, *22*, 4097–4105. [CrossRef]

11. Jung, S.; Dorrestijn, M.; Raps, D.; Das, A.; Megaridis, C.M.; Poulikakos, D. Are superhydrophobic surfaces best for icephobicity? *Langmuir* **2011**, *27*, 3059–3066. [CrossRef] [PubMed]

12. Oh, J.; Birbarah, P.; Foulkes, T.; Yin, S.L.; Rentauskas, M.; Neely, J.; Pilawa-Podgurski, R.C.N.; Miljkovic, N. Jumping-droplet electronics hot-spot cooling. *Appl. Phys. Lett.* **2017**, *110*, 123107. [CrossRef]

13. Wiedenheft, K.F.; Guo, H.A.; Qu, X.; Boreyko, J.B.; Liu, F.; Zhang, K.; Eid, F.; Choudhury, A.; Li, Z.; Chen, C.-H Hotspot cooling with jumping drop vapor chambers. *Appl. Phys. Lett.* **2017**, *110*, 141601. [CrossRef]

14. Miljkovic, N.; Enright, R.; Nam, Y.; Lopez, K.; Dou, N.; Sack, J.; Wang, E.N. Jumping-droplet-enhanced condensation on scalable superhydrophobic nanostructured surfaces. *Nano Lett.* **2013**, *13*, 179–187. [CrossRef] [PubMed]

15. Liu, T.; Chen, S.; Cheng, S.; Tian, J.; Chang, X.; Yin, Y. Corrosion behavior of super-hydrophobic surface on copper in seawater. *Electrochim. Acta* **2007**, *52*, 8003–8007. [CrossRef]

16. Huang, Y.; Sarkar, D.K.; Gallant, D.; Chen, X.-G. Corrosion resistance properties of superhydrophobic copper surfaces fabricated by one-step electrochemical modification process. *Appl. Surf. Sci.* **2013**, *282*, 689–694. [CrossRef]

17. Wang, P.; Zhang, D.; Qiu, R.; Wan, Y.; Wu, J. Green approach to fabrication of a super-hydrophobic film on copper and the consequent corrosion resistance. *Corros. Sci.* **2014**, *80*, 366–373. [CrossRef]

18. Liu, Y.; Li, S.; Zhang, J.; Liu, J.; Han, Z.; Ren, L. Corrosion inhibition of biomimetic super-hydrophobic electrodeposition coating on copper substrate. *Corros. Sci.* **2015**, *94*, 190–196. [CrossRef]

19. Enright, R.; Miljkovic, N.; Dou, N.; Nam, Y.; Wang, E.N. Condensation on superhydrophobic copper oxide nanostructures. *J. Heat Transf.* **2013**, *135*, 091304. [CrossRef]

20. Nam, Y.; Ju, Y.S. A comparative study of the morphology and wetting characteristics of micro/nanostructured Cu surfaces for phase change heat transfer applications. *J. Adhes. Sci. Technol.* **2013**, *27*, 2163–2176. [CrossRef]

21. Chavan, S.; Cha, H.; Orejon, D.; Nawaz, K.; Singla, N.; Yeung, Y.F.; Park, D.; Kang, D.H.; Chang, Y.; Takata, Y.; et al. Heat transfer through a condensate droplet on hydrophobic and nanostructured superhydrophobic surfaces. *Langmuir* **2016**, *32*, 7774–7787. [CrossRef] [PubMed]

22. Fadeev, A.Y.; McCarthy, T.J. Self-Assembly is not the only reaction possible between alkyltrichlorosilanes and surfaces: Monomolecular and oligomeric covalently attached layers of dichloro- and trichloroalkylsilanes on silicon. *Langmuir* **2000**, *16*, 7268–7274. [CrossRef]

23. Chen, H.-H.; Anbarasan, R.; Kuo, L.-S.; Hsu, C.-C.; Chen, P.-H.; Chiang, K.-F. Fabrication of hierarchical structured superhydrophobic copper surface by in-situ method with micro/nano scaled particles. *Mater. Lett.* **2012**, *66*, 299–301. [CrossRef]

24. Yang, H.; Pi, P.; Cai, Z.-Q.; Wen, X.; Wang, X.; Cheng, J.; Yang, Z.-R. Facile preparation of super-hydrophobic and super-oleophilic silica film on stainless steel mesh via sol–gel process. *Appl. Surf. Sci.* **2010**, *256*, 4095–4102. [CrossRef]

25. Xu, W.; Liu, H.; Lu, S.; Xi, J.; Wang, Y. Fabrication of superhydrophobic surfaces with hierarchical structure through a solution-immersion process on copper and galvanized iron substrates. *Langmuir* **2008**, *24*, 10895–10900. [CrossRef] [PubMed]

26. Fransolet, A.M.; Tarte, P. Infrared spectra of analyzed samples of the amblygonite-montrebasite series: A new rapid semi-quantitative determination of fluorine. *Am. Mineral.* **1977**, *62*, 559–564.

27. Salh, R. Defect related luminescence in silicon dioxide network: A review. In *Crystalline Silicon-Properties and Uses*; Basu, S., Ed.; InTech: Rijeka, Croatia, 2011.

28. Radev, L.; Mostafa, N.Y.; Michailova, I.; Salvado, I.M.M.; Fernandes, M.H.V. In vitro bioactivity of collagen/calcium phosphate silicate composites, cross-linked with chondroitin sulfate. *Int. J. Mater. Chem.* **2012**, *2*, 1–9. [CrossRef]

29. Nimittrakoolchai, O.U.; Supothina, S. Preparation of stable ultrahydrophobic and superoleophobic silica-based coating. *J. Nanosci. Nanotechnol.* **2012**, *12*, 4962–4968. [CrossRef] [PubMed]

30. Saleema, N.; Sarkar, D.K.; Gallant, D.; Paynter, R.W.; Chen, X.G. Chemical nature of superhydrophobic aluminum alloy surfaces produced via a one-step process using fluoroalkyl-silane in a base medium. *ACS Appl. Mater. Interface* **2011**, *3*, 4775–4781. [CrossRef] [PubMed]

31. Chen, H.-H.; Anbarasan, R.; Kuo, L.-S.; Tsai, M.-Y.; Chen, P.-H.; Chiang, K.-F. Synthesis, characterizations and hydrophobicity of micro/nano scaled heptadecafluorononanoic acid decorated copper nanoparticle. *Nano Micro Lett.* **2010**, *2*, 101–105. [CrossRef]

32. Gomez-Vidal, J.C.; Morton, E. Castable cements to prevent corrosion of metals in molten salts. *Sol. Energy Mater. Sol. Cells* **2016**, *153*, 44–51. [CrossRef]

33. Su, F.; Yao, K. Facile fabrication of superhydrophobic surface with excellent mechanical abrasion and corrosion resistance on copper substrate by a novel method. *ACS Appl. Mater. Interface* **2014**, *6*, 8762–8770. [CrossRef] [PubMed]

 *coatings*

Review

On the Durability and Wear Resistance of Transparent Superhydrophobic Coatings

Ilker S. Bayer

Smart Materials, Istituto Italiano di Tecnologia, Via Morego 30, 16163 Genoa, Italy; ilker.bayer@iit.it;
Tel.: +39-380-387-6699

Academic Editor: Mariateresa Lettieri
Received: 24 November 2016; Accepted: 9 January 2017; Published: 18 January 2017

Abstract: Transparent liquid repellent coatings with exceptional wear and abrasion resistance are very demanding to fabricate. The most important reason for this is the fact that majority of the transparent liquid repellent coatings have so far been fabricated by nanoparticle assembly on surfaces in the form of films. These films or coatings demonstrate relatively poor substrate adhesion and rubbing induced wear resistance compared to polymer-based transparent hydrophobic coatings. However, recent advances reported in the literature indicate that considerable progress has now been made towards formulating and applying transparent, hydrophobic and even oleophobic coatings onto various substrates which can withstand certain degree of mechanical abrasion. This is considered to be very promising for anti-graffiti coatings or treatments since they require resistance to wear abrasion. Therefore, this review intends to highlight the state-of-the-art on materials and techniques that are used to fabricate wear resistant liquid repellent transparent coatings so that researchers can assess various aptitudes and limitations related to translating some of these technologies to large scale stain repellent outdoor applications.

Keywords: transparent superhydrophobic; oleophobic; wear abrasion; hydrophobic nanoparticles; anti-graffiti coatings

---

1. Introduction

One of the most challenging aspects of preparing non-wettable coatings is to maintain a good degree of transparency while retaining robust and scratch resistant self cleaning and stain free characteristics [1–3]. Transparency and surface roughness are generally contradictory properties. Hydrophobicity of low surface energy coatings increases with surface roughness but coating transparency often decreases because of Mie scattering from the rough surface. When the roughness dimension is much smaller than the light wavelength, the film/coating becomes increasingly transparent due to refractive index change between air and the coating, which effectively reduces the intensity of refraction at the air (or water)/film interface and increases the optical quality. In other words, it is necessary to control the roughness below 100 nm to effectively lower the intensity of Mie scattering while maintaining non-wettable characteristics [4].

A forthright large-scale application of transparent non-wetting coatings is the prevention of unwanted markings in public areas or on public transportation vehicles known as graffiti. Graffiti prevention (anti-graffiti) treatments can be in the form of (a) transparent and self-adherent polymer films that can be peeled off and replaced from time to time; (b) in the form of polymeric paints (non-sacrificial) that are able to repel stains and graffiti tagging; or (c) they can be stain and dust repellent coatings obtained from ceramic precursors particularly suitable for special applications such as solar panels [5,6]. It is also important to take into account how the graffiti is applied. Most of the commercially available anti-graffiti paints are siloxane/silicone-based formulations and they can repel a majority of water-based paints and markers. However, if the graffiti is applied from a solvent or oil based paint, silicone chemistry may not protect against graffiti tagging due to solid surface energy and liquid surface tension match between the siloxane polymers and oil-based paint vehicle [7,8]. In other words, the coating should practically be oleophobic or superomniphobic in order to repel oil-based paints. Although beyond the scope of this review, another important aspect to consider is the type of the surface on which permanent marks (unwanted staining) are induced. In other words, a window like smooth transparent surface or highly porous concrete, brick, limestone, slate, wood and masonry walls. For instance, due to the high porosity of such surfaces, the graffiti is absorbed into the texture to a substantial degree, thereby making it difficult to remove or clean compared to smooth non-porous surfaces.

Mechanical robustness of non-wettable coatings signifies their resistance to wear as a result of rubbing induced abrasion [9,10]. In general, there are two approaches to creating a durable nonwetting surface: (a) limiting material removal so as to retain superhydrophobicity under wear for as long as possible and (b) developing a material that maintains superhydrophobicity as it wears away. For the latter type, such performance for surfaces under a single wear condition is known as "wear similarity". A simple example of wear similarity is sanded Teflon (polytetrafluoroethylene, PTFE) which can be rendered superhydrophobic by using fine grit sandpaper so that continued sanding would retain superhydrophobicity until the Teflon material is completely worn away. In almost all the cases reported in the literature, the most durable non-wettable surfaces or coatings that can withstand the abovementioned approaches are polymer-based nanocomposites [11,12]. However, recent progress indicates that transparent non-wettable coatings can also be produced with reasonable robustness, which can eventually resist continuous harsh abrasion conditions by one of the two abovementioned mechanisms.

In this review, article, we will distinctly analyze state-of-the-art on various transparent non-wetting coatings and treatments except for the sacrificial transparent films. We will review fabrication aspects of transparent nanoparticle films and coatings and ways to render them resilient against abrasion induced wear. Next, we will present and discuss transparent, flexible and non-wettable materials obtained by various mold transfer processes including nanoimprint lithography which utilize hydrophobic transparent elastomer precursors or others with UV-induced polymerization. Finally, we will present fabrication aspects and characteristics of ceramic precursor based transparent non-wettable coatings that are deemed suitable for windows or solar energy conversion unit surfaces. In each category, we will address inherent limitations, potentials for immediate outdoor applicability and future directions in order to render such transparent non-wetting coatings more resilient against wear abrasion.

2. Wetting Theories, Surface Roughness and Robust Metastable Superhydrophobicity

The most customary superhydrophobic surface roughness model is that of lotus leaf. On its surface both microscale and nanoscale protruding structures are found. It has been understood that in fact many plant leaves do not possess low-energy surface compounds that are commonplace in the fabrication of synthetic non-wettable materials. The most hydrophobic constituent is the epicuticular wax, and the apparent contact angle of water on waxes is only slightly above 90°, which is certainly not sufficient to explain extreme non-wettablity on certain plant surfaces. Herminghaus [13] also showed that it is possible to maintain non-wettability when instead of a protruding micro/nano texture, closed pore-like micro/nano texture is present even though hydrophobic chemistry is associated with the wetting of wax surface.

In its most simplistic form, wetting and surface roughness are correlated with two theories known as Wenzel and Cassie-Baxter [14]. As described by Marmur [14], wetting on rough surfaces may assume either of two regimes: homogeneous wetting (corresponding to Wenzel theory of hydrophobicity), where the liquid completely penetrates the roughness grooves, or heterogeneous wetting (corresponding to Cassie-Baxter theory of superhydrophobicity), where air (or another fluid) is trapped underneath the liquid inside the roughness grooves. In a way, durability or "robustness" of a superhydrophobic surface relies on how long it can resist the transition between these regimes [15].

In other words, metastable states (minimum liquid-solid contact) in which minimum droplet-surface contact points are maintained, have to be long lived. Evolution of non-wetting surfaces in nature also followed this approach as shown by calculations [14]. This strategy avoids the need for high steepness protrusions that may be more prone to erosion, wear and breakage. In addition, due also to the specific shape of the Lotus leaf protrusions, this strategy lowers the sensitivity of the superhydrophobic state to the protrusion distance.

As such, the complex relationship between surface roughness and durable superhydrophobicity can be summarized as follows:

- Microscale roughness features should resist damage under abrasion induced wear [10];
- Deeply embedded nanoscale features (away from the exposed microscale protrusions) can ensure longer superhydrophobic resistance against wear abrasion [16];
- A self-similar hierarchical texture throughout the non-wettable surface, film or coating bulk should be constructed in order to resist material removal due to wear abrasion [9];
- From tribology standpoint, energy or stress dissipating ceramic nanoparticles or organics should be built-in to the coatings such as rubbery domains [17].

Coating surface mechanical durability requires resistance to smoothing out in two roughness scales [16]. Once transparency is also introduced into the equation, resistance to durability becomes more problematic since transparent surface texture roughness should not enhance Mie and Rayleigh scattering [18]. Mie scattering occurs when the diameter of the surface features are close to the wavelength of the incident photon, and Rayleigh scattering occurs when the size of the surface features are much smaller than the wavelength of the incident photon. Note that herein we define "contact angle" as the apparent contact angle that is stable over time on a surface. In other words, the apparent contact angle on a superhydrophobic surface can still be greater than 150° but the droplet would not roll off or slide over the coating when tilted. On such surfaces, the real contact angle is established at submicron scales along the contact line, known as *sticky* superhydrophobicity.

3. Wear Damage Resistance Characterization Methods

Since recent reviews have sufficiently covered measurement and characterization of wear durability of non-wettable surfaces [10], it would be abundant to repeat them herein. However, some clarifications are needed in order to discuss "robustness" of non-wettable surfaces properly. Mechanical durability of non-wettable coatings is a broad terminology. It is used if the non-wettable surface is for instance stretch or bending resistant but also for abrasion induced wear. Some non-wettable coatings that were not properly tested against abrasion induced wear have been classified as durable if they showed good substrate adhesion [19]. Therefore, it is imperative to explicitly indicate the type of "robustness" the coating demonstrates or is tested against. For instance, there are many experimental tests that a non-wettable surface should pass in order to be classified as "robust" or mechanically durable that can be acceptable from an industrial application point of view. For instance, peel tests for substrate adhesion (including water immersion related lift off), underwater ultrasonic processing, both sand and rain (spray) based erosion damage, bending and flexibility, high pressure hydrostatic liquid build-up, scratch resistance and abrasion induced wear.

Of course, at the end of all these tests, the ideal Cassie-Baxter non-wetting state should prevail. In other words, droplets should freely roll off (not slide) the surfaces. In certain cases, due to wear abrasion, the non-wettable surface can transform into a superhydrophobic state but with low droplet mobility (high substrate tilt angles for droplet motion), generally referred to as rose petal effect [20,21], even though the hierarchical surface texture is preserved. This can still be considered as resistance to abrasion induced wear if the surface of the coating can be transformed back into a non-stick superhydrophobic state by a simple post treatment such as thermal annealing or a thin silane coating etc. [22].

In this review, we will focus on coating resistance against abrasion induced wear on transparent non-wettable coatings. It is believed that wear abrasion is the most relevant durability parameter that can be convincing for potential industrial applications as well as for outdoor anti-graffiti installations. Wear abrasion of transparent non-wettable coatings has not been adequetly addressed so far and more R&D efforts are needed in order to allow commercialization.

4. Transparent Non-Wettable Coatings from Nanoparticle Assembly

Rahmawan et al. [23] reviewed and discussed different self-assembly methods, including sol–gel processes, micro-phase separation, templating, and nanoparticle assembly, to create transparent, superhydrophobic surfaces. The review of literature indicates that one of the most commonly used nanoparticles in the fabrication of non-wettable coatings is SiO_2 nanoparticles [24]. They are extensively used in super-repellent polymer nanocomposite coatings but also in the fabrication of transparent non-wettable coatings. Irzh et al. [25] described a novel one-step or two short-step method for the production of superhydrophobic SiO_2 layers using microwave plasma. The one-step reaction was applied for the production of a transparent superhydrophobic layer that was not very UV-stable or a superhydrophobic UV-stable layer that was not very transparent. The two-step method but short process was applied for the production of a superhydrophobic transparent and UV-stable SiO_2 layer. Only a few seconds were required to produce these layers via a plasma polymerization mechanism. The authors used TEOS (silane-SiO_2 precursor) for roughness, decane (hydrocarbon precursor), perfluorodecaline ($C_{10}F_{18}$) or perfluorononane (C_9F_{20}) (fluorocarbon precursors). The microwave plasma reactor was loaded with the silica precursor and any one of the organic hydrophobic precursors to form a rough SiO_2 nanostructured surface functionalized with waxy or fluorinated macromolecules. Although the authors used many different substrates for deposition they did not report any performance related to mechanical abrasion resistance of the layers.

Bravo et al. [26] utilized the layer-by-layer assembly method to control the placement and level of aggregation of differently sized SiO_2 nanoparticles within the resultant multilayer thin film. They fully exploited the advantages of layer-by-layer processing to optimize the level of roughness needed to obtain superhydrophobicity and a low contact angle hysteresis and at the same time minimize light scattering (Figure 1a). Besides high transmittance (Figure 1b) and superhydrophobicity, very low levels of reflectivity was also achieved at specified visible wavelengths.

They created transparent superhydrophobic films by the sequential adsorption of silica nanoparticles and poly(allylamine hydrochloride). The final assembly was rendered superhydrophobic with silane treatment. Optical transmission levels above 90% throughout most of the visible region of the spectrum were realized in optimized coatings. Advancing water droplet contact angles as high as 160° with low contact angle hysteresis (<10°) were obtained for the optimized multilayer thin films. Because of the low refractive index of the resultant porous multilayer films, they also exhibited antireflection properties.

Wong and Yu [27] reported a simple protocol to prepare superhydrophobic and hydrophobic glass by treating standard microscope slides with methyltrichlorosilane and octadecyltrichlorosilane, respectively as shown in Figure 2a. Octadecyltrichlorosilane formed a closely packed, methyl-terminated, self-assembled monolayer that changed the glass surface from hydrophilic to hydrophobic (Figure 2b). Treatment with methyltrichlorosilane resulted in 3-dimensional polymethylsiloxane networked nanostructures, which produced a superhydrophobic surface. In both cases, the glass slides maintained optical transparency despite remarkable changes in the surface wettability (Figure 2a).

Tuvshindorj et al. [3] investigated the effect of different levels of surface topography on the stability of the Cassie state of wetting (droplets freely roll off the surfaces) in large-area and transparent superhydrophobic coatings. Three different organically modified silica (ormosil) coatings, (i) nanoporous hydrophobic coating (NC); (ii) microporous superhydrophobic coating (MC); and (iii) double-layer superhydrophobic coating with nanoporous bottom and microporous top layers (MNC), were prepared on glass surfaces (Figure 3a). The stability of the Cassie state of coatings against the external pressure was examined by applying compression/relaxation cycles to water droplets sitting on the surfaces (Figure 3b).

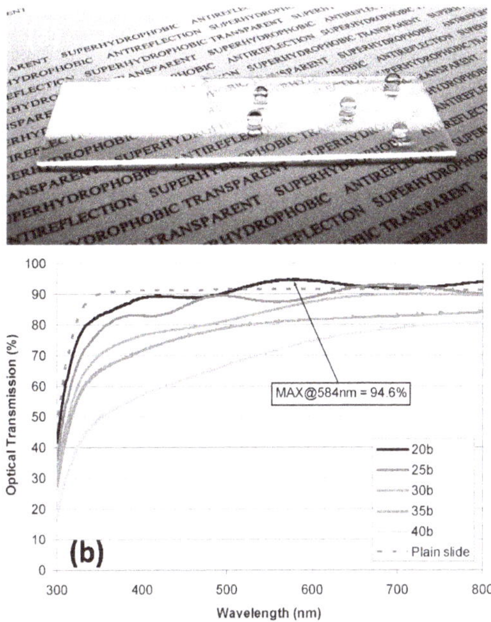

Figure 1. (**a**) Photograph of a glass slide coated with a transparent, superhydrophobic multilayer with antireflection properties: [PAH (7.5)/(50 + 20 nm) SiO_2 (9.0)] 20 + top-layers film. The right side of the glass slide is coated with the multilayers; (**b**) Transmittance of multilayer films with 20 to 40 bilayers of [PAH (7.5)/(50 + 20 nm) SiO_2 (9.0)] × and top layers. The transmittance of a plain glass slide (substrate for the films) is also plotted [26].

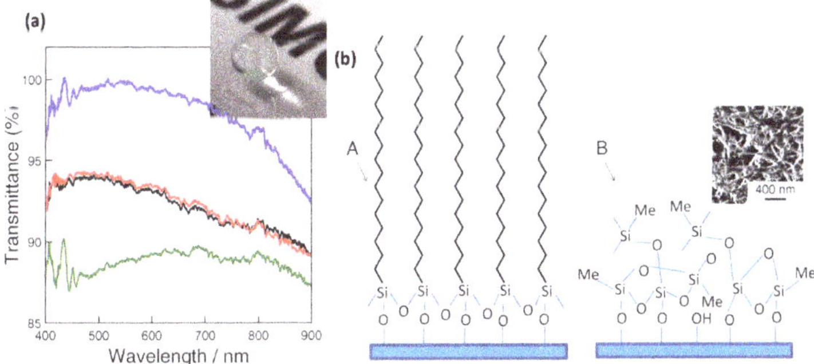

Figure 2. (**a**) UV-Vis spectra showing the transmittance of glass slides: untreated (**black trace**); hydrophobic (**red line**); superhydrophobic prepared at room temperature (**green line**); superhydrophobic glass prepared at 4 °C (**blue line**). The inset image shows a water droplet on a transparent superhydrophobic slide; (**b**) Schematic representation of a normal hydrophilic glass surface after treatment with (A) octadecyltrichlorosilane that results in the formation of a highly ordered self-assembled monolayer (hydrophobic) and (B) methyltrichlorosilane that results in the formation of a 3D polymethylsiloxane network (superhydrophobic). The inset shows a SEM image of the superhydrophobic surface [27].

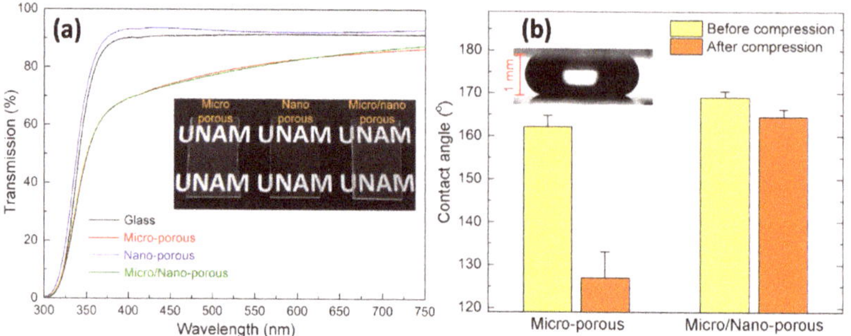

Figure 3. (**a**) Transmission spectra of ormosil coatings and an uncoated glass substrate. All of the coatings are highly transparent at the visible wavelengths. The inset shows a photograph of transparent coatings; (**b**) Droplet squeezing between two identical surfaces. Contact angles of surfaces before and after compression. The inset shows the squeezed water droplet between two MNC surfaces. There is only a slight decrease in the contact angle of the MNC after compression. The large decrease in the MC shows loss of the superhydrophobic property [3].

The changes of the apparent contact angle, contact-angle hysteresis, and sliding-angle values of the surfaces before and after compression cycles were studied to determine the Cassie/Wenzel transition behavior of the surfaces. In addition, water droplets were allowed to evaporate from the surfaces under ambient conditions, and the changes in the contact angles and contact-line diameters with increasing Laplace pressure were analyzed. They showed that, upon combination of coatings with different levels of topography it was possible to fabricate transparent superhydrophobic surfaces with extremely stable Cassie-Baxter states of wetting on glass surfaces. The stability of the Cassie state (droplets freely roll away) of the coatings against the external pressure was investigated. It was observed that a droplet sitting on a single-layer coating can easily transform to a sticky Wenzel state (hydrophobic but droplets can no longer roll away or slide over the surface) from a roll-off Cassie state under external pressures as low as approximately 80 Pa. On the other hand, Cassie to Wenzel state transition was observed at around 1600 Pa for a double-layer micro/nanoporous coating, which is almost 4-fold higher than the transition pressure for a Lotus leaf (slightly above 400 Pa). The extreme stability of the Cassie-Baxter state in a micro/nanoporous coating was attributed to its double-layer porous structure. With increasing external pressure, the contact line of a droplet in the Cassie-Baxter state gradually slipped down from the walls of a microporous top layer, and after a critical pressure, the contact line touched the nanoporous bottom layer. After removal of the pressure, the contact line partially recovered in the sense that dewetting of the nanoporous bottom layer took place. Although not indicated by the authors explicitly, this recovery was possibly an intermediate wetting state [28]. The authors did not report on the resilience of such coatings against abrasion or other similar means of wear effect.

Xu et al. [29] studied the non-wettable properties and transparency of coatings made from the assembly of fluorosilane modified silica nanoparticles (F-SiO$_2$ NPs) via one-step spin-coating and dip-coating without any surface post-passivation steps as shown in Figure 4. When spin-coating the hydrophobic NPs (100 nm in diameter) at a concentration $\geq$0.8 wt.% in a fluorinated solvent, the surface exhibited superhydrophobicity with an advancing water contact angle greater than 150° and a water droplet (5 µL) roll-off angle less than 5°. In comparison, superhydrophobicity was not achieved by dip-coating the same hydrophobic NPs. Scanning electron microscopy (SEM) and atomic force microscopy (AFM) images revealed that NPs formed a nearly close-packed assembly in the superhydrophobic films, which effectively minimized the exposure of the underlying substrate while offering sufficiently trapped air pockets (Figure 4c,d). In the dip-coated films, however, the surface coverage was rather random and incomplete (Figure 4a,b).

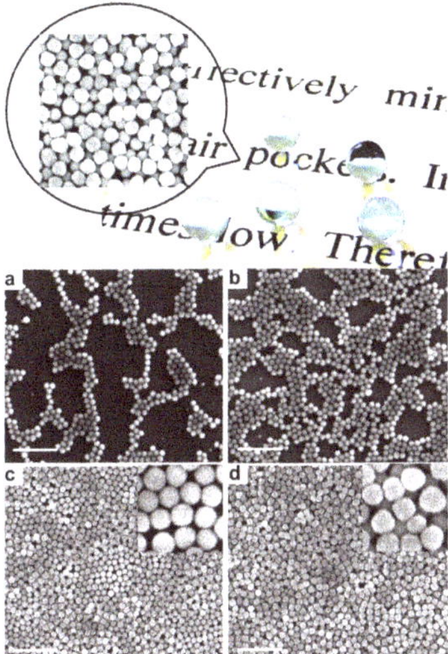

Figure 4. Top Panel: Photograph of the transparent F-SiO$_2$ NP (fluorinated silica nanoparticle film) on a glass slide. SEM images of spin-coated 100 nm F-SiO$_2$ NPs with different concentrations on TESPSA-functionalized Si wafers: (**a**) 0.1, (**b**) 0.4, (**c**) 0.8, and (**d**) 1.2 wt.%. The insets in c and d are high-magnification images. Scale bars: 1 μm [29].

Therefore, the underlying substrate was exposed and water was able to impregnate between the NPs, leading to smaller water contact angle and larger water contact angle hysteresis. The spin-coated superhydrophobic film was also highly transparent with greater than 95% transmittance in the visible region. They demonstrated that the one-step coating strategy could be extended to different polymeric substrates, including poly(methyl methacrylate) and polyester fabrics, to achieve superhydrophobicity [29].

Xu and He [30] described a simple and effective method to fabricate transparent superhydrophobic coatings using 3-aminopropytriethoxysilane (APTS)-modified hollow silica nanoparticle sols. The sols were dip-coated on slide glasses, followed by thermal annealing and chemical vapor deposition with 1H,1H,2H,2H-perfluorooctyltrimethoxysilane (POTS). The largest water contact angle (WCA) of coating reached as high as 156° with a sliding angle (SA) of ≤2° and a maximum transmittance of 83.7%. The highest transmittance of coated slide glass reached as high as 92% with a WCA of 146° and an SA of ≤6°. A coating simultaneously showing both good transparency (90.2%) and superhydrophobicity (WCA: 150°, SA: 4°) was achieved through regulating the concentration of APTS and the withdrawing speed of dip-coating. Scanning electron microscopy (SEM), transmission electron microscopy (TEM), and atomic force microscopy (AFM) were used to observe the morphology and structure of nanoparticles and coating surfaces. Optical properties were characterized by a UV-visible spectrophotometer. Surface wettability was studied by a contact angle/interface system. The effects of APTS concentration and the withdrawing speed of dip-coating were also discussed on the basis of experimental observations. They did not perform any abrasion induced wear-resistance experiments on the transparent non-wettable coatings.

Zhu et al. [31] produced a transparent superamphiphobic coating using carbon nanotubes (CNTs) as a template as shown in Figure 5a–f. The optical transmittance of the resulting coating was greater than 80% throughout a broad spectrum of ultraviolet and visible wavelengths. Meanwhile, water and numbers of extremely low surface tension liquids, such as dodecane (25.3 mN/m), did not wet the coatings and could roll off easily (Figure 5g). They did not report any abrasion induced wear-resistance properties of the coatings but the superamphiphobic property of the obtained transparent coating was kept even after thermal treatment at 400 °C. Separate experiments demonstrated that CNTs-directed surface texture enhanced the oleophobicity significantly by promoting high CA and low CAH with liquids tested, when compared to the pure SiO_2- and carbon black-directed surface texture.

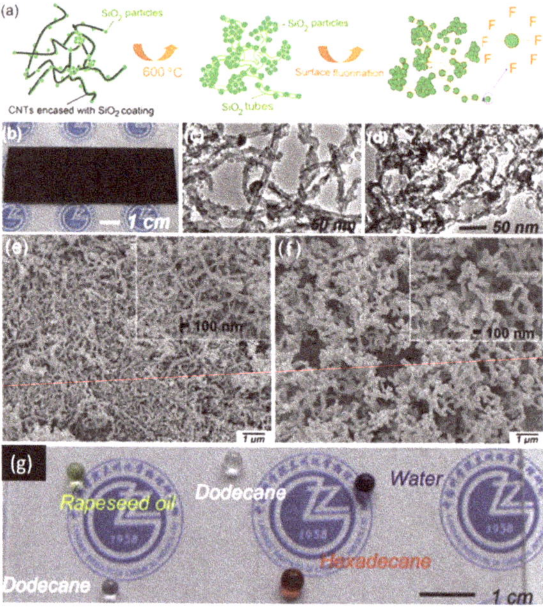

Figure 5. (**a**) Schematic diagram illustrating the fabrication procedure of the transparent superamphiphobic coating; (**b**) digital image of the CNTs–SiO$_2$ coating sprayed onto glass slide; TEM images of the sprayed CNTS–SiO$_2$ coating before (**c**) and after (**d**) thermal treatment; FESEM images of the sprayed CNTS–SiO$_2$ coating before (**e**) and after (**f**) thermal treatment; (**g**) transparent state of the coating after thermal treatment to burn away the CNTs. Various liquid droplets are also visible in their non-wetting state [31].

Ling et al. [32] produced a superhydrophobic surface with a static water contact angle WCA > 150° using a simple dip-coating method of 60-nm SiO$_2$ nanoparticles onto an amine-terminated (NH$_2$) self-assembled monolayer (SAM) glass/silicon oxide substrate, followed by chemical vapor deposition of a fluorinated adsorbate (Figure 6). For comparison, they also fabricated a close-packed nanoparticle film, formed by convective assembly, which gave WCA~120°. The mechanical stability (adhesion to substrate) of the superhydrophobic coating was enhanced by sintering of the nanoparticles in an O$_2$ environment at high temperature (1100 °C). A sliding angle of <5° indicated the self-cleaning properties of the surface. The dip-coating method was applied to glass substrates to prepare surfaces that are superhydrophobic and transparent. They indicated that for potential commercial application of a self-cleaning superhydrophobic surface, the mechanical integrity of the nanoparticle film must be sufficient to withstand wear. However, they tested their surfaces using a standard tape peel test rather than a real wear abrasion test. Adhesive tape with a pressure-sensitive adhesive (PSA) was applied on the surface and removed from the surface prior to gas phase deposition of 1*H*,1*H*,2*H*,2*H*-perfluorodecyltriethoxysilane (PFTS) because PFTS prevents the adhesion of scotch tape on the surface. The nanoparticle films before and after the peel tests were examined by SEM (Figure 6A,B).

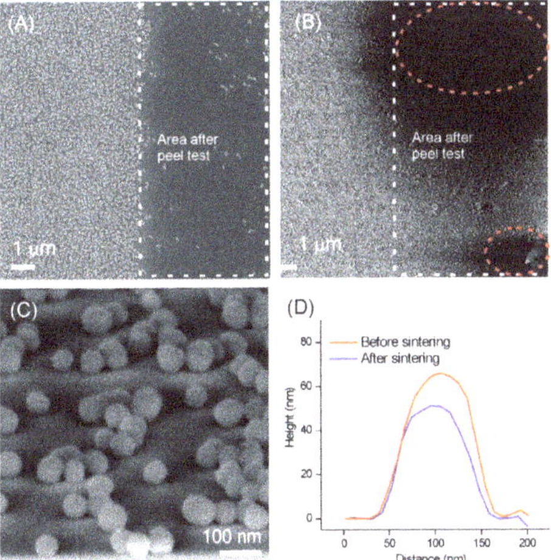

Figure 6. SEM images of the peel test on dip-coated nanoparticle films before (**A**) and after a sintering process (**B**); white boxes indicate the areas after peel tests, while red-dot ellipsoids show the area where scotch tape adhesives remained on the substrate after the test; (**C**) Titled SEM image of the nanoparticles after sintering, and (**D**) averaged height profiles of the nanoparticles before and after sintering (as measured by AFM) [32].

Figure 6A shows such nanoparticle film after the peel test. In the area highlighted by the white square, where the peel test was applied, almost all the nanoparticles were removed from the surface, indicating a rather poor adhesion of the nanoparticles on the NH_2-SAM. In order to improve the stability and to obtain a stable superhydrophobic surface, the substrates were subjected to a sintering process at high temperature. Sintering was performed under O_2 at 1100 °C for 2 h. Figure 6B shows the SEM image of the as-sintered substrate after the peel test. The areas with and without peel test showed an equally dense coverage of nanoparticles, indicating a good stability of the nanoparticle layer. This is attributed to the chemical and thermal bonding of nanoparticle onto the SiO_2 substrates. Some tape adhesives were seen on top of the nanoparticles (red ellipsoids in Figure 6B), which refers to a partial adhesive failure between the PSA and the sintered and hydrophilic surface. Lowering the sintering time to 30 min and the sintering temperature to 900 or 1000 °C was possible without apparent loss of adhesion improvement. After deposition of PFTS, the static water contact angle was 150°, indicating the formation of a stable superhydrophobic surface on the sintered substrate. After sintering at 1100 °C for 30 min, the nanoparticles were only slightly melted onto the substrates (Figure 6C). No obvious deformation was observed. The height profiles of 20 nanoparticles were averaged on samples before and after sintering (Figure 6D), showing that the aspect ratios (height/fwhm) of the nanoparticles before and after sintering were 1.1 and 0.9, respectively.

Ebert and Bhushan [33] performed a systematic study in which transparent superhydrophobic surfaces were created on glass, polycarbonate, and poly(methyl methacrylate) (PMMA) substrates using surface-functionalized SiO_2, ZnO, and indium tin oxide (ITO) nanoparticles. The contact angle, contact angle hysteresis, and optical transmittance were measured for samples using all particle-substrate combinations. To examine wear resistance, multiscale wear experiments were performed using an atomic force microscope (AFM) and a water jet apparatus. Dip coating was used as the means of fabrication. They used commercial and silane-modified SiO_2 nanoparticles as well as ZnO and ITO particles which were not surface-modified and used as received. To hydrophobize them, they were treated in solution by octadecylphosphonic acid (ODP) and the prepared samples did not require post-treatment with low surface energy substances.

Figure 7 shows the AFM induced wear results on coatings deposited on glass. Changes in roughness due to wear are given as roughness profile above each AFM topography panel. The nanoparticles showed different tendencies in the way they deposited onto substrates from dip coating, which was attributed to differences in primary particle size. This caused variation in coating thickness and morphology between particles, which was reflected as differences in wettability and transmittance between samples. ITO samples had slightly lower WCA and slightly higher CAH than SiO_2, ZnO, which is likely the result of a comparatively lower liquid-air fractional area. Roughness and coating thickness influenced transmittance more than inherent optical properties of particles, which were attributed to the proximity of roughness and thickness values to the 100 nm threshold for visible transparency. Samples on PMMA substrates performed modestly better than those on PC and glass in terms of wettability and transmittance. However, all samples exhibited a superhydrophobic CA (>150°), low CAH (<10°), and high transmittance of visible light (>90% in most cases). In addition, all surfaces showed wear resistance for potential commercial use in AFM wear and water jet experiments indicating strong bonding of the silicone resin and sufficient hardness of nanoparticles and resin. They concluded that transparent superhydrophobic surfaces with wear resistance can be fabricated with a broad range of materials to expand potential engineering applications. However, primary particle size, roughness, and coating morphology appear to be at least as important a factor in transparency as inherent optical properties of nanoparticles when coating thickness is on the order of 100 nm.

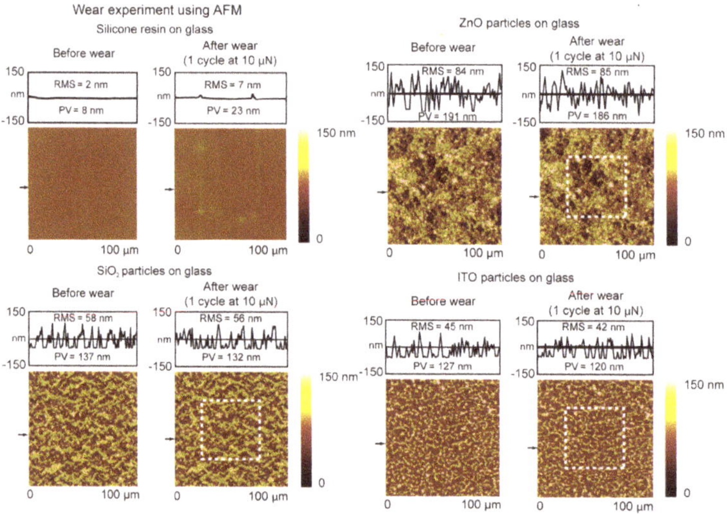

Figure 7. Surface height maps and sample surface profiles (locations indicated by arrows) before and after AFM wear experiment using 15 µm radius borosilicate ball at load of 10 µN for glass samples with silicone resin alone, SiO_2 nanoparticles, ZnO nanoparticles, and ITO nanoparticles. rms roughness and PV distance values for surface profiles are displayed within surface profile boxes. Results shown are typical for all substrates [33].

Deng et al. [34] presented a simple method to fabricate a superhydrophobic coating based on porous silica capsules (Figure 8). The superhydrophobic coating showed a static contact angle of 160° and sliding angle less than 5°. Moreover, it was thermally stable up to 350 °C. On the other hand, it also acted as anti-fouling layer for solar cells. The superhydrophobic coating did not diminish the efficiency of organic solar cells, also due to its excellent transparency. Further, the coating retained its superhydrophobicity under adhesion tape peeling and sand abrasion tests, which are demonstrated in Figure 8a,b (tape peel) and Figure 8c (sand abrasion).

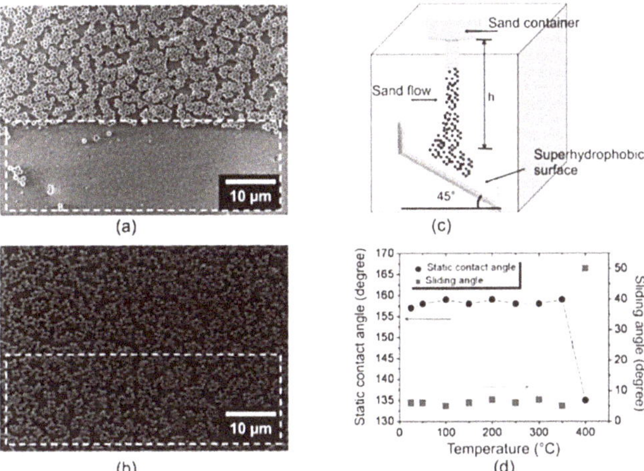

Figure 8. SEM images of superhydrophobic surfaces that were partially exposed to double sided tape (white boxes indicate the exposed areas). (**a**) If the particles stick to the surface by van der Waals interaction only, they can easily be removed; (**b**) Particles cannot be removed by double sided tape after binding them chemically to the surface by silica bridges; (**c**) Sketch of the setup used to determine the stability of the surface against sand impact; (**d**) Static contact angle and sliding angle measured after annealing the samples for 10 h at different temperatures. The surface remains superhydrophobic until annealing at 350 °C [34].

To improve the mechanical stability of the nanoparticle films, chemical vapor deposition (CVD) of tetraethoxysilane in the presence of ammonia was performed. To quantify the mechanical stability of the surfaces two separate tests were performed. Firstly, double sided adhesive tape is pressed with approximately 10 kPa to the surfaces, both, before and after performing CVD of tetraethoxysilane. If the capsules are attached to the surface by van der Waals interactions only a sharp boundary would be visible, separating areas that are and are not exposed to tape. After peeling the tape off, the area underneath it is almost particle free, substantiating the poor adhesion of the particles to the substrate (white box in Figure 8a). Contrary, if CVD of TES is performed beforehand peeling the tape off does not change the particle coverage (Figure 8b). In a second test, sand gains are impacted on a superhydrophobic surface and the minimal height is determined at which the porous silica particles burst. Bursting led to an increase of the sliding angle and finally to loss of superhydrophobicity. When 100 to 300 μm sized sand grains were impacted on a superhydrophobic surface the shells remained intact for impact heights up to 30 cm. After the sand abrasion the surface remained superhydrophobic, i.e., water droplets placed on the surface can bounce and slide off easily.

Inspired by mussels, Si et al. [35] designed a novel green superhydrophobic nanocoating with good transparency and stability through a facile reaction at room temperature with trimethyl silyl modified process. The nano-coating was coated from ethanol onto various substrates via a simple spray process without requiring toxic substances. The application also rendered the coatings self-healing. The nanocoating demonstrated rapid self-healing superhydrophobicity induced by exposure to organic solvents which can be applied at industrial levels (see Figure 9). At first, silica sol was prepared via a typical tetraethoxysilane (TEOS) hydrolysis reaction in an alkaline environment. After dopamine (DOPA) was added into the silica sol system forming an opaque DOPA–silica gel due to the presence of the –OH groups after ageing. The DOPA–silica gel was reacted with 1,1,1,3,3,3-hexamethyl disilazane (HMDS) to convert the rest of the –OH groups on the surface of the gel to $-OSi(CH_3)_3$. In order to obtain a transparent superhydrophobic nanocoating, the DOPA–silica gel was dried at 60 °C and ground to get an hydrophobic powder Using environmentally friendly nontoxic ethanol as a solvent, a transparent superhydrophobic DSTM gel nanocoating with varying mass ratios was prepared which was sprayed on various engineering material substrates without any pre-treatment regardless of the composition and morphology. No abrasion related wear tests were conducted on the prepared coatings, the self-healing properties were demonstrated against damage by solvent soaking only.

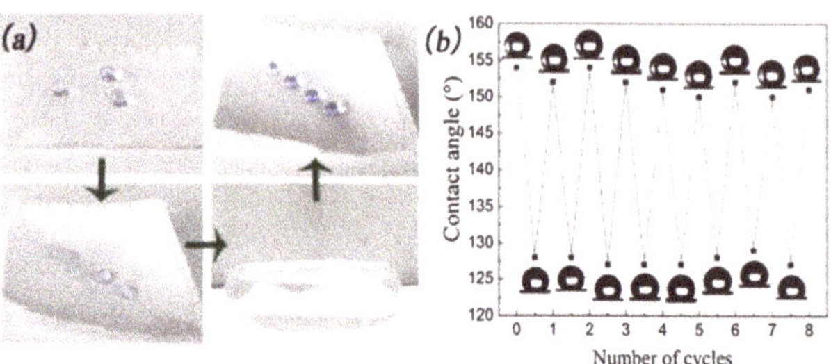

Figure 9. (**a**) Demonstration of self-healing superhydrophobicity of the coated cotton fabric induced by acetone; (**b**) WCAs on the coated copper wire mesh in the eight cycles of 1 M HCl treatment [35].

Deng et al. [36] designed an easily fabricated, transparent, and oil-rebounding superamphiphobic coating (Figure 10B) using porous deposit of candle soot that was coated with a 25-nanometer-thick silica shell. The black soot-colored coating became transparent after calcination at 600 °C. After silanization, the coating was superamphiphobic and remained so even after its top layer was damaged by sand impingement (Figure 10C). The coating consisted of a fractal-like assembly of nanospheres (Figure 10D). With increasing duration of CVD of TES or annealing above 1100 °C, the necks between particles were filled with silica and more rod-like shapes evolved, which reduced the superamphiphobicity. This was attributed to the notion that convex small-scale roughness can provide a sufficient energy barrier against wetting thus rendering superamphiphobicity possible. The coating was sufficiently oil-repellent to cause the rebound of impacting drops of hexadecane. Even low-surface-tension drops of tetradecane rolled off easily when the surface was tilted by 5°, taking impurities along with them. The surface kept its superamphiphobicity after being annealed at 400 °C.

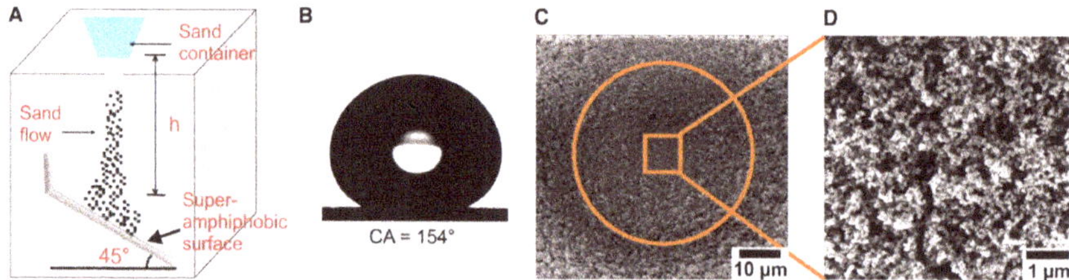

Figure 10. Mechanical resistance quantified by sand abrasion. (**A**) Schematic drawing of a sand abrasion experiment; (**B**) Hexadecane drop deposited on the coating after 20 g of sand abrasion from 40 cm height. The 100- to 300-µm-sized grains had a velocity of 11 km/hour just before impingement. After impingement, the drops rolled off after the substrate was tilted by 5°; (**C**) SEM image of a spherical crater (orange circle) after sand abrasion; (**D**) SEM image of the surface topography inside the cavity [36].

Yokoi et al. [37] produced a superhydrophobic polyester mesh (Figure 11a) possessing both mechanical stability and high transparency in the visible light range by coating 1*H*,1*H*,2*H*,2*H*-perfluorodecyltrichlorosilane (PFDTS)-treated fibers, after chemical etching, with SiO$_2$ nanoparticles functionalized with 1*H*,1*H*,2*H*,2*H*-perfluorooctyltriethoxysilane (PFOTS). It was found that the fabricated mesh maintained its superhydrophobicity and low water sliding angle because of the PFDTS surface treatment, although the SiO_2 nanoparticles modified with PFOTS are removed by the abrasion (Figure 11b). The vacant space between the mesh fibers allowed the penetration of visible light and protected the nanoroughness provided by the SiO_2 nanoparticles. They claimed that it was unnecessary to control the refractive index of the materials for improving transparency and to contain strong chemical or physical bonding between the particles or the particles and the substrate for improving the abrasion resistance. Therefore, compared with the traditional technology, the combination of the see-through mesh and the SiO_2 nanoparticle hierarchical structure appears to be an effective and simple method for improving the abrasion resistance and transparency of these superhydrophobic films (Figure 11c,d).

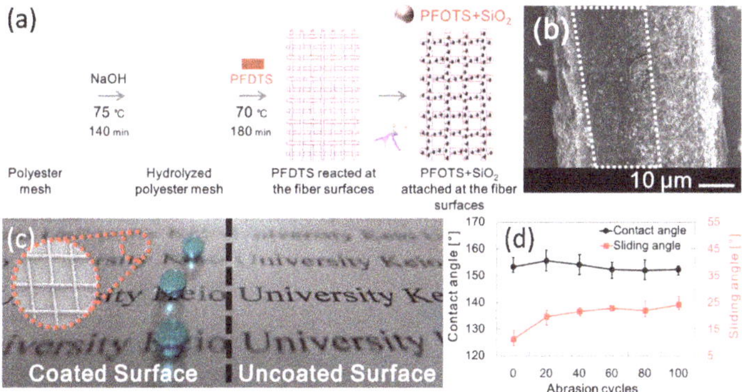

Figure 11. (**a**) Schematic of the fabrication procedure for the superhydrophobic polyester mesh. First, a polyester mesh undergoes an alkaline treatment with NaOH. Second, PFDTS is reacted on the fiber surfaces using the chemical vapor deposition method. Third, the mesh is treated with SiO_2 nanoparticles modified with PFOTS using a spray method; (**b**) FE-SEM images of the superhydrophobic polyester meshes after 100 cycles of abrasion with a pressure of ~10 kPa. Dashed zone is shows loss of nanoparticle from the fiber surface; (**c**) Photograph of blue-colored water on the coated polyester mesh surface (left) and a photograph of an uncoated surface on white paper for comparison; (**d**) Contact angle and sliding angle measurements of the superhydrophobic meshes using aqueous solutions with a pH range of 2–14 [37].

5. Transparent Non-Wettable Coatings from Pattern Transfer to Polymers

Gong et al. [38] recently reported a convenient and efficient duplicating method, being capable to form a transparent PDMS surface with superhydrophobicity in mass production. They claimed that the fabrication process had extensive application potentials. They used a femtosecond laser processing to fabricate mirror finished stainless steel templates (Figure 12a) with surface structures combining microgrooves with microholes array. Then liquid PDMS was poured for the duplicating process to introduce a particular structure composed of a microwalls array with a certain distance between each other and a microprotrusion positioned at the center of a plate surrounded by microwalls (Figure 12b,c). The parameters such as the side length of microwalls and the height of a microcone were optimized to achieve required superhydrophobicity at the same time as high-transparency properties (Figure 12d,e). The PDMS surfaces showed superhydrophobicity with a static contact angle of up to $154.5 \pm 1.7°$ and sliding angle lower to $6 \pm 0.5°$, also with a transparency over 91%, a loss less than 1% compared with flat PDMS by the measured light wavelength in the visible light scale (Figure 12f). They tested wear resistance by sandpaper over 100 cycles of abrasion. The superhydrophobic PDMS surfaces were also tested for thermal stability up to 325 °C.

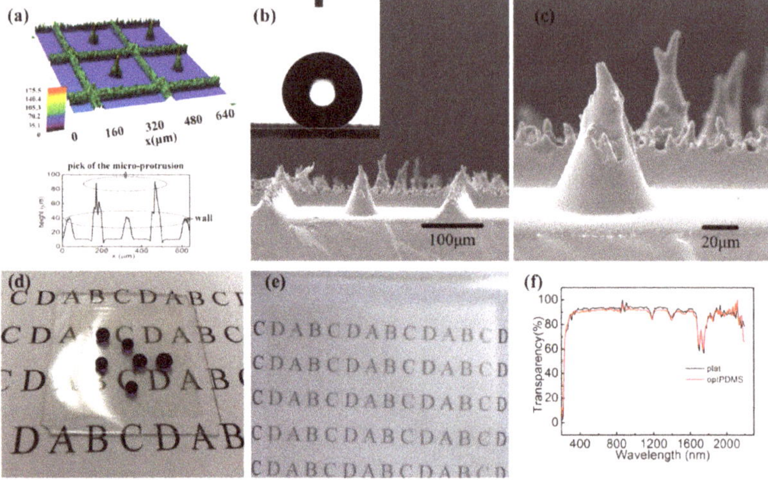

Figure 12. Surface microstructure of PDMS as a result of molding into a stainless steel substrate (**a**) laser microscope and SEM (**b**) 500× and (**c**) 2000×. The SCA of this optimal surface is 154.5° ± 1.7° and SA 6° ± 0.5°; (**d**) Sample of optimally designed PDMS coated on the paper; and (**e**) it was held up to keep a certain distance (10 cm) from the paper; (**f**) Transparency of the optimal designed surface is over 91% in the visible light wavelength [38].

Im et al. [39] demonstrated a robust superhydrophobic and superomniphobic surface based on transparent polymer microstructuring (Figure 13). On a large-size template of the transparent polydimethylsiloxane (PDMS) elastomer surface, perfectly ordered microstructures with an inverse-trapezoidal cross section were fabricated with two consecutive PDMS replication processes and a three-dimensional diffuser lithography technique. Figure 13 demonstrates a schematic representation of these surface patterns. The hydrophobicity and transparency were improved by additional coating of a fluoropolymer layer. The robustness of superhydrophobicity was confirmed by the water droplet impinging test. Additionally, the fabricated superhydrophobic surface was also superomniphobic, which shows a high contact angle with a low surface tension (γ_{la}) liquid, such as methanol (γ_{la} = 22.7 mN·m^{-1}).

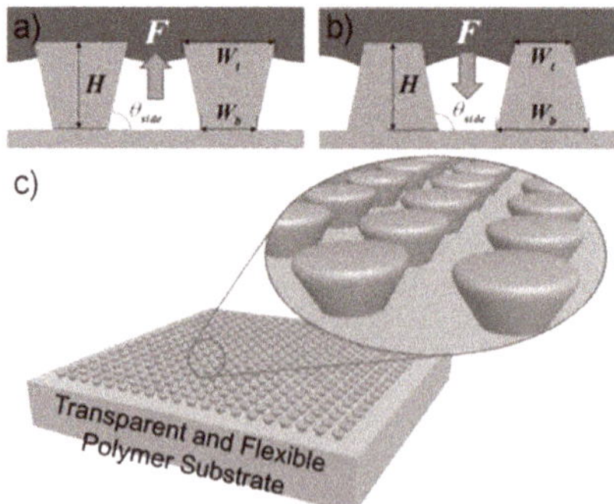

Figure 13. Water meniscus and the net force on: (**a**) overhang structures ($W_t > W_b$); (**b**) structures like truncated pyramids with inclined sidewalls ($W_t < W_b$); (**c**) Schematic illustration of the PDMS trapezoids surface [39].

The authors did not present any durability or mechanical wear or abrasion results but the intrinsic properties of PDMS rubber should dictate a certain degree of resilience against rubbing induced wear or abrasion.

Kim et al. [40] demonstrated highly transparent super-hydrophobic surface fabrication approach using nanoimprint lithography with a flexible mold (see Figure 14). They also applied a PDMS-based coating to achieve both a highly transparent super-hydrophobic surface and an anti-adhesion layer coating for high-resolution nanoimprint lithography by intrinsic low surface energy and easy release of PDMS. The PDMS-coated flexible mold was used repeatedly, more than 10 times, without losing the anti-adhesion property of PDMS and endured severe chemical cleaning processes, such as sonication, which were performed periodically to wash the mold after several consecutive imprinting.

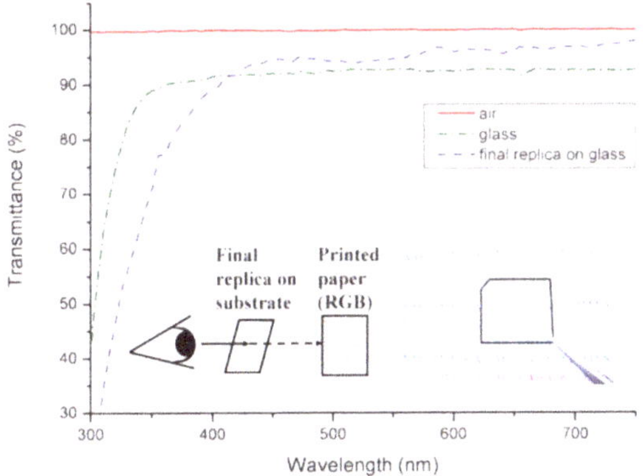

Figure 14. The transmittance of the highly transparent super-hydrophobic surface and a photograph of letters underneath the transparent superhydrophobic film [40].

They claimed that this process could be suitable for various applications that require both super-hydrophobic and anti-reflective surface coatings. Further, they claimed that the process could be easily extended to a large area patterning; therefore, this method could be suitable for mass production of nanopatterned polymeric optical substrates, and could be applicable towards solar cell surface contamination and plastic optics which require dust-free and self-cleaning surfaces with high transmission. They did not report any performance data on abrasion induced wear resistance on these materials.

Dufour et al. [41] conducted a systematic study on mushroom shaped (with sharper microscale reentrant shapes compared to Figure 13) PDMS molds functionalized with 1*H*,1*H*,2*H*,2*H*-perfluorodecyltrichlorosilane in vapor phase. They used low surface tension liquids (down to 28 mN/m) in their wetting analysis. In summary, on these mushroom-like PDMS surfaces, they concluded that since the contact line was always pinned around the cap contour, variation of apparent static contact angle with the liquid surface tension was negligible. Consequently, contact angle hysteresis was relatively large even for water. These microstructures were able to sustain a composite interface (an intermediate wetting state) with liquids having surface tension down to 27.9 mN/m, but as soon as the state became unstable, the drop completely spread into the lattice. The transition from Cassie-Baxter to Wenzel state was discontinuous, and there was no intermediate metastable configuration (partial impalement). Thus, the superomniphobic property of PDMS microstructures was attributed to their ability to prevent local Cassie Baxter-Wenzel transition since lateral spreading occurred immediately after the composite interface disappeared.

6. Transparent Non-Wettable Coatings from Ceramic-Based Nanostructures

The design and development of robust self-cleaning coatings for use in solar panels or solar cells is especially important given that nearly half of the overall power conversion efficiency of solar panels can be lost due to dust or dirt accumulation every year. In fact, some solar panels are known to be protected well with certain transparent anti-graffiti coatings. In addition to water- and dust-repellent property requirements, superhydrophobic coatings for photovoltaics must be highly transparent to both visible and near-IR light as well as being UV-resistant and durable. One promising approach [42] fabricated highly transparent porous silica coatings on glass substrates through layer-by-layer (LbL) assembly of raspberry-like polystyrene@silica (PS@SiO$_2$) microparticles followed by calcination at high temperature. Initially the coatings were superhydrophilic but became superhydrophobic after chemical vapor deposition of 1H,1H,2H,2H-perfluorodecyltriethoxysilane. Superhydrophobic porous silica coating had a water contact angle greater than 160° with a sliding angle of 7° and a transmittance of 85%. The authors claimed that transparency can be increased by lowering the LbL cycles but did not demonstrate superhydrophobic coatings with higher transparency.

Gao et al. [43] fabricated highly transparent and UV-resistant superhydrophobic arrays of SiO$_2$-coated ZnO nanorods that were prepared in a sequence of low-temperature (<150 °C) steps on both glass and thin sheets of polyester, PET (2 × 2 in.^{2}), and the superhydrophobic nanocomposite is shown to have minimal impact on solar cell device performance under AM1.5G illumination (See Figure 15a). They argued that such flexible plastics can serve as front cell and backing materials in the manufacture of flexible displays and solar cells. To demonstrate the minimal impact of the presence of the superhydrophobic nanorod arrays on solar cell performance, they prepared bulk-heterojunction (BHJ) devices with the polymer donor PBDTTPD and the fullerene acceptor PC71BM. The configuration of the OPV device including the SiO$_2$/ZnO nanocomposite is shown in Figure 15a; Figure 15b shows perfectly spherical droplets positioned on the front of the superhydrophobic device (static angle, 157°; sliding angle, 13°). The superhydrophobic cell (red curve, circles) and bare reference cell (blue curve, squares) under AM1.5G solar illumination (100 mW·cm^{-2}) (Figure 15c) exhibit equivalent current-voltage characteristics with comparable power conversion efficiencies (PCEs) of 6.9% and 6.8%, respectively, values comparable within the limits of experimental accuracy. In parallel, their external quantum efficiency (EQE) spectra (Figure 15d) show comparably broad and efficient EQE responses, with values >60% in the 370–630 nm range, and peaking at ca. 70% at 550 nm, which confirms the minimal impact of the presence of the superhydrophobic nanorod arrays on solar cell performance. They did not present any results on the durability of these coatings against abrasion.

Although no abrasion induced wear resistance tests were conducted, bending resistance of the transparent (Figure 14c,d) non-wettable coatings were demonstrated by the authors as seen in Figure 14c. They claimed that such robustness is important for flexible solar energy converters.

Yadav et al. [44] reported the emergence of superhydrophobic wetting behavior and enhanced UV stability of indium oxide (IO) nanorods due to their vertical alignment (Figure 16). Both randomly distributed and vertically aligned IO nanorods were synthesized (Figure 16a,b) via chemical vapor deposition (CVD) method. Their results showed that the static water contact angle (WCA) demonstrated a significant dependence on the alignment of the nanorods. The randomly distributed IO nanorods had 133.7° ± 6.8° wetting whereas for vertically aligned IO nanorods WCA was found to be 159.3° ± 4.8°. Continuous UV light illumination for 30 min exhibited the change in contact angle (ΔWCA) of about 41° for vertically aligned IO nanorods whereas randomly distributed IO nanorods become hydrophilic with a dramatic change in WCA value of 108°. The superhydrophobicity of vertically aligned IO nanorods and their enhanced UV stability were discussed by comparing the effective solid fraction at solid-liquid interface and the reactivity of surface crystallographic planes. They argued that the superhydrophobic surface of aligned vertically standing IO nanorods along with its resistance against photoinduced wetting transition can make them suitable for electronic devices with reduced surface discharge even at relatively high humidity levels. They did not test and measure any results related to mechanical durability of their coatings.

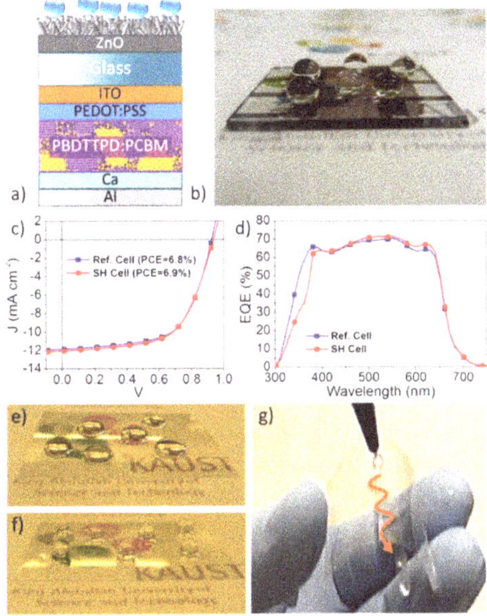

Figure 15. (**a**) Schematic of a BHJ polymer solar cell including the SiO_2/ZnO nanocomposite; (**b**) Water droplets positioned on the front of the superhydrophobic device remain perfectly spherical; (**c**) *J–V* characteristic of a superhydrophobic (SH) cell (red circle) superimposed on that of a bare reference (Ref.) cell (blue square); AM1.5G solar illumination (100 mW·cm^{-2}); (**d**) EQE spectra of the SH cell (red circle) and the Ref. cell (blue square); (**e**) Water droplets positioned on a bare transparent sheet of PET (2 × 2 in.^{2}); (**f**) Droplets positioned on superhydrophobic PET (static angle: 160°); (**g**) PET retains its superhydrophobicity upon repeated bending (×350) [43].

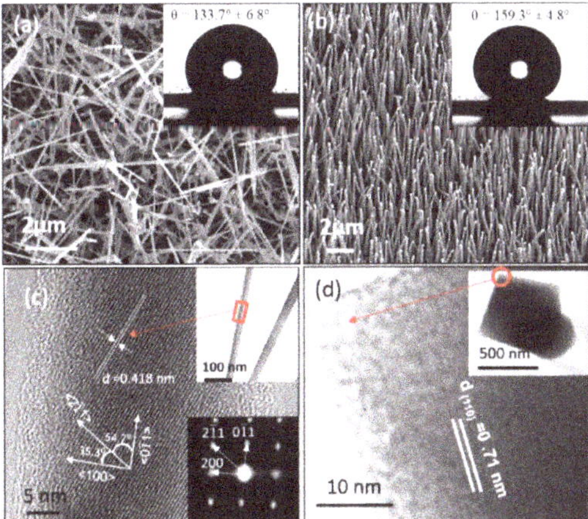

Figure 16. FESEM images of (**a**) randomly distributed and (**b**) vertically aligned IO nanorods. The insets show the contact angle images of the respective samples. HRTEM images of (**c**) randomly distributed and (**d**) vertically standing IO nanorods. The insets show the TEM and SAED pattern of respective nanorods [44].

Aytug et al. [45] reported the formation of low-refractive index antireflective glass films that embody omni-directional optical properties over a wide range of wavelengths, while also possessing specific wetting capabilities. The coatings were made up of an interconnected network of nanoscale pores surrounded by a nanostructured silica framework. These structures came from a method that exploited metastable spinodal phase separation in glass-based materials. Their approach not only enabled design of surface microstructures with graded-index antireflection characteristics, where the surface reflection was suppressed through optical impedance matching between interfaces, but also enabled self-cleaning ability through modification of the surface chemistry. Based on near complete elimination of Fresnel reflections (yielding >95% transmission through a single-side coated glass) and corresponding increase in broadband transmission, the fabricated nanostructured surfaces were found to promote a general and an invaluable ~3%–7% relative increase in current output of multiple direct/indirect bandgap photovoltaic cells. Moreover, these antireflective surfaces also demonstrated superior resistance against mechanical wear and abrasion. Their antireflective coatings were essentially monolithic, enabling simultaneous realization of graded index anti-reflectivity, self-cleaning capability, and mechanical stability within the same surface.

Figure 17a shows typical indenter generated scanning probe microscopy images of scratch tracks for a sputter deposited film, a fully processed film (i.e., heat treated and etched), and the underlying borosilicate substrate. The scratches, produced after the application of load ramped up to the equipment peak of 10,000 μN, showed no evidence of cracking, brittle fragmentation or tendency toward delamination. Moreover, for loads at one-half and at full equipment maximum the average penetration depth (calculated from the lateral and vertical displacements) of the indenter is nearly the same for both the reference substrate and the coated sample. This result suggests that the scratch resistance behavior of the as-deposited films is similar to that of the underlying substrate material. Accordingly, the coefficient of friction profiles for a representative template revealed similar behavior, as displayed in Figure 17b. Such transparent (Figure 17c), robust and low friction coefficient coatings do have potential as preventive treatments for anti-graffiti applications particularly for public transportation vehicle windows.

Metal oxides, in general, are known to exhibit significant wettability towards water molecules because of the high possibility of synergetic hydrogen-bonding interactions at the solid-water interface. Very recently, Sankar et al. [46] showed that nano sized phosphates of rare earth materials (Rare Earth Phosphates, REPs), $LaPO_4$ in particular, exhibited without any chemical modification, unique combination of intrinsic properties including notable hydrophobicity that could be retained even after exposure to extreme temperatures and harsh hydrothermal conditions (Figure 18a). Nanostructured film surface is depicted in Figure 18b. Transparent nanocoatings of $LaPO_4$ as well as mixture of other REPs on glass surfaces were shown to display hydrophobicity with water contact angle (WCA) value of 120° (Figure 18c) while sintered and polished monoliths manifested WCA greater than 105°. Significantly, these materials in the form of coatings and monoliths also exhibited complete non-wettability and inertness towards molten metals like Ag, Zn, and Al well above their melting points. They proposed that these properties, coupled with their excellent chemical and thermal stability, ease of processing, machinability and their versatile photo-physical and emission properties, render $LaPO_4$ and other REP ceramics utility in diverse applications.

The crystalline structure of the films ($LaPO_4$, $GaPO_4$, $NdPO_4$) is depicted in Figure 18d. They did not conduct any kind of mechanical wear abrasion tests on these interestingly hydrophobic inorganic coatings. However, due to their ceramic nature these films are expected to have satisfactory resistance to abrasion induced wear or scratches. Nonetheless, such tribological experiments should also be reported.

Fukada et al. [47] developed semi-transparent superomniphobic surfaces and applied as coating to integrate antifouling properties in solar cell devices. The coatings featured mesh and stripe structures (see Figure 19a,b). The stripe structure was more suitable for generating transparent and highly oleophobic coatings (Figure 19c), thereby displaying excellent antifouling properties. The stripe-based coating also displayed good thermally stability. Total transmittance was a key factor influencing the conversion efficiency. By applying an antifouling treatment, reduction in the performance of the solar cell devices was inhibited. For instance, the stripe-coated solar cells maintained a high conversion efficiency of 92% of the non-coated solar cell even in the presence of oil contaminant. In contrast, in the absence of the stripe substrate, conversion efficiency decreased to 64%.

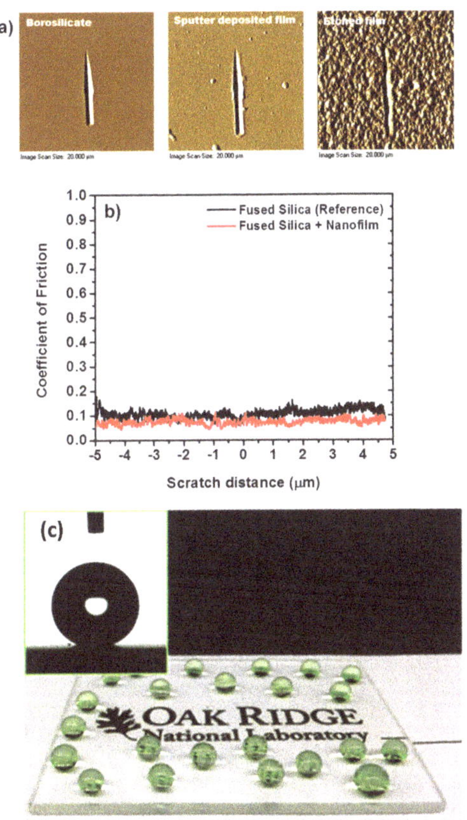

Figure 17. (**a**) Topographical images of the scratches made on an as-deposited and a nanotextured glass thin film. Image for the underlying uncoated borosilicate substrate is also included for comparison. No debris is observed on any sample, aside from several small and large sized surface-bound particles on the sputtered film. Note that the as-sputtered borosilicate films are generally somewhat rougher than the underlying substrates; (**b**) Comparison of the coefficient of friction profiles as a function of scratch distance for fused silica substrates with and without a dense borosilicate film; (**c**) Photograph of blue dyed water droplets on a borosilicate substrate coated with nanoporous antireflective glass film. The film surface is modified by a covalently bonded organosilane chemistry. Inset shows the profile of a 5 µL water droplet resting on a similarly processed film, displaying a static contact angle of 165° [45].

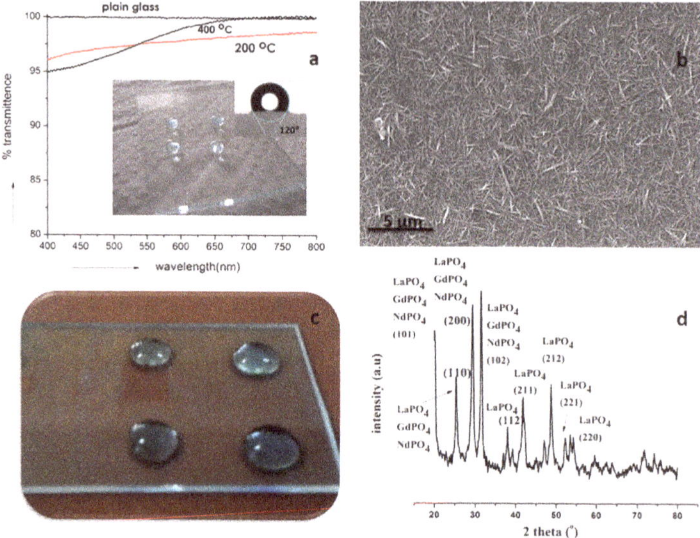

Figure 18. Hydrophobicity achieved for LaPO$_4$ coated thin film. (**a**) Water droplets sitting over LaPO4 coated glass plate and the high optical transparency recorded for the films; (**b**) SEM image showing spike-like arrangement of LaPO$_4$ nanorods over the glass surface; (**c**) Mixed REP sol coated glass slide showing hydrophobic character; (**d**) Diffraction pattern obtained for mixed rare earth phosphates [46].

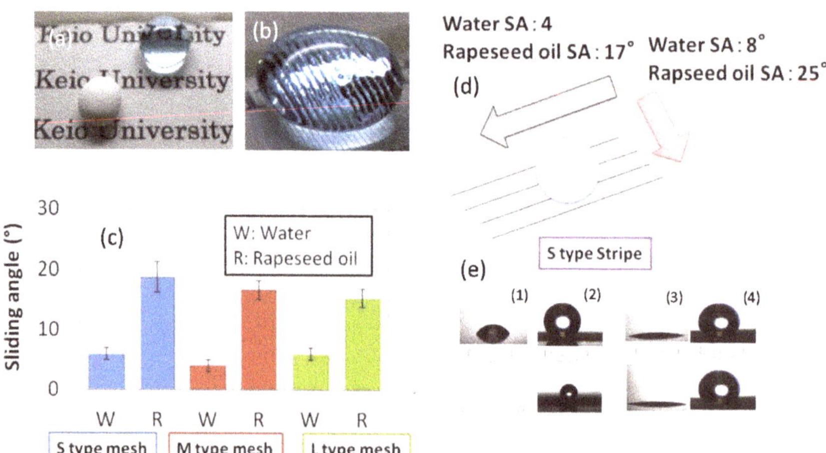

Figure 19. (**a**) Methylene blue-stained water and white ink stained rapeseed oil on a stripe substrate and (**b**) enlarged stripe structure in the presence of water; (**c**) Sliding angle measurements; (**d**) The droplet of water or rapeseed oil was placed on the mesh/stripe substrate. The substrate was then tilted slowly and the angle was measured when the liquid moved. (**e**) Thermal stability of mesh and stripe substrates, examined at 80 °C for 10 min. (1)Water (10 µL) on glass, the CA was 44°; (2) Water (10 µL) on L-type mesh, CA was changed from 151.6° to 146.0°; (3) Rapeseed oil (10 µL) on glass, CA was changed from <5.0° to <5.0°; (4) Rapeseed oil (10 µL) on L-type mesh, CA was changed from 141.9° to 141.0° [47].

7. Recommendations for Future Directions

Over the last decade and particularly in the last five years, many efficient methods of producing transparent non-wettable surfaces have been demonstrated. Most of the methods reviewed above are still not suitable for sustained mechanical durability against abrasion induced wear. As a future direction, the most important aspect would be to focus on substrate adhesion improvements of nanoparticle films. These could be done by applying transparent primers or adhesive layers or by welding the nanoparticles into transparent polymeric substrates by thermal annealing methods. The second important point would be to maintain "wear independent texture similarity" discussed earlier. In other words, the surface of the coatings or the films may wear away but the newly exposed layers can still have hierarchical texture with hydrophobic chemistry. In the case of dynamic oleophobicity (non-stick oil droplets), more environmentally friendly chemicals should be used rather than C-8 based fluorinated oils or silanes.

Furthermore, researchers should use established mechanical durability testing methods such as ASTM standards of paint industry instead of homemade tests such as running sandpaper over the non-wettable coating or film. Type of the abradant used as well as the downward pressure should always be indicated otherwise cross-comparisons may not be possible. A standard wear abrasion test should be the primary indicator to label non-wettable coatings as *mechanically durable* but secondary tests such as substrate adhesion, scratch resistance; powder and/or rain/spray erosion tests should also be conducted and reported. All in all, studies to date still lack these features and as such it is still not possible to attract industrial attention to produce robust commercial transparent non-wettable coatings for outdoor installations.

8. Conclusions and Outlook

In this review, we attempted to present the state-of-the-art in fabricating and testing transparent hydrophobic, superhydrophobic and oleophobic coatings against mechanical durability in terms of wear and abrasion resistance. Review of recent literature indicates that most of the published works have not presented any performance results related to resilience against abrasion induced wear [48]. Reports on fluorinated oil infused transparent nanotextured surfaces also exist [49], which claim excellent outdoor performance and substrate independent universal fabrication, but with no abrasion induced wear resistance characterization. However, dynamically oleophobic surfaces deserve mention here [50]. Dynamically oleophobic surfaces (mostly transparent) do not present hierarchical surface textures, in general, but low surface tension hydrocarbon liquid droplets bead up instead of spreading and can still slide away and clear such surfaces. This approach could be good against oil-based paint stain resistance on outdoor installations. Moreover, some of these environmentally friendly coatings have been designed to display excellent thermal stability, dynamic/thermoresponsive oleophobicity, and hydrolytic stability [51]. Perhaps, solid-air interfaces with dynamic oleophobicity that are generally grown from precursors can be similarly grown on wear abrasion resistant transparent coating textures so that oil-based stains can also be prevented on wear resistant transparent coatings. We identified three major fabrication methods namely, nanoparticle assembly and films, transparent elastomer surface micro/nano structuring by molding and formation of nanostructured ceramic surfaces functionalized with hydrophobic macromolecules, suitable for solar energy conversion devices. The relevance of this review to anti-graffiti applications stems from the fact that transparency maintains see-thru properties after the coatings are applied and in certain cases the oleophobicity results in not only water repellence but also grease/dirt repellence.

As summarized above, combining transparency with mechanical abrasion resistance (important for anti-graffiti applications) is very challenging and is not described frequently. However, latest reports indicate considerable progress towards achieving both in one layer coating. It is also important to note that most of these fabrication methods should eventually translate into and applied over large areas commonly encountered for graffiti prevention; in other words, public transport vehicle windows, walls, or building windows, to name a few. Although such an immediate transformation is premature currently, future approaches should also address this need so that more industrial interest can be attracted towards joint development efforts.

Conflicts of Interest: The author declares no conflict of interest.

References

1. Tserepi, A.D.; Vlachopoulou, M.E.; Gogolides, E. Nanotexturing of poly (dimethylsiloxane) in plasmas for creating robust super-hydrophobic surfaces. *Nanotechnology* **2006**, *17*, 3977–3984. [CrossRef]
2. Wang, D.; Zhang, Z.; Li, Y.; Xu, C. Highly transparent and durable superhydrophobic hybrid nanoporous coatings fabricated from polysiloxane. *Appl. Mater. Interfaces* **2014**, *6*, 10014–10021. [CrossRef] [PubMed]
3. Tuvshindorj, U.; Yildirim, A.; Ozturk, F.E.; Bayindir, M. Robust cassie state of wetting in transparent superhydrophobic coatings. *Appl. Mater. Interfaces* **2014**, *6*, 9680–9688. [CrossRef] [PubMed]
4. Yabu, H.; Shimomura, M. Single-step fabrication of transparent superhydrophobic porous polymer films. *Chem. Mater.* **2005**, *17*, 5231–5234. [CrossRef]
5. Liu, Y.; Das, A.; Xu, S.; Lin, Z.; Xu, C.; Wang, Z.L.; Rohatgi, A.; Wong, C.P. Hybridizing ZnO nanowires with micropyramid silicon wafers as superhydrophobic high-efficiency solar cells. *Adv. Energy Mater.* **2012**, *2*, 47–51. [CrossRef]
6. Giolando, D.M. Transparent self-cleaning coating applicable to solar energy consisting of nano-crystals of titanium dioxide in fluorine doped tin dioxide. *Sol. Energy* **2016**, *124*, 76–81. [CrossRef]
7. Bayer, I.S.; Megaridis, C.M.; Zhang, J.; Gamota, D.; Biswas, A. Analysis and surface energy estimation of various model polymeric surfaces using contact angle hysteresis. *J. Adhes. Sci. Technol.* **2007**, *21*, 1439–1467. [CrossRef]
8. Milionis, A.; Bayer, I.S.; Loth, E. Recent advances in oil-repellent surfaces. *Int. Mater. Rev.* **2016**, *61*, 101–126. [CrossRef]
9. Steele, A.; Davis, A.; Kim, J.; Loth, E.; Bayer, I.S. Wear independent similarity. *Appl. Mater. Interfaces* **2015**, *7*, 12695–12701. [CrossRef] [PubMed]
10. Milionis, A.; Loth, E.; Bayer, I.S. Recent advances in the mechanical durability of superhydrophobic materials. *Adv. Colloid Interface Sci.* **2016**, *229*, 57–79. [CrossRef] [PubMed]
11. Milionis, A.; Languasco, J.; Loth, E.; Bayer, I.S. Analysis of wear abrasion resistance of superhydrophobic acrylonitrile butadiene styrene rubber (ABS) nanocomposites. *Chem. Eng. J.* **2015**, *281*, 730–738. [CrossRef]
12. Milionis, A.; Dang, K.; Prato, M.; Loth, E.; Bayer, I.S. Liquid repellent nanocomposites obtained from one-step water-based spray. *J. Mater. Chem. A* **2015**, *3*, 12880–12889. [CrossRef]
13. Herminghaus, S. Roughness-induced non-wetting. *Europhys. Lett.* **2000**, *52*, 165–170. [CrossRef]
14. Marmur, A. The lotus effect: Superhydrophobicity and metastability. *Langmuir* **2004**, *20*, 3517–3519. [CrossRef] [PubMed]
15. Bico, J.; Thiele, U.; Quéré, D. Wetting of textured surfaces. *Colloids Surf. A* **2002**, *206*, 41–46. [CrossRef]
16. Xiu, Y.; Liu, Y.; Hess, D.W.; Wong, C.P. Mechanically robust superhydrophobicity on hierarchically structured Si surfaces. *Nanotechnology* **2010**, *21*, 155705. [CrossRef] [PubMed]
17. Sidorenko, A.; Ahn, H.S.; Kim, D.I.; Yang, H.; Tsukruk, V.V. Wear stability of polymer nanocomposite coatings with trilayer architecture. *Wear* **2002**, *252*, 946–955. [CrossRef]
18. Schaeffer, D.A.; Polizos, G.; Smith, D.B.; Lee, D.F.; Hunter, S.R.; Datskos, P.G. Optically transparent and environmentally durable superhydrophobic coating based on functionalized SiO_2 nanoparticles. *Nanotechnology* **2015**, *26*, 055602. [CrossRef] [PubMed]
19. Mates, J.E.; Bayer, I.S.; Palumbo, J.M.; Carroll, P.J.; Megaridis, C.M. Extremely stretchable and conductive water-repellent coatings for low-cost ultra-flexible electronics. *Nat. Commun.* **2015**, *6*, 8874. [CrossRef] [PubMed]
20. Bhushan, B.; Nosonovsky, M. The rose petal effect and the modes of superhydrophobicity. *Philos. Trans. R. Soc. Lond. A* **2010**, *368*, 4713–4728. [CrossRef] [PubMed]
21. Bormashenko, E.; Starov, V. Impact of surface forces on wetting of hierarchical surfaces and contact angle hysteresis. *Colloid Polym. Sci.* **2013**, *291*, 343–346. [CrossRef]

22. Liu, S.; Liu, X.; Latthe, S.S.; Gao, L.; An, S.; Yoon, S.S.; Liu, B.; Xing, R. Self-cleaning transparent superhydrophobic coatings through simple sol–gel processing of fluoroalkylsilane. *Appl. Surf. Sci.* **2015**, *351*, 897–903. [CrossRef]
23. Rahmawan, Y.; Xu, L.; Yang, S. Self-assembly of nanostructures towards transparent, superhydrophobic surfaces. *J. Mater. Chem. A* **2013**, *1*, 2955–2969. [CrossRef]
24. Si, Y.; Guo, Z. Superhydrophobic nanocoatings: From materials to fabrications and to applications. *Nanoscale* **2015**, *7*, 5922–5946. [CrossRef] [PubMed]
25. Irzh, A.; Ghindes, L.; Gedanken, A. Rapid deposition of transparent super-hydrophobic layers on various surfaces using microwave plasma. *Appl. Mater. Interfaces* **2011**, *3*, 4566–4572. [CrossRef] [PubMed]
26. Bravo, J.; Zhai, L.; Wu, Z.; Cohen, R.E.; Rubner, M.F. Transparent superhydrophobic films based on silica nanoparticles. *Langmuir* **2007**, *23*, 7293–7298. [CrossRef] [PubMed]
27. Wong, J.X.; Wong, H.; Yu, H.-Z. Preparation of transparent superhydrophobic glass slides: Demonstration of surface chemistry characteristics. *J. Chem. Educ.* **2013**, *90*, 1203–1206. [CrossRef]
28. Erbil, H.Y.; Cansoy, C.E. Range of applicability of the Wenzel and Cassie-Baxter equations for superhydrophobic surfaces. *Langmuir* **2009**, *25*, 14135–14145. [CrossRef] [PubMed]
29. Xu, L.; Karunakaran, R.G.; Guo, J.; Yang, S. Transparent, superhydrophobic surfaces from one-step spin coating of hydrophobic nanoparticles. *Appl. Mater. Interfaces* **2012**, *4*, 1118–1125. [CrossRef] [PubMed]
30. Xu, L.; He, J. Fabrication of highly transparent superhydrophobic coatings from hollow silica nanoparticles. *Langmuir* **2012**, *28*, 7512–7518. [CrossRef] [PubMed]
31. Zhu, X.; Zhang, Z.; Ren, G.; Men, X.; Ge, B.; Zhou, X. Designing transparent superamphiphobic coatings directed by carbon nanotubes. *J. Colloid Interface Sci.* **2014**, *421*, 141–145. [CrossRef] [PubMed]
32. Ling, X.Y.; Phang, I.Y.; Vancso, G.J.; Huskens, J.; Reinhoudt, D.N. Stable and transparent superhydrophobic nanoparticle films. *Langmuir* **2009**, *25*, 3260–3263. [CrossRef] [PubMed]
33. Ebert, D.; Bhushan, B. Transparent, superhydrophobic, and wear-resistant coatings on glass and polymer substrates using SiO_2, ZnO, and ITO nanoparticles. *Langmuir* **2012**, *28*, 11391–11399. [CrossRef] [PubMed]
34. Deng, X.; Mammen, L.; Zhao, Y.; Lellig, P.; Müllen, K.; Li, C.; Butt, H.-J.; Vollmer, D. Transparent, thermally stable and mechanically robust superhydrophobic surfaces made from porous silica capsules. *Adv. Mater.* **2011**, *23*, 2962–2965. [CrossRef] [PubMed]
35. Si, Y.; Zhu, H.; Chen, L.; Jiang, T.; Guo, Z. A multifunctional transparent superhydrophobic gel nanocoating with self-healing properties. *Chem. Commun.* **2015**, *51*, 16794–16797. [CrossRef] [PubMed]
36. Deng, X.; Mammen, L.; Butt, H.J.; Vollmer, D. Candle soot as a template for a transparent robust superamphiphobic coating. *Science* **2012**, *335*, 67–70. [CrossRef] [PubMed]
37. Yokoi, N.; Manabe, K.; Tenjimbayashi, M.; Shiratori, S. Optically transparent superhydrophobic surfaces with enhanced mechanical abrasion resistance enabled by mesh structure. *Appl. Mater. Interfaces* **2015**, *7*, 4809–4816. [CrossRef] [PubMed]
38. Gong, D.; Long, J.; Jiang, D.; Fan, P.; Zhang, H.; Li, L.; Zhong, M. Robust and stable transparent superhydrophobic polydimethylsiloxane films by duplicating via a femtosecond laser-ablated template. *Appl. Mater. Interfaces* **2016**, *8*, 17511–17518. [CrossRef] [PubMed]
39. Im, M.; Im, H.; Lee, J.H.; Yoon, J.B.; Choi, Y.K. A robust superhydrophobic and superoleophobic surface with inverse-trapezoidal microstructures on a large transparent flexible substrate. *Soft Matter* **2010**, *6*, 1401–1404. [CrossRef]
40. Kim, M.; Kim, K.; Lee, N.Y.; Shin, K.; Kim, Y.S. A simple fabrication route to a highly transparent super-hydrophobic surface with a poly (dimethylsiloxane) coated flexible mold. *Chem. Commun.* **2007**, *22*, 2237–2239. [CrossRef] [PubMed]
41. Dufour, R.; Perry, G.; Harnois, M.; Coffinier, Y.; Thomy, V.; Senez, V.; Boukherroub, R. From micro to nano reentrant structures: Hysteresis on superomniphobic surfaces. *Colloid Polym. Sci.* **2013**, *291*, 409–415. [CrossRef]
42. Shang, Q.; Zhou, Y. Fabrication of transparent superhydrophobic porous silica coating for self-cleaning and anti-fogging. *Ceram. Int.* **2016**, *42*, 8706–8712. [CrossRef]
43. Gao, Y.; Gereige, I.; El Labban, A.; Cha, D.; Isimjan, T.T.; Beaujuge, P.M. Highly transparent and UV-resistant superhydrophobic SiO_2-coated ZnO nanorod arrays. *Appl. Mater. Interfaces* **2014**, *6*, 2219–2223. [CrossRef] [PubMed]

44. Yadav, K.; Mehta, B.R.; Singh, J.P. Superhydrophobicity and enhanced UV stability in vertically standing indium oxide nanorods. *Appl. Surf. Sci.* **2015**, *346*, 361–365. [CrossRef]

45. Aytug, T.; Lupini, A.R.; Jellison, G.E.; Joshi, P.C.; Ivanov, I.H.; Liu, T.; Wang, P.; Menon, R.; Trejo, R.M.; Lara-Curzio, E.; et al. Monolithic graded-refractive-index glass-based antireflective coatings: Broadband/omnidirectional light harvesting and self-cleaning characteristics. *J. Mater. Chem. C* **2015**, *3*, 5440–5449. [CrossRef]

46. Sankar, S.; Nair, B.N.; Suzuki, T.; Anilkumar, G.M.; Padmanabhan, M.; Hareesh, U.N.S.; Warrier, K.G. Hydrophobic and metallophobic surfaces: Highly stable non-wetting inorganic surfaces based on lanthanum phosphate nanorods. *Sci. Rep.* **2016**, *6*. [CrossRef] [PubMed]

47. Fukada, K.; Nishizawa, S.; Shiratori, S. Antifouling property of highly oleophobic substrates for solar cell surfaces. *J. Appl. Phys.* **2014**, *115*. [CrossRef]

48. Cao, L.; Gao, D. Transparent superhydrophobic and highly oleophobic coatings. *Faraday Discuss.* **2010**, *146*, 57–65. [CrossRef] [PubMed]

49. Ma, W.; Higaki, Y.; Otsuka, H.; Takahara, A. Perfluoropolyether-infused nano-texture: A versatile approach to omniphobic coatings with low hysteresis and high transparency. *Chem. Commun.* **2013**, *49*, 597–599. [CrossRef] [PubMed]

50. Park, J.; Urata, C.; Masheder, B.; Cheng, D.F.; Hozumi, A. Long perfluoroalkyl chains are not required for dynamically oleophobic surfaces. *Green Chem.* **2013**, *15*, 100–104. [CrossRef]

51. Masheder, B.; Urata, C.; Hozumi, A. Transparent and hard zirconia-based hybrid coatings with excellent dynamic/thermoresponsive oleophobicity, thermal durability, and hydrolytic stability. *Appl. Mater. Interfaces* **2013**, *5*, 7899–7905. [CrossRef] [PubMed]

MDPI

St. Alban-Anlage 66

4052 Basel

Switzerland

Tel. +41 61 683 77 34

Fax +41 61 302 89 18

www.mdpi.com

Coatings Editorial Office

E-mail: coatings@mdpi.com

www.mdpi.com/journal/coatings

9 783039 217380